The Archaic

The Archaic takes as its major reference points C. G. Jung's classic essay 'Archaic Man' (1930), and Ernesto Grassi's paper on 'Archaic theories of history' (1990). Moving beyond the confines of a Jungian framework to include other methodological approaches, this book explores the concept of the archaic.

Defined as meaning 'old-fashioned', 'primitive' or 'antiquated', the archaic is, in fact, much more than something very, very old: it is timeless, inasmuch as it is before time itself. *Archē*, *Urgrund*, *Ungrund*, 'primordial darkness', 'eternal nothing' are names for something essentially nameless, yet whose presence we nevertheless intuit.

This book focuses on the reception of myth in the tradition of German Idealism or Romanticism (Creuzer, Schelling, Nietzsche), which not only looked back to earlier thinkers (such as Jacob Boehme) but also laid down roots for developments in twentieth-century thought (Ludwig Klages, Martin Heidegger). *The Archaic* also includes:

- studies of the Germanic dimension of the archaic (Charles Bambach, Alan Cardew)
- a discussion of the mytho-phenomenological approach to the archaic (Robert Josef Kozljanič)
- a series of articles on Jung's understanding of the archaic (Paul Bishop, Susan Rowland, Robert A. Segal).

This book will be of interest to psychoanalysts, anthropologists and phenomenologists, as well as students of psychology, cultural studies, religious studies, and philosophy, as it seeks to rehabilitate a concept of demonstrable and urgent relevance for our time.

Paul Bishop is Professor of German at the University of Glasgow. His previous publications include *Analytical Psychology and German Classical Aesthetics*, 2 vols (Routledge, 2007–2008) and Jung's *Answer to Job: A Commentary* (Routledge, 2002).

The Archaic

The Past in the Present

Edited by Paul Bishop

Routledge
Taylor & Francis Group

LONDON AND NEW YORK

First published 2012 by Routledge
27 Church Road, Hove, East Sussex, BN3 2FA

Simultaneously published in the USA and Canada
by Routledge
711 Third Avenue, New York NY 10017

Routledge is an imprint of the Taylor & Francis Group, an Informa business

British Library Cataloguing in Publication Data
A catalogue record for this book is available
from the British Library

Library of Congress Cataloging in Publication Data
The archaic : the past in the present : a collection of papers / edited
by Paul Bishop.
        p. cm.
   Includes bibliographical references and index.
   ISBN 978–0–415–54755–0 (hardback) — ISBN 978–0–415–54756–7
   (paperback)
   1. Cosmology—Philosophy.   2. Phenomenology. I. Bishop, Paul, 1967–
   BD494.A73 2012
   113—dc22

                                                    2011011102

ISBN: 978–0–415–54755–0 (hbk)
ISBN: 978–0–415–54756–7 (pbk)
ISBN: 978–0–203–80342–4 (ebk)

Typeset in Times
by RefineCatch Ltd, Bungay, Suffolk
Paperback cover design by Andrew Ward
Printed by TJ International Ltd, Padstow, Cornwall

He that has a secret to hide should not only hide it, but hide that he has it to hide.

Thomas Carlyle, *The French Revolution: A History*, vol. 2, 2.4.7

Contents

Acknowledgements

This volume has been the fruit of much collaboration, and it could not have been realized without the help of its contributors.

The idea for such a volume was first suggested to me by Susan Rowland, following the publication of our papers in *The Journal of Analytical Psychology* on the significance of C. G. Jung's essay 'Archaic Man'. Subsequently, Robert A. Segal agreed with alacrity to the republication of his own paper, which had sparked the exchange, and the nucleus of the second half of this volume was formed. I should like to record my thanks to Susan for her original suggestion and for her support of this project. I am also grateful to Warren Coleman, the editor of the *JAS*, for supporting my request to republish these papers, and to Wiley-Blackwell for granting permission to use previously published material.

Likewise, I am grateful to my colleagues in Essex (Alan Cardew), Munich (Robert Josef Kozljanič), and Texas (Charles Bambach), for accepting the invitation to contribute to this volume, which seeks to broaden the debate out from Jung's approach. In particular, I should like to thank Robert Kozljanič for suggesting the inclusion of a text by Ernest Grassi in this volume, as well as for his encouragement in sustaining a *philosophische Germanistik* in the form of electronic mail between Munich and Glasgow. And I should like to thank Alan Cardew for his supportive erudition, and for reminding me that, amid the institutional darkness, one should follow Lord Byron's advice and learn to love despair. Thanks to Alan's enthusiasm for it, I read *Heidegger's Roots: Nietzsche, National Socialism, and the Greeks* (Cornell University Press, 2003) shortly after it was published, and I am delighted and honoured in equal measure to include a contribution by Charles Bambach in this volume.

For permission to reprint R. F. C. Hull's translation of Jung's paper 'Archaic Man', I am grateful to the Heirs of C. G. Jung, and for permission to translate Grassi's paper on 'Archaic Theories of History', originally published in German in *Synthesis philosophica*, I am grateful to Emilio Hidalgo-Serna. For steering the project from the proposal stage through to publication, I am grateful to Kate Hawes and all her colleagues at Routledge.

In a special way, this volume is dedicated to Helen Bridge and to her help and support, which allows me to say, to borrow a phrase, that 'ours is a home where the past still lives'.

Contributors

Charles Bambach is Professor of Philosophy at the University of Texas, Dallas. His books include *Heidegger's Roots: Nietzsche, National Socialism, and the Greeks* (Cornell University Press, 2003) and *Heidegger, Dilthey, and the Crisis of Historicism* (Cornell University Press, 1995). He has also written a variety of articles on hermeneutics, phenomenology, ethics, and the history of German philosophy. Bambach's current book project, *Thinking the Poetic Measure of Justice: Heidegger, Hölderlin, Celan, and the Greek Experience of Dikē*, deals with the tragic aporia between ethics and justice in modern German philosophy, specifically Heidegger's dialogue with the poetry of Friedrich Hölderlin and Paul Celan.

Paul Bishop is Professor of German at the University of Glasgow. His research focuses on lines of filiation between the German eighteenth and nineteenth centuries and twentieth-century thought, with particular reference to Goethe, Schiller, Herder, Feuerbach, Nietzsche, Freud, Jung, Klages, and Steiner. His publications include *Reading Goethe at Midlife: Ancient Wisdom, German Classicism, and Jung* (Spring Books, 2011), and *Analytical Psychology and German Classical Aesthetics*, 2 vols (Routledge, 2007–2008). He is the editor of *A Companion to Friedrich Nietzsche: Life and Works* (Camden House, 2012), *Nietzsche and Antiquity* (Camden House, 2004), and *Jung in Contexts: A Reader* (Routledge, 1999).

Alan Cardew is currently Director of the Enlightenment in the Department of Literature, Film, and Theatre Studies at the University of Essex. In 2005–2006 he was the Director of the Centre for American (US) Studies, where he currently teaches courses on conspiracy theory and the history of New York. For some years he taught in the Department of Language and Linguistics. He is a member of the Centre for Theoretical Studies, and also teaches in the Centre for Psychoanalytic Studies in the area of Refugee Care. He is particularly interested in interdisciplinary studies, and has recently written on Jung and Heidegger, Cassirer, hermeneutics, and Nietzsche. He is currently pursuing research on the interpretation of myth, as well as on the place of God in the Enlightenment.

Robert Josef Kozljanič has lectured at the TU Chemnitz and teaches at the Münchener Volkshochschule, Germany. Currently he is engaged in an inter-disciplinary project at the University of Mainz, taking a cultural-hermeneutical approach to the symbolic phenomenon of 'The Historico-Cultural Landscape of the Upper Middle Rhine Valley'. He is editor of the *Jahrbuch für Lebensphilosophie* (*Yearbook for the Philosophy of Life*), and his research focuses on phenomenological and hermeneutical philosophy, as well as cultural and religious studies. Selected publications include *Freundschaft mit der Natur: Naturphilosophische Praxis und Tiefenökologie* (Drachen Verlag, 2008), *Der Geist eines Ortes: Kulturgeschichte und Phänomenologie des Genius Loci*, 2 vols (Albunea Verlag, 2004), *Lebensphilosophie: Eine Einführung* (Kohlhammer, 2004), *Ernesto Grassi: Leben und Denken* (Fink, 2003), and *Kunst und Mythos* (Igel Verlag, 2001).

Susan Rowland is Professor of English and Jungian Studies at the University of Greenwich, UK. She is the author of four books on Jung, literary theory and gender, including *Jung as a Writer* (Routledge, 2005) and *Jung: A Feminist ReVision* (Polity, 2002). Her latest book, *C. G. Jung in the Humanities* (Routledge, 2010), looks at Jung, the arts, alchemy, science, myth, history, culture and religion. She is currently working on Jung and ecocriticism.

Robert A. Segal is Sixth Century Chair in Religious Studies at the University of Aberdeen, Scotland. Among the books he has written or edited are *The Poimandres as Myth* (de Gruyter 1986), *Joseph Campbell* (rev. edn, Plume, 1990), *The Gnostic Jung* (Routledge, 1992), *The Allure of Gnosticism* (Open Court, 1995), *Jung on Mythology* (Routledge, 1998), *The Myth and Ritual Theory* (Blackwell, 1998), *Theorizing About Myth* (University of Massachusetts Press, 1999), *Hero Myths* (Blackwell, 2000), *Myth: A Very Short Introduction* (OUP, 2004), *The Blackwell Companion to the Study of Religion* (Blackwell, 2006), and *Myth: Critical Concepts* (4 vols, Routledge, 2007).

Part I

Theorizing the archaic

Chapter 1

Introduction

A brief history of the archaic

Paul Bishop

In the beginning was . . .

The word 'archaic' derives from the Greek *arkhaios* (ἀρχαῖος), which in turn is related to the word ἀρχή (or *archē*), meaning 'principle', 'origin', or 'cause'.[1] As the word is used, it covers a wide range of meanings, from the political to the metaphysical.[2] For instance, *arkhaois* is used in Plato's *Laws* when Megillus refers to Homer's 'legendary account' which ascribes 'the primitive habits of the Cyclopes to their savagery', or when the Athenian stranger talks about 'an ancient tale, told of old'.[3] Then again, in *Theaetetus*, Socrates speaks about 'the ancients, who concealed their meaning from the multitude by their poetry',[4] while in *Cratylus* he refers to 'the earliest times' and to 'the ancient word'.[5] Elsewhere *arkhaois* acquires the sense of 'original' or 'primordial', as when Aristophanes tells Eryximachus in the *Symposium* that 'the way to bring happiness to our race is to give love its true fulfilment: let every one find his own favourite, and so revert to his *primal* estate'.[6] Conversely, in his *Politics* Aristotle uses the term in a negative sense, when he remarks that 'old customs are exceedingly simple and barbarous', and observes that 'the remains of ancient laws which have come down to us are quite absurd'.[7] In his *Rhetoric*, however, he notes that 'what is long established seems akin to what exists by nature'.[8] To clarify this ambiguity about Plato and Aristotle's understanding of the *archaic*, we need to examine more closely the term 'archē'.

The notion of the *archē* is found in the works of the earliest philosophers whose works have come down to us, usually referred to as the pre-Socratics. For Thales of Miletus (c. 625–c. 545 BCE), hailed by Aristotle as the founder of natural philosophy, the 'archē', the 'first principle' or 'material principle' of all things, is – water.[9] For Anaximander of Miletus (c. 610–c. 545 BCE), the 'archē' is the *apeiron*, the 'limitless' or the 'infinite'; according to the sixth-century neo-Platonist Simplicius of Cilicia, Anaximander was the first to introduce the term 'archē' in its technical sense, holding that 'it is neither water nor any of the so-called elements but some different infinite nature',[10] and the term is also found later in Philolaus of Croton (c. 470–c. 390 BCE), associated with the school of Pythagoras.[11] Behind the abstract concept of the archaic *apeiron*, the sole surviving fragment of

Anaximander conjures up something momentous and awe-inspiring: 'And the things from which existing things come into being are also the things into which they are destroyed, in accordance with what must be. For *they give justice and reparation to one another for their injustice in accordance with the arrangement of time*'.[12] So when Anaximander's younger contemporary, Anaximenes of Miletus (c. 585–c. 525 BCE) speaks of the 'archē' as air,[13] we might think of this less as the element we breathe and more as the life-giving breath of the soul.

If the school of Pythagoras of Samos (c. 570–c. 496 BCE), including Philolaus, pointed the way to a mathematization of the world through the importance attached to harmony and number, other thinkers were avowedly more mysterious. In the thought of Heraclitus of Ephesus (c. 540–c. 475 BCE), known as 'the Obscure', the 'archē' is not water, not air, but fire: fire, 'ever-living, kindling in measures and being extinguished in measures',[14] the universe itself being 'generated from fire and consumed in fire again, alternating in fixed periods throughout the whole of time':[15] a symbol of the principle of *becoming* that characterizes Heraclitean thought. Conversely, Parmenides of Elea (c. 540/515–c. 470/445 BCE) advanced as 'archē' the concept of *being*: evoked in his didactic poem 'The Way of Truth' in the fragment preserved by Simplicius:

> [. . .] Being, it is ungenerated and indestructible, whole, of one kind and unwavering, and complete. Nor was it, nor will it be, since now it is, all together, One, continuous. [. . .]
>
> And since there is a last limit, it is completed on all sides, like the bulk of a well-rounded ball, equal in every way from the middle. For it must not be at all greater or smaller here or there.
>
> For neither is there anything which is not, which might stop it from reaching its like, nor anything which is in such a way that it might be more here or less there than what it is, since it all is, inviolate. Therefore, equal to itself on all sides, it lies uniformly in its limits.[16]

Subsquent thinkers sought to mediate between the 'archē' as water, air, or fire, or the 'archē' as being or becoming. For Empedocles of Acragas (c. 495–c. 435 BCE), there are four 'archai' – earth, air, water, fire – held together by the twin powers of Love and Strife; for Anaxagoras of Clazomenae (c. 500–c. 428 BCE), an infinite number, called *chrēmata* ('primordial things') or *spermata* ('seeds'), held together by *nous* ('spirit'); while for Leucippus of Milet (c. 450 BCE) and Democritus of Abdera (c. 470–c. 370/360 BCE) there were only two things: the 'full' and the 'empty',[17] or 'atoms' and 'space'.[18] (For the atomists, what determined events was a spontaneous and indeterminate 'swerve' in the fall of the atoms – the 'clinamen', an idea taken up by Epicurus and, later, Lucretius.) If some Pythagoreans believed the specks of dust one sees floating in a beam of sunlight, others what moves the dust, to be souls, Democritus compared the motion of the atoms to 'the motes in the air which we see in shafts of light coming through windows',[19] and bound up with this materialist conception of the archaic-as-atoms-and-emptiness were two important ideas: the formulation of a law of causality ('Nothing happens in vain but everything for a reason and by necessity'),[20] and an ethics ('true happiness is the purpose of the soul', for 'happiness is procured by beautiful things').[21]

Among the early Greek thinkers, it was the Ionic natural philosophers of Miletus, the school of Pythagoras, and the Eleatic school – rather than the atomists, such as Leucippus and Democritus – who exercised the greatest influence on later Greek thought, represented by Aristotle and Plato. In the Platonic dialogue called *Parmenides*, Socrates engages directly with the thought of predecessors; in Plato's *Phaedo* and *Phaedrus*, 'archē' retains the more abstract sense of 'principle', and in the *Republic* it also means 'rule' or 'beginning'.[22]

In Aristotle's *Metaphysics*, no fewer than six different uses of 'archē' in the sense of 'beginning' are distinguished, of which the first defines the term as 'that part of a thing from which one would start first, for example, a line or a road has a beginning in either of the contrary directions' and the sixth as 'that from which a thing can first be known'.[23] His fourth definition understands 'archē' in a non-immanent sense as 'that from which a thing first comes to be, and from which the movement or the change naturally first begins, as a child comes from its father and its mother, and a fight from abusive language'; while, in the fifth definition, 'archē' is related not only to politics, but also to aesthetics, as 'that at whose will that which is moved is moved and that which changes changes, for example, the magistracies in cities, and oligarchies and monarchies and tyrannies, are called *archai* and so are the arts, and of these especially the architectonic arts'.[24] As we shall see, its relation to the aesthetic returns in more modern interpretations of the archaic, so Aristotle's observation that 'the nature of a thing is a beginning, and so is the element of a thing, and thought and will, and essence, and the final cause – for the good and the beautiful are the beginning both of the knowledge and of the movement of many things', is one we should remember.[25]

Elsewhere Aristotle refers to the pre-Socratic philosophers as the *arkhaioi*, and to the originative principles of pre-Socratic philosophy as *archai*;[26] in his *Physics*, he discusses whether the number of *archai* is two (identified as matter and form) or three (identified as matter, form, and privation);[27] and in his *Posterior Analytics*, he investigates the relation between the *archai* and science (*episteme*), positing each of the *archai* as ontologically and epistemology prior to the sciences, arguing that, in the case of some subjects, an individual 'archē' is specific to a particular *epistēmē*:

> I call the basic truths of every genus those elements in it the existence of which cannot be proved. As regards both of these primary truths and the attributes dependent on them the meaning of the name is assumed. The fact of their existence as regards the primary truths must be assumed; but it has to be proved of the remainder, the attributes. [. . .] Of the basic truths used in the demonstrative sciences some are peculiar to each science, and some common,

but common only in the sense of analogous, being of use only insofar as they fall within the genus constituting the province of the science in question.[28]

Thus, as Richard McKirahan observes, whilst 'archē' is, in one sense, a 'relative term', it is, in another sense, 'absolute', for 'a fact is basic whether or not anyone happens to recognize it as such'.[29]

Now Aristotle's conception of the *archai* was bequeathed to his successors, including the Stoics and the neo-Platonists. For the former, the dual *archai* of which Aristotle had spoken were identified as matter (or a passive principle) and God (considered as an active principle); while for the neo-Platonists, and in particular Plotinus (205–c. 269/70), the *archai* are (as in Plato and Aristotle) identified with the early philosophers or 'the ancients'.[30] Nevertheless, in Plotinus the 'archē' acquires the specific meaning of the 'original' form of the soul,[31] 'the nobility and the ancient privilege of the soul's essential being',[32] and 'the ancient staple of the soul'.[33]

At the same time, in Plotinus the 'archē' acquires a new and cosmic dimension as the One. According to Parmenides' account of the One in the Platonic dialogue that bears his name, 'the One is both all things and nothing whatsoever, alike with reference to itself and to the others'.[34] Correspondingly, for Plotinus 'the One is all things and no one of them', and 'it is precisely because there is nothing within the One that all things are from it: in order that being may be brought about, the source must be no being but being's generator, in what is to be thought of as the primal act of generation'.[35] Of the three first principles or *archai*, also known as hypostases – namely, the One, the Intellect, and the Soul – it is the One, the 'archē', that is both the source of human beings and their goal, their *telos*, towards which they (and, indeed, all things) strive.[36]

In Plotinus we can see that the archaic, in the sense of 'archē', has taken on mysterious, even mystic, overtones. But the archaic had already begun to acquire the sense of a special atmosphere towards the end of the period of classical Rome, before the pagan cults were eclipsed by Christianity, which would fuse the One with the monotheistic deity of Judaism. For instance, in a passage in his *Metamorphoses* – discussed in further detail by Robert J. Kozljanič in his contribution to this volume – Ovid (43 BCE–17 CE) describes a physical landscape in terms that evoke its timeless and its numinous (and, in this sense, its archaic) atmosphere:

> There was a valley clothed in hanging woods
> Of pine and cypress, named Gargaphie,
> Sacred to chaste Diana, huntress queen.
> Deep in its farthest combe, framed by the woods,
> A cave lay hid, not fashioned by man's art,
> But nature's talent copied artistry,
> For in the living limestone she had carved

A natural arch; and there a limpid spring
Flowed lightly babbling with a grassy sward.
Here, tired after the hunt, the goddess loved
Her nymphs to bathe her with the water's balm.[37]

Then again, in one of the leading representatives of late Stoicism, the philosopher Seneca (c. 4 BCE–65 CE), who lived on the very cusp of the pagan and the Christian eras, sought to persuade his correspondent, Lucilius, of the existence of divinity by evoking the sense of its archaic – in the sense of 'originary', 'primordial', or, again, 'numinous' – presence in nature:

If you're confronted by some dense grove of aged and giant trees shutting out every glimpse of sky with screen upon screen of branches, the towering stems, the solitude, the sense of strangeness in a dusk so deep and unbroken where no roof is, will make deity real to you. Again, the cavern that holds a hill-side poised on its deep-tunnelled galleries of rock, hollowed into that roomy vastness by nature's tools, not man's, will strike some hint of sanctity into your soul. We render homage to the sources of great rivers. The vast stream that bursts suddenly from hiding has its altars; hot springs are worshipped; there are lakes hallowed by their dark inscrutable waters and unplumbed depths.[38]

As Kozljanič suggests, what Seneca is talking about here is what, some nineteen centuries later, Rudolf Otto (1869–1937) formulated as 'the sacred' (*das Heilige*), the sense of 'the holy' with its different aspects of the numinous, of *mysterium tremendum*, and of *fascinosum*.[39] The response to the archaic is typically described as a 'shudder'; in the Mothers scene in *Faust*, Part Two, one finds the famous lines, when Faust tells Mephistopheles:

Yet not in torpor would I comfort find;
Awe is the finest portion of mankind;
However scarce the world may make this sense –
In awe one feels profoundly the immense.

Doch im Erstarren such ich nicht mein Heil,
Das Schaudern ist der Menschheit bestes Teil;
Wie auch die Welt ihm das Gefühl verteure,
Ergriffen, fühlt er tief das Ungeheure.[40]

According to the French intellectual historian Pierre Hadot, what Goethe expresses is something new, the expression of a specifically modern sensibility that one subsquently finds in Schelling, Nietzsche, Rilke, and Heidegger, but not earlier, in Plato, or Epicurus, or the Stoics, or Spinoa.[41] Yet this *frisson*, this *nuance d'angoisse*, as Hadot describes it, does appear to date back at least as far as Seneca.

If the concept of the archaic intersects with the sense of the numinous or the *mysterium tremendum* that Otto called 'the sacred', it is in the writings of the Gnostics, chronologically contemporary with the neo-Platonism of Plotinus, that the archaic acquires an overtly religious dimension. Within the cosmology of Valentinus, for instance, the 'archē' becomes synonymous with *abyssus*, or *vorago*, or βυθός:

> Within invisible and unnameable heights there was – they say – a preexistent, perfect eternity; this they call also prior source, ancestor, and the deep [= προαχην, προπατορα, βυθον χαλουσιν; *proachen, propatorem*, Bythum *vocant*]. And it existed uncontained, invisible, everlasting, and unengendered. Within infinite eternal realms it was in great stillness and rest. And with it coexisted thought, which they also call loveliness [Χαριν] and silence [Σιγην]. And eventually the aforementioned deep took thought to emit a source of the entirety. And it deposited this emanation that it had thought to emit, like sperm, in the womb of the silence that coexisted with it. And the latter received this sperm, conceived, and brought forth intellect, which was like and equal to its emitter and was the only being that comprehended the magnitude of its parent. And this intellect they call also only-begotten, parent, and source of the entirety.[42]

Now, according to Arthur Schopenhauer (1788–1860), the source of mystical thought – specifically, Jacob Boehme in his *Gründliches Bericht vom irdischen und himmlischen Mysterio* (see below) – was indeed Valentinian Gnosticism; for here, as he put it, the *Ungrund* was 'at home'.[43] And other commentators have discerned a clear link between, on the one hand, the neo-Platonic concept of the archaic (the 'archē' as the One) and, on the other, its sense of 'beginning' or 'origin', which developed in the medieval period into the mystical concept of the *origo* (Latin) or the *ursprunc* (Middle High German), now referred to in Modern German (and particularly its philosophical tradition) as the *Ursprung*.[44]

The archaic as mystical 'ground' or 'abyss'

The concept of the 'ground' (*Grund*) or its absence, an 'abyss' (*Ungrund*), informs the mysterious (and strangely moving) texts of the medieval thinkers, fascinating to Hegel and Rudolf Steiner alike: the mystics.[45] In one of his sermons, one of the greatest mystics of the German Middle Ages, Meister Eckhart (c. 1260–1327), describes God as 'the one natural spring' (*éin natiurlîcher ursprunc*), out of which 'the single natural outlet of the Son' (*éin natiurlîcher ûzvluz des sunes*) proceeds.[46] Just as the Son represents an 'outflow' from God the Father, so God flows into the soul of the believer, there – in the soul – to be (re)born, a process equated by Eckhart with 'the light of grace'.[47] Set against his crucial distinction between God (*Gott*) and divinity (*Gottheit*), Eckhart defines divinity as the 'ungrounded ground' (*grunt ane grunt*), the 'originator without origin' (*principium sine principio*).[48] On

Adolf Lasson's account of Meister Eckhart, the 'ungrounded ground' of divinity is like – or actually *is* – a 'silent desert' (*stille Wüste*) or an 'impenetrable "darkness" ' (*Finsternis*).[49] In his sermon on a text from the Book of Wisdom, – 'For while all things were in quiet silence, and the night was in the midst of her course, thy Almighty word leapt down from heaven from thy royal throne'[50] – Eckhart comments that this silence is 'in the purest thing that the soul is capable of, in the noblest part, the ground [*grund*] – indeed, in the very essence of the soul which is the soul's most secret part', which he terms 'the silent "middle" '; and it is here, in the 'silent "middle" ', the 'ground of the soul', that God 'enters with his all'.[51] For Eckhart, this is the moment when 'God is known by God in the soul', and 'with this Wisdom [. . .] she knows the power of the Father in fruitful travail, and essential Being [*ur-sein*] in simple unity void of all distinctions';[52] this is the true moment of incarnation: 'When the word speaks in the soul and the soul replies in the living Word, then the Son is alive in the soul'.[53]

A similar vocabulary and a similar set of ideas can be found in the later figure of Johannes Tauler (c. 1300–1361) who, in a sermon on Pentecost, contrasts the response to the outpouring of the Holy Spirit of the foolish and the wise man. The former enjoys the comfort of the Paraclete, but by falling in love with this enjoyment, he loses the 'true ground' (*dem woren grunde*). The wise man, however, gives himself up utterly to the 'origin' (*ursprung*), he achieves 'enlightened purification' (*verklerte lúterunge*), and sees only God.[54] On the other hand, for Tauler God Himself is a form of abandonment, He is the 'divine desolation' of which Tauler speaks in highly mystic terms in a sermon on the hidden nature of God: 'Then Man can see the quality of divine isolation in silent emptiness, in which no word in the essence is ever spoken in an essential way, nor is any work ever done. For there everything is so silent, so secret, and so empty' (*Denne mag der mensche an sehen die eigenschaft der goetlichen wuestunge in der stillen einsamkeit, do nie wort in dem wesende nach weselicher wise inne gesprochen enwart noch werk gewúrkt enwart; denne do ist es so stille, so heimelich und so wuest*).[55] And another contemporary of Eckhart and Tauler, Heinrich Seuse (or Suso) (c. 1293–1366), depicted God in *The Little Book of Eternal Wisdom* as telling his servant that 'the highest path of all beings away from their first origin is taken (in accordance with the natural order) through the noblest beings to the lowest; but the return to the origin is taken from the lowest to the highest', and that 'no one should seek the groundless ground of my secrecy and my hiddenness, in which I arrange all things according to my eternal providence, for no one can understand it.'[56]

Over two centuries later, in the mature mysticism of Jacob Boehme (1575–1624), known as the *theosophus teutonicus* ('Teutonic theosopher') or (in deference to his trade) the Shoemaker of Görlitz, the notion of the 'groundless ground', the *Ungrund*, is explored in its fullest theological implications.[57] In *De incarnatione verbi, or Of the Incarnation of Jesus Christ* (1620), Boehme writes:

> In eternity, as in the *Ungrund* outwith nature, there is nothing other than a silence without being; there is nothing to give rise to anything, it is an eternal

silence, and there is nothing like it, an *Ungrund* without beginning or end: there is no goal nor resting-place, nor any searching or finding, or anything that could be possibly be. [*In der Ewigkeit, als im Ungrund ausser der Natur, ist nichts als eine Stille ohne Wesen; es hat auch nichts, das etwas gebe, es ist eine ewige Ruhe, und keine Gleiche, ein Ungrund ohne Anfang und Ende: Es ist auch kein Ziel noch Staette, auch kein Suchen oder Finden, oder etwas, da eine Moeglichkeit waere.*][58]

And Boehme has a striking image for this *Ungrund* (which may, in turn, have had an influence on Schopenhauer):[59]

This *Ungrund* is like an eye, for it is its own mirror, it has no being, neither light nor darkness, and it is above all a *magia*, and has a will after which we should neither strive nor enquire, for it disturbs us. With the same will we understand the ground of the divinity, which has no origin, for it contains itself in itself, about which we ought properly remain silent; for it is outwith nature. [*Derselbe Ungrund ist gleich einem Auge, denn er ist sein eigener Spiegel, er hat kein Wesen, auch weder Licht noch Finsterniß, und ist vornehmlich eine* Magia, *und hat einen Willen, nach welchem wir nicht trachten noch forschen sollen, denn es turbiret uns. Mit demselben Willen verstehen wir den Grund der Gottheit, welcher keines Ursprungs ist, denn er fasset sich selber in sich, daram wir billig stumm sind; denn er ist ausser der Natur.*][60]

Boehme's subtle, even paradoxical thought regards the ground of divinity as having no origin, as being nothing, yet being full of desire, or so he explains in *Mysterium Pansophicum, Or Complete Account of the Earthly and Heavenly Mysteries* (1620):

The *Ungrund* is an eternal nothing, but it creates an eternal beginning as a desire. For the nothing is a desire for something: and since there is nothing that could give rise to something, the desire is itself what gives rise to something that is nothing other than a longing desire. [*Der Ungrund ist ein ewig Nichts, und machet aber einem ewigen Anfang, als eine Sucht; Dann das Nichts ist eine Sucht nach Etwas: Und da doch auch Nichts ist, das Etwas gebe; sondern die Sucht ist selber das Geben dessen, das doch auch ein Nichts ist, als blos eine begehrende Sucht.*][61]

However aw(e)fully abstract Boehme's thinking may appear, its source lay (or so it is said) in Boehme's sensory experience of sunlight – not, like Democritus, watching the particles of dust at play in it, but observing its reflection on a metal dish: 'Sitting one day in his room, his eye fell upon a burnished pewter dish, which reflected the sunshine with such marvellous splendour that he fell into an inward ecstasy, and it seemed to him as if he could now look into the principles and deepest foundations of things'; believing it to be 'only a fancy, and in order to banish it from his mind, he went out upon the green', but 'he remarked that he gazed into the very heart of things, the very herbs and grass, and that actual nature harmonized with what he had inwardly seen'.[62]

According to Schopenhauer, the gnostic thought of Valentinus that had re-emerged in Boehme was subsequently transmitted to the Dutch philosopher Benedict Spinoza (1632–1677),[63] for whom God was the 'cause of itself' (*causa sui*) as well as 'the efficient cause of all things' (*Deum omnium rerum esse*);[64] in other words, an entirely self-grounded being. But Boehme, Schopenhauer believed, also acted as a source of gnostic thoughts for such 'neo-Spinozists' (as he termed them) as Friedrich Wilhelm Joseph von Schelling (1775–1854),[65] who spoke of 'that within God which is not *God himself*' (*was in Gott selbst nicht Er Selbst ist*),[66] but is the ground of God – 'the "primal ground" or, rather, the "groundless" ' (*als den Urgrund oder vielmehr* Ungrund).[67]

The Romantic archaic

There can be no doubt about Schelling's admiration for Jacob Boehme. In his lectures *On the History of Recent Philosophy* (1833/1834) Schelling hailed Boehme as 'the most remarkable individual' among all the theosophists,[68] and in *Introduction to the Philosophy of Revelation* (1842/1843) he described Boehme as 'a miracle in the history of humankind, and particularly in the history of the German mind'.[69] As Charles Bambach discusses in greater detail in his contribution in this volume, the sense in which some believed there to be a specifically German dimension to the concept of the archaic forms an important aspect of its history.[70] In *Philosophical Inquiries into the Nature of Human Freedom* (1809), Schelling defined the 'ungrounded', or the *Ungrund*, as follows:

> As it [i.e., *der Ungrund*] precedes all antitheses these cannot be distinguishable in it or be present in any way at all. It cannot then be called the identity of both, but only the absolute indifference as to both. [. . .] The essence of the basis, or of existence, can only be precedent to all basis, that is, the absolute viewed directly, the groundless. [*Das Wesen des Grundes, wie das des Existierenden, kann nur das vor allem Gründe Vorhergehende sein, also das schlechthin betrachtete Absolute, der Ungrund.*][71]

In the course of what Schelling called 'dialectical exposition' (*dialektische Erörterung*), he related the 'ungrounded' – the *Ungrund* – to the very source of life itself, in its most concrete (individual and interpersonal) aspects:

> Indifference is not a product of antitheses, nor are they implicitly contained in it, but it is a unique being, apart from all antitheses, in which all distinctions break up. It is naught else than just their non-being, and therefore has no predicates except lack of predicates, without its being naught or a non-entity.

[. . .] But, as has been shown, it [i.e., *der Ungrund*] cannot be this [i.e., *das Absolute*] in any other way than by dividing into two equally eternal beginnings, not that it is both *at the same time* but that it is in both *in the same way*, as the whole in each, or a unique essence. But the groundless divides itself into the two equally eternal beginnings only in order that the two which could not be in it as groundless at the same time, or there be one, should become one through love; that is, it divides itself only that there may be life and love and personal existence. [*Der Ungrund theilt sich aber in die zwei gleich ewigen Anfänge, nur damit die zwei, die in ihm, als Ungrund, nicht zugleich oder Eines sein konnten, durch Liebe eins werden, d.h. er teilt sich nur, damit Leben und Lieben sei und persönliche Existenz.*][72]

'This is a real masterpiece', quipped Schopenhauer, dismissively;[73] but the category of the 'ungrounded ground', the *Ungrund*, in this text from Schelling's middle period can be related to a much later text, entitled 'Another Deduction of the Principles of Positive Philosophy', based on a lecture held in Munich in 1839. As his starting point Schelling takes 'what unconditionally exists, prior to all thought' (*das allem Denken zuvor, das unbedingt Existirende*), and tries to think it.[74] Is this unconditional being or existence, by virtue of being prior to everything, really 'the monad' (*die Monas*), in other words, 'what remains, the principle that stands over everything' (*das Bleibende, das über allem stehende Princip*)? What is the status of this 'unprethinkable being' (*das unvordenkliche Seyn*)? And does this 'unprethinkable being' in no way 'allow for an opposite by which it could be changed and against which it could consequently be considered contingent'?[75] In his discussion Schelling, following the mystical distinction between God and divinity, disassociates this 'unprethinkable being' from God, describing it as 'a being which, however early we arrive, is already there' (*das Seyn, das, so früh wir kommen, schon da ist*).[76] The archaic, one might say, is, to borrow postmodern parlance (see below), that which is *always already there*. Ultimately, Schelling's philosophy must be seen in the context of the response to the transcendental philosophy of Kant offered by German Idealism as a whole; a context which forms the background to Alan Cardew's contribution in this collection of papers.

In the critical philosophy, Immanuel Kant (1724–1804) had sought to shift the entire problem of the *archē*, in the sense of 'origin' or 'ground', away from the realm of transcendence (whether real or apparent, i.e., whether ontological or epistemological/historical) and into the sphere of what he called the 'transcendental', i.e., belonging to knowledge that is 'occupied not so much with objects but rather with our *a priori* concepts of objects in general'.[77] On the one hand, he interprets the *archē* or *Ursprung* in a (psycho- or anthropo)logical sense, interrogating in the *Critique of Pure Reason* the source, the origin, the foundations of our knowledge. On the other hand, he remains alert to the (potentially) metaphysical dimensions of such problems as the origin of evil,[78] or the very idea of the origin itself, i.e., the *Ens originarium*,[79] an 'absolutely necessary being' (*schlechterdings notwendigen Wesens*),[80] or God, whether defined as 'the highest being' (*das höchste Wesen*) and as 'the original ground of all things' (*Urgrund aller Dinge*),[81] as the original ground of nature (*Urgrund der Natur*),[82] or as 'the original ground of the world' (*Urgrund der Welt*) as 'the sum total of all objects of experience'.[83]

Kant was not alone in interpreting the *archē* or *Ursprung* in a psychological or anthropological sense, as shown by the titles of works by many eighteenth-century thinkers, such as Rousseau's *Discours sur l'origine et les fondements de l'inégalité parmi les hommes* (1755), Edmund Burke's *A Philosophical Inquiry into the Origin of our Ideas of the Sublime and the Beautiful* (1757), Condillac's *Essai sur l'origine des connaissances humaines* (1746), or Herder's *Über den Ursprung der Sprache* (1772).[84] Nor was he alone in establishing a critical philosophy; indeed, Kant was – as Tobias Trappe has argued –[85] surpassed in his ambitions (at least, in the eyes of its author) by the *Fundamental Critique of Pure Understanding*, published in 1788 – or, to give it its full title, *Attempt at Enlightenment of the Optics of the Eternal Light of Nature, From the First Ground of All Grounds, To the Most Profound Fundamental Critique of Pure Understanding* (*Aufklärungs-Versuch der Optik des ewigen Natur-Lichts bis auf den ersten Grund aller Gründe, zur tieffsten Grund-Critik des reinen Verstandes*). Its author was Jacob Hermann Obereit (1725–1798), a writer, philosopher, and medical doctor, who belongs, like Franz Paul von Herbert (1759–1811), Johann Benjamin Erhard (1766–1827), or Friedrich Carl Forberg (1770–1848), to the forgotten figures of German Romanticism, although he is a member, along with Johann Gottfried Herder (1744–1803), of an older generation of Kant critics.[86] Obereit developed a philo- sophy of the Universal Absolute, of the Universal Ground (*Allgrund*), that was not just an *Allgrund* but an *Urgrund*, an *Urmittel*, an *Urzweck*. Beginning with an *archē*, with 'the simplest of all beings, the eternal itself, something essentially and infinitely free, something entirely sufficient in itself' (*vom simpelsten aller Wesen, dem Ewigen von selbst, dem wesentlich unendlich Freien, in sich selbst allein Allgenugsamen*), Obereit transposed the implicitly Trinitarian structure of St Paul's letter to the Romans into the intellectual discourse of German Idealism: 'Of Him, and through Him, and to Him, are all things',[87] and thus all things are 'of Him, as the highest, freely efficient primal ground [*Urgrund*]; through Him, as the highest, freely active primal means [*Urmittel*]; to Him, as the highest, freely fulfulling primal purpose [*Urzweck*]'.[88] The *Urgrund*, or *Urmittel*, or *Urzweck*, is 'the single, absolute *Allgrund*', it is:

> The pure Being-There of what is eternally independent of itself, the first ground, just as it is the final goal *of all thought*, what is absolutely positive in itself, of, through, for itself, the single ground of *all that can be thought*, the final endpoint *of all that can be represented*, the single primal ground of unlimited affirmation, in and of which no negation is possible, the single goal of absolute agreement, foundation, sufficiency without limits, thus the single, absolute foundational and teleological object above everything [. . .] the Absolute in itself, the Infinite itself with the essential law and with freedom given *a priori, per se* [. . .] the One, the eternal resting-point, *the*

unconditioned condition of everything, the eternal beginning of everything, the primal ground [. . .].[89]

For Johann Gottlieb Fichte (1762–1814), the critical project – as proposed by Kant, rather than by Obereit, that is – was in itself insufficient, requiring supplementation in the form, first, of a fourth (proto-)critique, the *Attempt at a Critique of All Revelation* (*Versuch einer Kritik aller Offenbarung*) (1792; 1793); second, of a doctrine of science, or a science-of-science, a *Wissenschaftslehre*, that Fichte presented in numerous versions; and third, of such concepts as the transcendental ego (*tranzendentales Ich*).[90] For Fichte, the very possibility of knowledge – and at the same time, he appears to argue, of existence – is bound up with what he termed 'the interiority of the origin' (*die Innerlichkeit des Ursprunges*), or 'the inwardness of the *archē*', which underpins the possibility of 'pure knowledge' (*reines Wissen*):

> Pure knowledge thought of as an *origin* for itself, and its opposite as the nonbeing of knowledge, for otherwise it could not arise, is *pure being*. [*Das reine Wissen gedacht, als* Ursprung *für sich, und seinen Gegensatz als Nichtseyn des Wissens, weil es sonst nicht entspringen könnte, ist* reines Seyn.] (Or let us say, if we want to understand properly, the absolute creation, as the act of creation, not as something created, is the standpoint of absolute knowledge; the latter creates itself out of its pure possibility, as what alone is its precondition, and precisely this is pure being.)[91]

Such 'pure thought' is self-originating thought, it is thought that takes place in an act of unlimited freedom, and yet this thought is, for Fichte, intimately bound up with our very act of being:

> The origin is in itself an absolute origin, starting from which and beyond which it is impossible to go further – or so we said [*Der Ursprung ist für sich ein absoluter, aus dem und über welchen nicht hinausgegangen werden kann – sagten wir*]. In its being-for-itself it would be unchanging, and yet it [i.e., the origin] presupposes it [i.e., being-in-itself]. But it [i.e., the origin] is not in it [i.e., being-for-itself], except inasmuch as it is accomplished with absolute formal freedom (which, as we know, can either be present or not); it [i.e., the origin] cannot be perceived, unless it makes itself perceptible; it does not make itself [perceptible] without precisely being perceived: which distinction, between the subject and the object, must strictly speaking be abolished in the unity of the subject – in other words, in the interiority of the origin; – and it [i.e., the origin] is not perceived, except inasmuch as this freedom as such is for-itself, and perceived as arising within itself (accomplishing itself) [*– welcher Unterschied des Subjects und Objects hier jedoch der Strenge nach zu einer Einheit des Subjects vernichtet werden muss – eben zu einer Innerlichkeit des Ursprunges; – und er ist nicht angeschaut, ausser inwiefern*

diese Freiheit als solche eben selbst für sich ist, als in sich entspringend (sich vollziehend) angeschaut wird].[92]

Thus in German Idealism, at least of the Fichtean variety, the notion of the 'archē' as origin is deployed as a central concept, hard as it is (and Fichte realized this) to grasp.[93] The primary innovation in Fichte's earlier statements of his *Wissenschaftslehre* is the notion of 'intellectual intuition' (*intellektuelle Anschauung*), the supreme philosophical (as well as the defining human) act: 'the immediate consciousness that I act and of what I do when I act: it is because of this that it is possible for me to know something because I do it' (*das unmittelbare Bewusstseyn, dass ich handle, und was ich handle: sie ist das, wodurch ich etwas weiss, weil es ich thue*).[94] According to Fichte, intellectual intuition is 'something everyone has to discover immediately within himself' (*jeder muss es unmittelbar in sich selbst finden*), for it is 'not something that can be demonstrated by means of concepts' (*es [. . .] lässt sich nicht durch Begriff demonstriren*), yet 'it contains within itself the source of life, and apart from it there is nothing but death' (*in ihr ist die Quelle des Lebens, und ohne sie ist der Tod*).[95]

Intellectual intuition – this scintillating concept, despite or precisely because its possibility had been rejected by Kant, had also caught the attention of the young Schelling, who defined this form of intuition as an 'originating action' (*ursprüngliches Handeln*),[96] a mode of perception (*Anschauung*) that is simultaneously an act (*Handlung*), by means of which there arises 'pure self-consciousness' (*ein reines SelbstBewusstseyn*).[97] Over and against this use of the term in a precise, Fichtean philosophical sense, however, intellectual intuition was also described by Schelling in a far more mystical way. In his *Philosophical Letters on Dogmatism and Criticism* (1795), where much of the argument is conducted with Spinozism in the background,[98] Schelling talks about how 'in us all there dwells a secret, wonderful ability to withdraw from the changes and chances of time into our innermost self, stripped of everything that has attached to it from outside, and there, under the form of unchangeability, to contemplate eternity within us' (*uns allen nämlich wohnt ein geheimes, wunderbares Vermögen bei, uns aus dem Wechsel der Zeit in unser innerstes, von allem, was von außenher hinzukam, entkleidetes Selbst zurückzu-ziehen, und da unter der Form der Unwandelbarkeit das Ewige in uns anzuschauen*).[99] (Or, as Spinoza had put it, 'we feel and know by experience that we are eternal'.)[100]

Later, in his lectures on the *Philosophy of Revelation* (that is, on the same topic that had launched Fichte's career back in 1792), Schelling remarked that 'Fichte gave the first impetus to return to the real *archais* [*Fichte gab den ersten Anlaß, zu den wirklichen* archais *wieder zu gelangen*], when he opposed I and not-I, an opposition that obviously includes more than thought and extension. But inasmuch as Fichte understood by the I nothing other than the human I, which is already something extremely concrete, it could not be said that in this distinction there lay a true principle of being. But it showed the path to it. It was

Naturphilosophie that uncovered the pure *archas* [*Die Naturphilosophie kam zuerst wieder auf die reinen* archas)'.[101]

In his *Philosophical Inquiries into the Nature of Human Freedom* of 1809, Schelling, no doubt because of the influence of Boehme,[102] abandoned the terminology of intellectual intuition and spoke instead of the foundational act *per se* as 'the longing which the eternal One feels to give birth to itself' (*die Sehnsucht, die das ewige Eine empfindet, sich selbst zu gebären*),[103] and he developed a neo-Gnostic, ontological drama of depth and revelation, light and darkness, Gott and *Ungrund*:

> Following the eternal act of self-revelation, the world as we now behold it is all rule, order and form; but the unruly lies ever in the depths as though it might again break through, and order and form nowhere appear to have been original [*als wären Ordnung und Form das Ursprüngliche*], but it seems as though what had initially been unruly had been brought to order. This is the incomprehensible basis of reality in things, the irreducible remainder which cannot be resolved into reason by the greatest exertion but always remain in the depths [*Dieses ist an den Dingen die unergreifliche Basis der Realität, der nie aufgehende Rest, das, was sich mit der größten Anstrengung nicht in Verstand auflösen läßt, sondern ewig im Grunde bleibt. Aus diesem Verstandlosen ist im eigentlichen Sinne der Verstand geboren*]. Out of this which is unreasonable, reason in the true sense is born. Without this preceding gloom, creation would have no reality; darkness is its necessary heritage. Only God – the Existant himself – dwells in pure light; for he alone is self-born [*Ohne dies vorausgehende Dunkel gibt es keine Realität der Kreatur; Finsternis ist ihr notwendiges Erbteil. Gott allein – Er selbst der Existierende – wohnt im reinen Lichte, denn er allein ist von sich selbst*].[104]

In one of his mature works preceding the *Philosophical Inquiries*, however, his *System of Transcendental Idealism* (1800), Schelling had redefined 'intellectual intuition' as 'aesthetic intuition'; or, more precisely, 'aesthetic intuition is precisely intellectual intuition become objective' (*die ästhetische Anschauung ist die objektiv gewordene intellektuelle*).[105] Thus 'intellectual intuition' becomes mystical, but also aesthetic, although already Fichte, in his *Foundations of Natural Law according to Principles of the Wissenschaftslehre* (1796), had hinted at its proto-aesthetic dimension:

> Let me draw attention to the fact that the eye itself and for the human being is no mere dead, passive mirror, like the surface of calm water, artificial mirrors, or the eye of an animal. It is a powerful organ, which of its own accord encapsulates, sketches, and reproduces the tree; which of its own accord marks out the shape that is to emerge from the raw marble or on the canvas, before the chisel or the brush is even touched [*Ich mache darauf aufmerksam, dass das Auge selbst und an sich dem Menschen nicht bloss ein*

todter, leidender Spiegel ist, wie die Fläche des ruhenden Wassers, durch Kunst verfertigte Spiegel, oder das Thierauge. Es ist ein mächtiges Organ, das selbstthätig die Gestalt im Baume umläuft, abreisst, nachbildet; das selbst- thätig die Figur, welche aus dem rohen Marmor hervorgehen, oder auf die Leinwand geworden worden soll, vorzeichnet, ehe der Meissel, oder der Pinsel berührt ist]. [. . .] Through this life and movement unto infinity of its individual parts among themselves, what is earth-bound of their material is, as it were, sloughed off and cast out, the eye becomes transfigured into light, and becomes a visible soul [*Durch dieses Leben und Weben der Theile unter einander ins Unendliche, wird das, was sie irdisches vom Stoffe an sich hatten, gleichsam abgestreift und ausgeworfen, das Auge verklärt sich selbst zum Lichte, und wird eine sichtbare Seele*].[106]

Not for the first time, then, but following Aristotle, we see that, in the German Idealism of Fichte and Schelling, the *archē* (as the *Ursprung*) becomes associated *qua* foundational act with the aesthetic. Indeed, one wonders whether it is precisely the archaic that illustrates how the mystic is, in fact, the aesthetic. Only a few years after Schelling's death, in a note dating from the mid 1880s, Friedrich Nietzsche (1844–1900) was to speak of the world itself as an aesthetic product – 'the world as a work of art that gives birth to itself' (*die Welt als ein sich selbst- gebärendes Kunstwerk*).[107] And for Nietzsche, it was the world's status as a work of art that justified it: the justification of the world as an aesthetic phenomenon.[108] *Thus the following question mark hangs over the archaic: is it ultimately what offers us salvation? Or is it that from which we need to be saved? Or is it both?*

Nietzsche and the archaic

Nietzsche's approach to the archaic – both in the sense of the primordial, the ancient, the essence of antiquity, as well as its more specific, technical sense of the *archē* – is shot through with ambiguity. It is in Nietzsche that one first finds a radical scepticism towards the idea of the origin (which has itself become a given, a *sine qua non*, indeed a shibboleth of postmodernity), as well as one of the most significant advocates of the archaic – prophesying, heralding, and in a sense inau- gurating its return. Both a classical philologist, enanmoured of the world of the pre-Socratics and Greek tragedy, and a post-Kantian and post-Hegelian (as well as, more importantly, a post-Feuerbachian and neo-Stirnerian) philosopher, Nietzsche is uniquely placed to demonstrate the questionableness and the rele- vance alike of the archaic; hence it should come as no surprise that he occupies such an important place in Charles Bambach's account in this volume of the history of the archaic in German thought.

On the sceptical side of the Nietzschean equation, we should turn to the works that may be conveniently categorized as belonging to his 'middle period'.[109] In the very first section that opens the first volume of *Human, All Too Human* (1878) Nietzsche pours scorn on the notion of the origin or *Ursprung*, regarding it as a chief failing of 'metaphysical philosophy' that, confronted with the question: 'how can something originate in its opposite, for example, rationality in irrationality, the sentient in the dead, logic in unlogic, disinterested contemplation in covetous desire, living for others in egocentrism, truth in error?', it 'den[ies] that the one originates in the other and assum[es] for the more highly valued thing a miraculous source [*Wunder-Ursprung*] in the very kernel and being of the "thing in itself" '.[110] Subsequently, in an aphorism in *Daybreak* (1881) entitled 'Origin and significance', Nietzsche notes that – '*formerly*' – it was always 'presupposed' that 'the *salvation* of man must depend on *insight into the origin of things*', but that – 'now' – 'the more insight we possess into an origin the less significant does the origin appear', whereas '*what is nearest to us*, what is around us and in us, gradually begins to display colours and beauties and enigmas and riches of significance of which earlier mankind had not an inkling'.[111] Yet, for all that, in his view, 'mankind likes to put questions of origin [*Herkunft*] and beginnings [*Anfänge*] out of its mind', and although he wonders whether one must not 'be almost inhuman to detect in oneself a contrary inclination',[112] Nietzsche admits that 'the question of where our good and evil really *originated*' was a problem that pursued him as a child, and that his 'ideas on the *origin* [*Herkunft*] of our moral prejudices' lie at the heart of philosophical concerns.[113]

By contrast, and despite the fact that Nietzsche, as he tells us, soon ceased 'to look for the origin of evil *behind* the world', his early work had in fact been profoundly concerned with the question of origins: the question of the 'birth of tragedy' revolves around nothing less than the origin of art, indeed, the origin of the phenomenal world as appearance (*Schein*). 'In Dionysian art and its tragic symbolism', he wrote in 1872, nature calls out to us: 'Be as I am! Amid the ceaseless flux of phenomena I am the eternally creative primordial mother [*Urmutter*]!'[114] In turn, 'the *tragic myth* [. . .] leads the world of phenomena to its limits where it denies itself and seeks to flee back again into the womb of the true and only reality', and as the 'vast Dionysian impulse' of the 'tragic artist' 'devours his entire world of phenomena', so we 'sense beyond it, and through its destruction, the highest artistic primal joy [*Urfreude*], in the bosom of the primordially One [*im Schoße des Ur-Einen*]'.[115] And in a sketch for an expanded version of *The Birth of Tragedy*, Nietzsche wrote that humankind is nothing less than 'the continual birth of genius', in that 'from the perspective of that monstrous, ever-present viewpoint of the Primal One genius is in every moment attained, the entire pyramid of appearance is perfect to its tip'.[116] In particular, what he calls 'Dionysian genius' is defined as 'the human being – become, in utter self-forgetting, one with the primal ground [*Urgrunde*] of the world – who now creates from its primal pain [*Urschmerze*] the reflection of the same to the end of its redemption'.[117]

In reality, of course, Nietzsche's position is not the contradiction it (may) initially appear(s) to be. For whilst attacking metaphysics in general, Nietzsche was nevertheless a champion of what he called an *Artisten-Metaphysik*: a 'metaphysics of the artist', as he put it in his 'Attempt at a Self-Criticism' in the third edition of *The Birth of Tragedy* in 1886.[118] One of the chief ideas associated with his mature philosophy, the doctrine of the eternal recurrence of all things, links his earlier philological concerns with the pre-Socratics and the function of culture with his later ambition to inaugurate a new era precisely through a return not to the actuality, but to the values of the past. 'Every day we are becoming *more and more Greek*', he wrote in 1885: 'To begin with, as is proper, in our concepts and in our value judgements [. . .]: but at some stage, one hopes, also with our *body!* Here lies (and here has always lain) my hope for the Germans!'[119] Nietzsche's scepticism about origins in his middle period, rather than his concern with the *Urgrund* of tragedy in his earlier writings, were taken up by the many thinkers that wrote in his wake. And for this reason we must turn in our brief survey of the history of the notion of the archaic to such critics and thinkers of the concept as Walter Benjamin (1892–1940), Theodor W. Adorno (1903–1969), and Michel Foucault (1926–1984).

After(-) Nietzsche: Benjamin, Adorno, Foucault, Derrida

Nietzsche's first major publication was *The Birth of Tragedy*, published in 1872, which was greeted with incomprehension and controversy; it might even be said to mark the beginning of his departure from the academic world.[120] Walter Benjamin's *The Origin of German Tragic Drama* was submitted as his *Habilitationsschrift* in 1925, but it was not accepted; it marks *his* failure ever even to enter the academic world. Benjamin's study (eventually published in 1928) echoes the title of Nietzsche's, yet its approach is as fundamentally opposed as its subject matter. Whereas the object of tragedy is myth, and its *dramatis personae* are heroes, the baroque *Trauerspiel* is concerned with history, and the stature of its characters derives from their high rank.[121] (The distinction between tragedy and *Trauerspiel* is underpinned by a distinction between symbol and allegory –[122] opposing modes of representation, understood differently by Romanticism than by classicism.)[123] Thus whereas 'in the symbol', and in tragedy, 'destruction is idealized and the transfigured face of nature is fleetingly revealed in the light of redemption', by contrast 'in allegory', and in the mourning-play, 'the observer is confronted with the *facies hippocratica* [or "Hippocratic face", the changes produced by impending death] of history as a petrified, primordial landscape [*Urlandschaft*]'.[124] Championing the cause of the mourning-play over and against the tragedy, Benjamin inevitably seeks to distance himself from Nietzsche, especially his conception of myth as leading us 'back into the womb of the true and only reality' (see above).[125] Moreover, in his opening 'Epistemo-Critical Prologue', Benjamin offers a critique of the very notion of 'origin', denying its status as a logical category – Benjamin's target here is Hermann Cohen –[126] and insisting on its status as a historical category, albeit one entirely unrelated to how things come into being.

For Benjamin argues that 'origin' (*Ursprung*) has 'nothing to do with genesis [*Entstehung*]', since 'the term origin [*Ursprung*] is not intended to describe the process by which the existent came into being, but rather to describe that which emerges from the process of becoming and disappearance' (*Im Ursprung wird kein Werden des Entsprungenen, vielmehr dem Werden und Vergehen Entspringendes gemeint*), so that 'origin is an eddy in the stream of becoming, and in its current it swallows the material involved in the process of genesis' (*Der Ursprung steht im Fluß des Werdens als Strudel und reißt in seine Rhythmik das Entstehungsmaterial hinein*).[127] This denial of archaicity, of 'origin'-ality, is part and parcel of his championing of the baroque (and melancholy) view of the world as a heap of ruins.

In the case of Michel Foucault, Nietzche's project of genealogy is enthusiasti-cally taken up, but in an entirely de-transcendentalized form. In *The Archeology of Knowledge* (*L'archéologie du savoir*) (1969), Foucault announced 'a new form of history', one that would replace 'global history' with a 'general history'.[128] His target was 'this continuous chronology of reason, which is invariably traced back to an inaccessible origin [*l'inaccessible origine*], to a foundational beginning', but instead of 'tradition and the trace' Foucault insisted on 'the caesura and the limit'.[129]

For someone who once declared he was 'a Nietzschean',[130] Foucault's essay 'Nietzsche, Genealogy, History' (1971) demonstrates an appropriate sensivity to the need to distinguish between the different senses in which Nietzsche spoke about *Ursprung*, which he used to sustain his own distinction between history and genealogy.[131] Instead of concerning itself with 'monotonous finality', genealogy must record 'the singularity of events', since 'what is found at the historical beginning of things is not the inviolable identity of their origin; it is the dissension of other things' (or, as Foucault calls it, 'disparity').[132] Thus the genealogist, in Foucault's eyes, 'refuses to extend his faith in metaphysics' and 'listen to history', where 'he find that there is "something altogether different" [*"tout autre chose"*] behind things: not a timeless and essential secret, but the secret that they have no essence or that their essence was fabricated in a piecemeal fashion from alien forms'.[133] Hence Foucault urges the complete abandonment of 'an attempt to capture the exact essence of things, their purest possibilities, and their carefully protected identities', because 'this search assumes the existence of immobile forms that precede the external world of accident and succession', and is 'directed to "that which was already there" [*ce qui était déjà*], the image of a primordial truth fully adequate to its nature [*le "cela même" d'une image exactement adéquate à soi*]', which 'necessitates the removal of every mask to ultimately disclose an original identity'.[134]

It should be noted that, just as Foucault spoke of madness in a voice that itself bore all the hallmarks of classical rationality and elegance,[135] so he announced the end of conventional history and the beginning of its 'new form' in such a way that betrays little doubt about whether such a caesura in the history of history was in itself problematic, or whether such a contrast between the 'old' and the 'new' itself would make sense only within the context of a linear, uni-directional concep-tion of history. Nevertheless, Foucauldian genealogy is entirely in line with two other enemies of the primordial, the archaic, and the search for origins: the Frankfurt School and deconstruction.

The Frankfurt School was as attracted to the idea of the archaic as it was repelled by it. In his first, rejected *Habilitationsschrift* on the unconscious in Kant and Freud (1926), Theodor W. Adorno offered a critique of the 'irrational' metaphysics of Schopenhauer and *Lebensphilosophie*,[136] rehearsing in this early work his constantly trenchant criticisms of the archaic. But the discussion of the 'Arcades' project in the correspondence between Adorno and Benjamin reveals a surprising degree of ambivalence,[137] at least on Benjamin's part.[138] Writing on 5 April 1934, Adorno told Benjamin: 'I have come to realize that just as the modern is the most ancient, so too the archaic itself is a function of the new: it is thus first produced historically as the archaic, and to that extent it is dialectical in character and not "pre-historical", but rather the exact opposite', for 'it is precisely nothing but the site of everything whose voice has fallen silent because of history: something which can only be measured in terms of that historical rhythm which alone "produces" it as a kind of primal history.'[139] Alarmed by Benjamin's interest in Jung and Klages, and whilst conceding a proximity between their own interests and Klages' doctrine of the *Bild* ('image'), Adorno sought to argue that 'it is exactly here that the decisive distinction between archaic and dialectical images really lies'.[140] Yet it required several letters to persuade Benjamin of the error of his Jungian, Klagesian, and 'mythic-archaic' ways.[141] Nevertheless, the later Adorno stuck to his neo-Marxist guns. In *Zur Metakritik der Erkenntnistheorie* (1956), for example, he argued that 'totalitarian systems' were the brutal implementation of 'what ideology had been preparing for millennia as the domination of the spirit', the word 'elementary' (*elementar*) covering the same scientific simplicity as the mythical 'originary' (*Ursprüngliche*). Adorno remarked caustically that 'the identity of originariness and domination means that whoever has power should be not just the first, but the very first to do so'; as a political programme, 'absolute identity becomes absolute ideology – in which no one believes any more'.[142] In *Negative Dialectics* (1966) Adorno restated this rejection of the origin:

> The category of the root, the origin, is a category of dominion. It confirms that a man ranks first because he was there first; it confirms the autochthon against the newcomer, the settler against the immigrant. The origin – seductive because it will not be appeased by the derivative, by ideology – is itself an ideological principle.
>
> [*Die Kategorie der Wurzel, des Ursprungs selbst ist herrschaftlich, Bestätigung dessen, der zuerst drankommt, weil er zuerst da war; des Autochthonen gegenüber dem Zugewanderten, des Seßhaften gegenüber dem Mobilen. Was lockt, weil es durchs Abgeleitete, die Ideologie, nicht sich beschwichtigen lassen will, Ursprung, ist seinerseits ideologisches Prinzip.*][143]

Instead, Adorno offered the lapidary insight: 'There is no origin except in ephemeral life' (*Kein Ursprung außer im Leben des Ephemeren*).

In *Dialectic of Enlightenment* (1947), Adorno and Max Horkheimer (1895–1973) proposed an analysis of Western philosophy as leading (more or less) directly from Plato to NATO, from the pre-Socratics to the culture industry (as it were: from Homer to *The Simpsons*). On their account of this dialectic, myth is initially proposed as a form of reason, which itself becomes a new myth; reason eventually *and of necessity* undermines reason, so that myth returns in the form of 'technical rationality', a horrifyingly irrational form of reason.

Like Lessing's Tellheim, Horkheimer and Adorno find themselves moved to pondered 'how rational is this reason',[144] yet in another sense they overlook the point of this remark. For the Enlightenment reason that becomes transformed into instrumental reason is not the only kind of reason; in his conversations with Eckermann, for example, Goethe evokes the idea of a 'higher' or even a 'highest' reason (*höchste Vernunft*),[145] while in *Human, All Too Human* Nietzsche stated his ambition to show 'how reason comes to its senses' (*wie Vernunft kommt zur Vernunft*),[146] and in a *Nachlass* note he explored the idea that the 'highest' reason is to be found in aesthetics:

> Happiness resides alone in reason, the rest of the world is gloomy [*Das einzige Glück liegt in der Vernunft, die ganze übrige Welt ist triste*]. But I see the highest reason [*die höchste Vernunft*] in the work of the artist, and he can knows it for what it is; there may be something which, if it could be brought forth with consciousness, would yield an even greater feeling of reason and happiness: for example, the course of the solar system, the conception and the education of a human being.[147]

Moreover, one can regard myth, as Bronislaw Malinowski (1884–1942) did, not as 'an explanation put forward to satisfy scientific curiosity', but rather as 'the re-arising of a primordial reality in narrative form'.[148] In his 'Prolegomena' to his and C. G. Jung's *Essays on a Science of Mythology* (1941), the philologist Carl Kerényi (1897–1973) attempted a rehabilitation of precisely the concepts that would come under attack from (Western) Marxists, deconstructionists, and postmodernists alike.

Complementary to the Frankfurt School's critique of Western philosophy, Jacques Derrida (1930–2004) offered an extensive critique of its metaphysical assumptions. In 'Structure, Sign and Play in the Discourse of the Human Sciences' in *Writing and Difference* (*L'Écriture et la différence*) (1967), Derrida undertakes a deconstruction of the thought of Claude Lévi-Strauss and outlines a logic of the 'supplement'. The very notion of origin or *archē* is an instance of what Derrida calls the 'metaphysics of presence', an assumption that sets up a binary opposition, privileging one term, subordinating the second, and deriving the latter from the former.[149] Thus the term origin becomes privileged over and against all that is said to derive from it, yet – paradoxically – it requires those secondary elements in order to establish and maintain its priority. This is the 'difference' to which Derrida later opposes *différance*, an 'originality' without origin, inasmuch as *différance* is 'neither a word nor a concept', 'not only the play of differences within language but also the relation of speech to language, the detour through which I must pass in order to speak, the silent promise must make', 'the play of the trace [. . .] which has no meaning and is not'.[150] *Différance* is (or is not) 'the non-full, non-simple, structured and differentiating origin of differences. Thus, the name "origin" no longer suits it' (*l' "origine" non-pleine, non-simple, l'origine structurée et différante des différences. Le nom d'"origine" ne lui convient donc plus*).[151]

(Drawing on Derrida's critique of the origin, Gilles Deleuze [1925–1995] and Félix Guattari [1930–1992] substituted for the logocentric notion of the 'root' the term 'rhizome', opposing the horizontality of their postmodern approach to the supposed verticality of classical metaphysics.[152] Intriguingly, however, the concept of the 'rhizome' is found not just in Deleuze, but also in Jung; not surprisingly, given the early Deleuze's fascination with Jung's *Symbols and Transformations of Libido*.[153] Oddly, however, Jung is rarely thought of as a postmodern thinker, despite the multiple, non-hierarchical, and pluralist assumptions of his theory of the archetypes.)[154]

Yet within Derrida's thinking, particularly its later development, several important shifts undermine this rejection of the very possibility of origin. First, already in 'Structure, Sign, and Play', Derrida distinguishes between 'two interpretations of interpretation, of structure, of sign, of play': one 'seeks to decipher, dreams of deciphering a truth or an origin which escapes play and the order of the sign', whilst the other, 'no longer turned toward the origin', 'affirms play and tries to pass beyond man and humanism', abandoning the dream of 'full presence, the reassuring foundation, the origin and the end of play'.[155] There is, Derrida concluded, 'a kind of question, let us still call it historical, whose *conception, formation, gestation*, and *labour* we are only catching a glimpse of today' – only a glimpse, for we (and Derrida) 'turn [our] eyes away when faced by *the as yet unnamable* [*l'encore innommable*] which is proclaiming itself and which can do so, as is necessary whenever a birth is in the offing, only under the species of the nonspecies, in the formless, mute, infant, and terrifying form of monstrosity [*sous la forme informe, muette, infante et terrifiante de la monstruosité*]'.[156] Derrida's very language here evokes what one might call *the persistence of the archaic*, and *our naissant, incipient awareness of it*.

Second, in 1983 Derrida turned, as part of his collaboration with the architects Peter Eisenman and Bernard Tschumi to create the gardens for the Parc de la Villette in Paris,[157] to a reading of Plato's *Timaeus*, in which an account is offered of the creation of the world by the divine Creator (*demiurge*). Finding the division between the 'intelligible' (the realm of the ideal and the eternal) and the 'sensible' (the phenomenal copy of this ideal) insufficient, Timaeus proposes a third realm or dimension of being, 'the receptacle, and in a manner the nurse, of all generation' – the *chora*.[158] The *chora*, as 'the universal nature which receives all bodies', is always 'the same', for, 'inasmuch as she always receives all things, she never departs at all from her own nature and never, in any way or at any time, assumes a form like that of any of the things which enter into her'; the *chora* is 'the natural recipient of all impressions, and is stirred and informed by them, and appears different from time to time by reason of them'.[159] In a rejection of Thales, Anaximenes, Anaximander, and the pre-Socratic search for the *archē*, Timaeus holds that 'the mother and receptacle of all created and visible and in any way sensible things is not to be termed earth or air or fire or water, or any of their compounds, or any of the elements from which these are derived, but is an invisible and formless being which receives all things and in some mysterious way partakes of the intelligible, and is most incomprehensible'.[160] For Derrida, the *chora*, itself unrepresentable, is a spacing that provides the necessary condition for everything to take place and be represented;[161] or, to be put it another way, *the chora is the archaic.*

Third, Derrida's turn to religion in his last philosophical works should alert us to an insufficiency in his earlier theses, especially deconstruction, and the fact that (in literary studies) the attack on the canon has itself become canonical, providing the basis for a new orthodoxy in the academy. Although there is no space here to discuss in detail Derrida's announcement of the 'return of religion',[162] nor the ensuing debate to which it has given rise,[163] it has understandably been interpreted (and welcomed) as a sign of a re-awakening of the numinous,[164] or perhaps more accurately as *a new awareness of the archaic.*

And yet: Heidegger, Klages, Jung, Grassi

For there exists a counter-tradition, albeit one that has been subdued and, all too often, deliberately suppressed (and it is this tradition that is alert to the existence – the presence, even – the persistence, indeed – of the archaic).[165] For instance, Kaarino Kailo has argued that 'the marginalization of Jungian literary and psycho-therapeutic approaches in the academic context' is rooted in the same 'eurocentric epistemological ideologies' that lead to the 'trivialization and criticism' of feminist theories.[166] And Robert J. Kozljanič has (accurately) described Ludwig Klages (1872–1956) as 'a vitalist philosopher who, for no good reason, has fallen into oblivion'; he recalls how Klages' philosophy has been 'occasionally ridiculed, sometimes consciously passed over in silence, frequently criticised, but above all *repressed*', and draws a connection with depth ecology's concept of a psychic agency it calls 'the ecological unconscious'.[167] So maybe it is no coincidence that Jung's theory of the anima aims at a (re)evaluation of the Feminine, just as Klages (see below), following Bachofen, reintroduced the ancient idea of the Magna Mater.

Martin Heidegger (1889–1976), whose work is discussed in greater detail in Charles Bambach's contribution in this volume, was a thinker whose thought turned around the problems of the origin (*Ursprung*) , of 'grounding' (*Grund*), and the beginning (*Anfang*).[168] In his retrospective on his philosophical development, 'My Pathway Hitherto' (*Mein bisheriger Weg*) (1937/1938), Heidegger emphasizes how his work seeks to ask a 'fundamental question' which is a 'question about fundamentals' (*Grundfrage*).[169] In an early work entitled *Zur Bestimmung der Philosophie* (1919), Heidegger developed the project for a philosophy conceived as 'a science of the origin' (*Wissenschaft vom Ursprung*) or an '*Ur*-wissenschaft', from which even theory would derive its origin.[170] In a sense, his chief work, published in 1927, *Being and Time* (*Sein und Zeit*), was the realization of this programme. In this work Heidegger placed considerable emphasis on *Ursprünglichkeit* – 'originality' or 'primordiality' – as a methodological interpretaton of 'being-in-the-world' (*Dasein*). For Heidegger this *Ursprünglichkeit* is bound up with time, hence with temporality, hence with death, and hence with being (*Dasein*) as 'being-towards-death' (*Sein zum Tode*).[171] And 'being-towards-death' proves to be the precondition for an authentic existence (*die Eigentlichkeit der Existenz*) as 'being-as-a-whole' (*das Ganzsein*), as demanded by conscience (*das Gewissen*) and manifested by care (*Sorge*). All these themes – being, time, death, authenticity, care – hang together in Heidegger's 'origin-al' (*ursprünglich*) interpretation of existence:

> By pointing out that Dasein has an *authentic potentiality-for-Being-a-whole*, the existential analytic acquires assurance as to the constitution of Dasein's *primordial* Being [*die existenziale Analytik der Verfassung des* ursprünglichen *Seins des Daseins*]. But at the same time the authentic potentiality-for-Being-a-whole becomes visible as a mode of care. And therewith the phenomenally adequate ground for a primordial Interpretation of the meaning of Dasein's Being has also been assured. [*Damit ist denn auch der phänomenal zureichende Boden für eine ursprüngliche Interpretation des Seinssinnes des Daseins gesichert.*][172]

Subsquently in *On the Essence of Reason* (*Vom Wesen des Grundes*) (1929), Heidegger related the principle of reason (*den Satz vom Grund*) to freedom, as 'the origin of the principle of reason' (*der Ursprung des Satzes vom Grunde*), as itself 'the reason of the reason' (*der Grund des Grundes*), but also as 'the abyss of existence' (*der Ab-grund des Daseins*).[173]

Following his so-called 'turn' (*Kehre*), in Heidegger's later writings the ontological theme of the origin (*Ursprung*) becomes the existential theme of the 'primordial leap' (*Ur-Sprung*), from which the question of the ground of existence – the question 'why is there something rather than nothing' – 'leaps forth'. For Heidegger, this is *the* question of philosophy, which philosophers (and everyone else) have forgotten:

> The question 'why is there something-that-exists rather than nothing?' is first in rank for us because it is the most far reaching, second because it is the deepest, and finally because it is the most fundamental of all questions [*die dem Range nach erste einmal als die weiteste, sodann als die tiefste, schließlich als die ursprünglichste Frage*].[174]

It is, Heidegger argues, in every sense the *fundamental* question:

> The question aims at the ground of what is insofar as it is. To seek the ground is to try to get to the bottom [*ergründen*]; what is put in question is thus related to the ground [*rückt in den Bezug zu Grund*]. However, since the question is a question, it remains to be seen whether the ground arrived at is really a ground, that is, whether it provides a foundation [*ob der Grund ein wahrhaft gründender, Gründung erwirkender*]; whether it is a primal ground [*Ur-grund*]; or whether it fails to provide a foundation and is an abyss [*Ab-grund*]; or whether the ground is neither one nor the other but presents only a perhaps necessary appearance of foundation [*einen vielleicht notwendigen Schein von Gründung*] – in other words, it is a non-ground [*Un-grund*].[175]

To ask this question is, in Heidegger's Kierkegaardian terminology, a 'leap' (*Sprung*) but, because it is a question/leap about the origin, it is the originary/ original question/leap (*Ur-Sprung*):

> We call such a leap, which opens up its own source, the original source or origin, the finding of one's own ground [*Einen solchen, sich als Grund er-springenden Sprung nennen wir gemäß der echten Bedeutung des Wortes einen Ur-sprung: das Sich-den-Grund-erspringen*]. It is because the question 'Why are there things-that-exist rather than nothing?' breaks open the ground for all authentic questions and is thus at the origin [*Ursprung*] of them all that we must recognize it as the most fundamental of all questions [*die ursprünglichste Frage*].[176]

Along with the Gnostic-mystico-Schellingian vocabulary of *Urgrund* and *Ungrund*, the theme of art returns in the later Heidegger, and the question of being (*Dasein*) and beings (*Seiendes*) becomes the question of 'the origin of the work of art'. Indeed, *Ursprung* is the very first word of Heidegger's famous essay (1934/1936) on this question.[177] In the concluding pages of this complex yet rhetorically resonant text, which pushes language to its stretching point, Heidegger states that art (*Kunst*) 'happens' as poetry (*Dichtung*), which in turn is founding (*Stiftung*) in a triple sense: as bestowing (*Schenkung*), 'grounding' (*Gründung*), and beginning (*Anfang*).[178] Heidegger's densely philological argument links the beginning with the 'leap', as a 'leap-ahead' which (always) already contains its own end, but he distinguishes it from the 'primitive' (in a negative sense):

> A genuine beginning, as a leap, is always a jump-ahead, in which everything to come is already leaped over, even if as something disguised. The beginning already contains the end latent within itself. A genuine beginning, however, has nothing of the beginner about it as the primitive does [*Der echte Anfang ist als Sprung immer ein Vorsprung, in dem alles Kommende schon übersprungen ist, wenngleich als ein Verhülltes. Der Anfang enthält schon*

verborgen das Ende. Der echte Anfang hat freilich nie das Anfängerhafte des Primitiven]. The primitive is, because it lacks the bestowing, grounding leap and jump-ahead, always futureless. It is not capable of releasing anything more from itself because it contains nothing other than that in which it is caught. [*Das Primitive ist, weil ohne den schenkenden, grünenden Sprung und Vorsprung immer zukunftslos. Es vermag nichts weiter aus sich zu entlassen, weil es nichts anderes enthält als das, worin es gefangen ist.*][179]

Rather, the beginning contains a plenitude, a fullness, 'the undisclosed fullness of the monstrous' [the unfamiliar and the extraordinary] (*die unerschlossene Fülle des Ungeheuren*) and consequently of the struggle with its opposite, with 'the familiar and the ordinary' (*des Streites mit dem Geheuren*): 'Art lets truth leap forth. Art, as a foundational preservation, allows the truth of what-is-in-existence to spring forth in the work. To allow something to spring forth, in the foundational leap out of the source of its essence into its being, this is what origin means' (*Die Kunst läßt die Wahrheit entspringen. Die Kunst erspringt als stiftende Bewahrung die Wahrheit des Seienden im Werk. Etwas erspringen, im stiftenden Sprung aus der Wesensherkunft ins Sein bringen, das meint das Wort Ursprung*).[180] Heidegger's essay concludes with a series of questions to which, one assumes, the closing quotation from Hölderlin provides a reassuring answer:

> Are we in our existence historically at the origin? Do we know, which means do we pay heed to the essence of the origin? Or, in our relation to art, do we merely make an appeal to a cultivated acquaintance with the past?
> [. . .]
>
> 'Reluctantly,
That which dwells near its origin parts.'
>
> [*Sind wir in unserem Dasein geschichtlich am Ursprung? Wissen wir, d.h. achten wir das Wesen des Ursprungs? Oder berufen wir uns in unserem Verhalten zur Kunst nur noch auf gebildete Kenntnisse des Vergangenen?*
> [. . .]
>
> "*Schwer verläßt
Was nahe dem Ursprung wohnet, den Ort.*"][181]

It will come as no surprise that Heidegger was fascinated with the work of Anaximander, on whose sole existing fragment he wrote an essay published in 1946.[182] As well as returning philosophy to the theme of the *archē*, and linking this theme with the question of art, Heidegger's highly technical, almost incantatory writings are shot through with an intuitive sense of the archaic, as when he evokes in his *Introduction to Metaphysics* the motion of the planets in their orbit and the flow of the sap through the plant.[183] Many years later, in his Heraclitus seminar, organized with Eugen Fink in the winter semester of 1966–1967 in Freiburg, Heidegger recalled a similar revelatory moment on a Greek island:

'I remember an afternoon during my journey in Aegina. Suddenly I saw a single bolt of lightning, after which no more followed. My thought was: Zeus.'[184] Like the philologist Ulrich von Wilamowitz-Moellendorff (1848–1931),[185] or the philologist and philosopher Walter F. Otto (1874–1958),[186] Heidegger seems to have believed that the Olympian gods in some sense actually exist.[187]

The question of whether or not he believed in the gods or in God – when asked outright during his famous interview with John Freeman, he dodged the question, answering: 'I don't need to believe, I know' –[188] has bedevilled much of the reception of the work of the Swiss psychoanalyst C. G. Jung (1875–1961). Of all the thinkers presented in this volume, Jung's engagement with the problem of the archaic was arguably the most enduring and comprehensive, and three papers included in this volume (by Robert A. Segal, Susan Rowland, and Paul Bishop) debate extensively his paper translated as 'Archaic Man' (1931).[189] In contradistinction to the art-historical sense of the archaic, in the sense of being 'pseudo-antique or copied, as in later Roman sculpture or nineteenth-century Gothic', something is truly archaic, Jung once wrote, when it exhibits 'qualities that have the character of *relics*'; hence he considers an image to be archaic when it 'possesses unmistakable mythological parallels'.[190]

But there is more to the archaic for Jung than the possession of certain qualities, mythological or otherwise. Above all, the state which Jung – borrowing, as Robert A. Segal demonstrates at some length, a term from the French anthropologist Lucien Lévy-Bruhl (1857–1939) – calls *participation mystique* can be described as archaic inasmuch as in it, 'the subject cannot clearly distinguish himself from the object but is bound to it by a direct relationship which amounts to a partial identity', that is, an identity that 'results from an *a priori* oneness of subject and object'.[191] As well as a fusion of the subject with the object, the fusion *within* the subject of the 'psychological functions' – 'thinking with feeling, feeling with sensation, feeling with intuition, and so on' – is also archaic,[192] as well as reminiscent of the reciprocal subordination of drives that German aesthetics, especially Schiller, associated with the aesthetic.[193] Given which, perhaps we can understand better why, for Jung, the *archē* takes the form of the *archē*-type: initially referred to (somewhat Germanically) as the *Urbild*, or 'primordial image', and later (in a more Græcizing fashion) as the *Archetyp* – the 'impression' (*tupos*) of the 'origin' (*archē*). Although, on the one hand, Jung could sometimes argue that because 'the word "type" is [. . .] derived from τύπος, "blow" or "imprint"', the existence of an archetype 'presupposes an imprinter', or God,[194] on other occasions he made it clear that the archetypal was pre-eminently the quality of a particular situation, in which the subject became caught up – or fused – with the objective world in a way analogous to various mythical situations.[195] Or as Joseph Campbell put it, 'the latest incarnation of Oedipus, the continued romance of Beauty and the Beast, stand this afternoon on the corner of Forty-second Street and Fifth Avenue, waiting for the traffic light to change.'[196]

In her account of Jung's life and thought, Marie-Louise von Franz (1915–1998) sets up a contrast between 'the traditions and religions which have become contents of collective consciousness' and 'that primordial experience which is the final source of these contents' – an experience which, according to her, resides in 'the encounter of the single individual with his own god or daimon, his struggle with the overpowering emotions, affects, fantasies and creative inspirations and obstacles which come to light from within.'[197] To put it another way, such a primordial experience has its roots in the archaic. Or, to use a term from Jung's later thought, such an experience was an example of the kind he now termed 'synchronistic'. For a 'synchronistic' experience is a 'spiritual' experience, but not in a religious sense;[198] it is 'spiritual' in the sense of *geistig*, as Hegel or Cassirer understood the concept of *Geist*; it is the experience of a moment, in which the interiority (of the Self) coincides (or synchronizes) with the exteriority (of the World), revealing the identity (and hence the truth) of both.

Klages

A thinker who must surely rival Jung in the intensity of his engagement with the archaic is Ludwig Klages, a figure who is not well-known outside the German-speaking world. (Even there, Klages has lost his former pre-eminence: once enormously popular, in the Twenties and Thirties, nowadays one of the few signs of his impact is the ready availability of copies of his works, in the famous large, yellow paperbacks published in Leipzig by Johann Ambrosius Barth, in *Antiquariate* or second-hand bookshops.)[199]

One of the many achievements of Klages's mammoth masterpiece, *Der Geist als Widersacher der Seele* (1929–1932), translatable as *The Spirit as the Adversary of the Soul* or *Mind as Opponent of Psyche*, is to offer – in a simple, almost throwaway paragraph – nothing less than a redefinition of the archaic as the 'cosmic' (*kosmisch*). In book 5, 'The Reality of the Images', Klages turns, in his chapter on 'The Elementary Souls', to a consideration of the four elements: the earth – *not*, be it noted, in any *Blut-und-Boden* sense, but as an element in which we recognize the 'primordial image' (*Urbild*) of matter itself; water, as an embodiment of the manifestation of matter; air, as (following Anaximenes) the opposite of matter; and fire, something so elemental it is almost no longer an element. So the Persians recognized in fire a cosmic principle, while for Heraclitus it is the *archē* itself. In this Heraclitean-archaic fire, Klages sees a symbol of cosmic life, whose polar opposite, water, is the symbol of materiality as the bearer of individual life.[200]

In the second part of book 5, Klages develops his 'pagan' conception of time as a circle.[201] In his chapter on the Magna Mater, which explicitly acknowledges its debt to the work of J. J. Bachofen, Klages investigates the symbol of the perpetual cycle (or eternal recurrence) of life and death.[202]

> In complete opposition to logical consciousness, which – feeling its way along the straight line of time – considers each past thing to be destroyed, but in the present sees only repetitions of it, the Pelasgians – bound up with the

circle of time – live, know, and teach the *eternal return of the origin*. [*Völlig entgegengesetzt dem logischen Bewußtsein, das entlang tastend an der Geraden der Zeit jedes Vergangene für vernichtigt hält, im Gegenwärtigen aber von ihm nur Wiederholungen sieht, lebt und weiß und lehrt das im Kreise der Zeit gebundene Pelasgertum die* ewige Wiederbringing des Ursprungs.][203]

Yet in the conclusion to his central work, it becomes clear that, far from calling for a return to the past, what he is seeking is a new way of understanding our relation to the past in the present: in other words, what Klages demands is not a 'going back' (*Rückkehr*), but a 'turning round' (*Umkehr*).[204] No wonder Klages was of no use to the likes of Alfred Rosenberg (1893–1946), one of the chief ideologists of National Socialism; what Klages insistently calls for is to adopt a *holistic* sense of the world, which he presents in the hypostatized form of the Primordial One (*das Ureine*):

> What we call day and night, arising and passing, waking and sleeping, motion and rest, man and woman are the decomposing halves of the self-impregnating Primordial One: of the alternately-colored world egg, in the language of the Orphics, which hides in its yolk the Eros who unleashes and binds. [*Was wir Tag und Nacht, Entstehen und Vergehen, Wachen und Schlaf, Bewegung und Ruhe, Mann und Weib benennen, sind die Zersetzungshälften des in sich selbst sich befruchtenden Ureinen: des zwei-farbigen Welteis nach der Sprache der Orphiker, das im Dotter den lösend-bindenden Eros birgt.*][205]

In turn, the archaic unity of the cosmos reveals itself in what Klages calls the world's *Urbildlichkeit* – its essence as a 'primordial image', but also its 'pictur-ability as an image', its mode of revelation as a 'reality of images' (*Wirlichkeit der Bilder*):

> In every moment each detail of the world can be completely possessed by the soul, it can be submerged in the colour of the essence. These are the moments when we gain a view of the world of eternity. [*Jeder Weltausschnitt kann in jedem Augenblick vollkommener Besitz der Seele werden, in die Farbe der Essenz sich tauchen. Dies sind die Augenblicke des Durchblicks in die Welt der Ewigkeit.*][206]

Grassi

For the fifth-century BCE historian Thucydides, the Persian Wars of the previous century had been 'the greatest event of the past', yet they had been swiftly decided by 'two battles on land and two at sea' and their magnitude paled in comparison with the recent Peloponnesian War of 431 to 404 BCE, with its rivalry between Athens and Sparta:

Never before had so many cities been captured and depopulated, either by barbarians or by Greeks at war with each other, and then, in some cases, resettled by new inhabitants. Never before had there been so many exiles and so much killing, some brought about the war itself and some by civil strife. Ancient events that were better established in legend than in experience now seemed less incredible, for there were now violent earthquakes spread through much of the world; eclipses of the sun, which now occurred much more frequently than ever before in memory; terrible regional doughts and the famines they caused; and last but not least, the plague, which caused great harm and great loss of life. All of these things were associated with this war.[207]

In his study of the notion of the *archai* reproduced in this volume, Ernesto Grassi (1902–1991) contrasts Thucydides' account of the Peloponnesian War as an event of originary significance, driven by concrete human circumstances, with the priority attached in their differing conceptions of history by Hesiod to *myth* and by Pindar to *poetic metaphor*. Much of Grassi's work, unfortunately still insufficiently known in the English-speaking world,[208] interrogates the relation between reality and language in a way that is strikingly different from Heidegger's account.

In his major study of myth and aesthetics, *Kunst und Mythos* (1957), Grassi undertakes to recover 'our eye for and understanding of the original, but hopelessly submerged primordial being [*Ur-Seiende*]'.[209] On a reading of Plato's *Phaidros*, *Ion*, *Theaetetus*, and *Symposium*, and Aristotle's *Metaphysics*, and a consideration of the terms *légein* ('select' or 'interpret'), *hermeneuein* ('interpret'), *téchne* ('art'), *aisthesis* ('sense'), *empeiria* ('experiment'), *poieisis* ('creation'), and *mechanè* ('instrument'), Grassi distinguishes between, on the one hand, a primordial nature (*Urnatur*) that persists as 'an integral and structured order' in the vegetative life of plants and the instinctual life of animals, and, on the other, human beings who find themselves, by virtue of their 'empirical' and 'technical' understanding, their propensity to dissect and to extract, excluded from this order.[210] Now myth, with its a-causal, a-logical, and essentially emotional world view,[211] through dance, festivities, and play, transforms random phenomena into a cosmos – or in Vico's terms, the first fire lit on the altars served to clear forests and establish the cities.[212]

For Grassi, the price for the genesis of art is the loss of myth, the break with an absolute world order, and a shift from an 'eternal' to a particular perspective.[213] Myth becomes mere fable, sacral reality and sacred ritual become spectacle, but whilst art sounds the death-knell of myth, it preserves its cosmic, world-creating potential. For metaphor permits a recovery of myth,[214] albeit in a completely different way from how Heidegger, Grassi's former doctoral supervisor, contends poetic language or symbolic images work.[215] According to Grassi, 'the "finding", the *invenire* of a word becomes "something invented", in which existence [*das Seiende*] uncovers itself no longer in its "abstractness", but in its "authenticity", that is, in its historicity'.[216] Thanks to the Renaissance Humanists, such as Dante (1266–1321), Albertino Mussato (1261–1329), and Leonardo Bruni (1370–1444), the transhistorical capacity of poetry (of rhetoric, of metaphor) to *create history* – language as 'archē' – is fully realized.

In his study on *Renaissance Humanism* (1986), Grassi explores the achievement of these and other thinkers, including Guarino Veronese (1374–1460), Juan Luis Vives (1492–1540), and Cristoforo Landino (1424–1498), in illuminating detail. For the Renaissance Humanists, poetry has 'the function of making manifest the miracle, the admirable, the unfoundable, and, therefore, un-fathomable in the "here" and "now," not in the contemplation of the abstract.'[217] Through poetry, he continues, 'that power is disclosed which opens up a space for human action and creates a "clearing" [*Lichtung*] in the primeval forest, and that clearing will be the stage for history.'[218] For the Humanists – Dante, Mussato, Bruni, Coluccio Salutati (1331–1406), Giovanni Pontano (1426–1503), and Lorenzo Valla (1405–1457) – 'the poetic, rhetorical word has its own acknowledged function, and even preeminence over rational thought and language':

> It is impossible to speak "rationally" about the primal, the non-deducible, the pre-eminent – and, as such, the unfathomable. The non-deducible can have its claim adequately responded to only in metaphors – in the realm of the "here" and "now" and through the word of "indication" rather than "proof," through mythical rather than rational language.[219]

Thus, according to Grassi, the 'specific problem' of Humanist philosophy is 'how, when, and through what mode of language "being-there" experiences the claim of Being':

> 'Being-there' reveals itself in the existential necessity of language; first of all in the silent response of 'being-there' to the claims of Being in the given situation, and then as the silence of Being breaks into language in response to the demand. That is, the 'truth' of the word understood not as the rational correspondence to being ('adaequatio rei et intellectus,' as expressed traditionally), but as *alethia*, as the spoken disclosure of the claim of Being as it reveals itself in the given situation. Only in the primal sphere does the resorting to signs, to indicating movements, even to silence, attain meaning: outside of the claim everything is unresponsive and indeterminate, as in an impenetrable forest without 'clearings'.[220]

In the essay included in this volume (where it is translated for the first time), ' "Archaic Theories of History": Thucydides, Hesiod, Pindar: The Originary Character of Language' (1992), Grassi returns to the definition of 'archē' as an originary principle, rather than as 'primal reality' (*Urwirklichkeit*), examining three different aspects of the 'originariness' (*Ursprünglichkeit*) of language in respect of various approaches to human historicity. For Thucydides, the *archē* is *ananke*, 'necessity', exemplified by the military action of the Peloponnesian War; for Hesiod, history is rooted in *erga*, 'works', arising from *eris*, 'conflict' and 'competition'; and for Pindar, whose work emphasizes the word in celebration of the *agon*, the 'contest' or the 'struggle'. The poetological primacy of language in its archaic mode lies at the core of Grassi's rehabilitation of the humanist tradition, which rejects all rationalizing, abstract approaches to human existence and seeks the source of its meaning in the work of language. Thus the archaic has a dimension that is intimately bound up with the concrete, historical questions of humankind. *When does history begin? What makes history? And what is really going on in history?* The answer to these questions is, in its own way, the answer to the question about the nature of the archaic.

Among Grassi's contemporaries was the Romanian thinker Mircea Eliade (1907–1986), whose book *The Sacred and the Profane* (1957) took Rudolf Otto's concept of *das Heilige* as its starting point; one of the most interesting characteristics of modern, desacralized society was its preservation, albeit vestigial, of a sense of the sacred.[221] Today, Eliade believed, 'properly speaking, there is no longer any world', instead there are only 'fragments of a shattered universe', described as 'an amorphous mass consisting of an infinite number of more or less neutral places in which man moves, governed and driven by the obligations of an existence incorporated into an industrial society.'[222] But for Eliade, a proponent (along with Gershom Scholem and Henry Corbin) of what has been called a 'religion after religion',[223] our 'crypto-religious behaviour' points to the persistence of sacred space – the archaic – the experience of which 'makes possible the "founding of the world" ', for 'where the sacred manifests itself in space, *the real unveils itself*, the world comes into existence'.[224]

By way of a conclusion

For Goethe, as we shall see, it was small fig trees; for Ernesto Grassi, it was his 'immediate experience' of the Brazilian rain forest, with its mixture of 'fascination' and terror';[225] for Paul Gaugin, it was the lush vegetation of Taihiti;[226] whilst for D. H. Lawrence (1885–1930) it was 'huge, straight fir trees' and 'big beech trees' that revealed the archaic. In *Fantasia of the Unconscious* (1928) he recalls his sensations in the Black Forest:

> I listen again for noises, and I smell the damp moss. The looming trees, so straight. And I listen for their silence – big, tall-bodied trees, with a certain magnificent cruelty about them – or barbarity – I don't know why I should say cruelty. Their magnificent, strong, round bodies! It almost seems you can hear the slow, powerful sap drumming in their trunks. Great full-blooded trees, with strange tree-blood in them, soundlessly drumming.
>
> Trees that have no hands and faces, no eyes; yet the powerful sap-scented blood roaring up the great columns. A vast individual life, and an overshadowing will – the will of a tree; something that frightens you.

Suppose you want to look a tree in the face? You can't. It hasn't got a face. You look at the strong body of a trunk; you look above you into the matted body-hair of twigs and boughs; you see the soft green tips. But there are no eyes to look into, you can't meet its gaze. You keep on looking at it in part and parcel.

It's no good looking at a tree to know it. The only thing is to sit among the roots and nestle against its strong trunk, and not bother. [. . .] This marvellous vast individual without a face, without lips or eyes or heart. This towering creature that never had a face. Here am I between his toes like a pea-bug, and him noiselessly over-reaching me, and I feel his great blood-jet surging. And he has no eyes. But he turns two ways: he thrusts himself tremendously down to the middle earth, where dead men sink in darkness, in the damp, dense undersoil; and he turns himself about in high air; whereas we have eyes on one side of our head only, and only grow upwards.

Plunging himself down into the black humus, with a root's gushing zest, where we can only rot dead; and his tips in high air, where we can only look up to. So vast and powerful and exaltant in his two directions. And all the time he has no face, no thought: only a huge, savage, thoughtless soul. Where does he even keep his soul? – Where does anybody?[227]

Moreover, what Goethe called *das Schaudern*, Lawrence described as being 'terrified', and Pierre Hadot glosses as a *frisson*, a *nuance d'angoisse*, can still be felt today. In her recent book on silence, Sara Maitland describes an experience that is congruent with the idea of the archaic. Not in a grove of trees (as in Seneca and Ovid), in a forest (as in Lawrence) or up a mountain (as in the Corycian cave as Pomponius Mela describes it), but on the island of Skye, Maitland recounts the following experience of a sense of primordial unity:

On one unusually radiant day, [. . .] I took a walk up the burn above the house [. . .] and climbed on up into the steep-sided corrie. It was sheltered there and magnificent – almost vertical mountains on both sides – [. . .] and below, tiny strands of water that looked like handfuls of shiny coins tossed casually down. I sat on a rock and ate cheese sandwiches. [. . .] And there, quite suddenly, I slipped a gear. There was not me and the landscape, but a kind of oneness: [. . .] as though the molecules and atoms I am made of had reunited themselves with the molecules and atoms that the rest of the world is made of. I felt absolutely connected to everything. It was very brief, but it was a total moment. I cannot remember feeling that extraordinary sense of connectedness since I was a very small child.[228]

Yet this 'beautiful' aspect of the archaic has its 'sublime' counterpart, a sensation that is best described as the sheer terror of the archaic: one of Pascal's *Pensées* simply states, 'the eternal silence of these infinite spaces fills me with dread' (*le silence éternel de ces espaces infinis m'effraient*),[229] and during a walk from Luib to Loch Slapin, on 'a strange day, very still with no wind', Maitland experiences the terror of the primordial:

> I came to a little loch with reeds standing in the perfectly clear water, which reflected the hills rising sharply either side – and at first I was enchanted [. . .] – and then, abruptly, suddenly I was 'spooked'. It is so hard to describe – the silence, the fact there was low cloud, or mist [. . .] and the opposite side there were wisps of cloud [. . .]. I found myself becoming increasingly uneasy, nervous. Gradually I became convinced I was being watched. There were two black shapes on the hill above me. I thought, or rather I felt, that they were alive. [. . .] I decided, firmly and rationally, that they were in fact rocks, but I never entirely convinced myself. [. . .] I felt that the silence was stripping me down, desiccating, denuding me. I could hear the silence itself screaming.[230]

Maitland assimilates her panic to the primal experience of 'panic', attributed by the ancients to a manifestation of the god Pan, and she recalls how the French adventurer Augustine Courtauld spent six months alone in a tent in the Arctic, where he 'recorded strange and inexplicable screaming noises', saying afterwards 'that it was the only thing that really frightened him'.[231] On seeing a photograph of Courtauld, Jung apparently described him as a man 'stripped of his *persona*, his public self stolen, leaving his true self naked before the world'.[232] Nor are Pascal, Maitland, and Courtauld alone. For this is precisely the manifestation of the archaic as *das Ungeheure*, a mixture of the sublime and the numinous, that Goethe had emphasized in his scientific writings when he was talking about Nature.

For it is when we are confronted with the *Urphänomen* (variouly translated as 'primal phenomenon' 'archetypal phenomenon', or 'originary phenomenon') that, according to Goethe, we experience this anxiety, or even *Angst*. 'The immediate perception of originary phenomena', he wrote, 'plunges us into a kind of anguish: we sense our inadequacy [. . .]'.[233] Or again, 'when archetypal phenomena stand unveiled before our senses, we feel a kind of nervousness, to the point of feeling anguish [*eine Art von Scheu, bis zur Angst*]'.[234] Even more concretely, Goethe is recorded as saying in a conversation with J. D. Falk in the summer of 1809: 'This fig tree, this little snake, this cocoon [. . .] all these things are signatures, heavy with meanings. Yes, he who could decipher their meaning exactly, would soon be able to do without all writing and all words. Yes, the more I think about it, the more it seems to me that there is something useless, idle, and even fatuous, I might say, in human discourse, so that we are terrified by Nature's silent seriousness, and by her silence [*daß man vor dem stillen Ernste der Natur und ihrem Schweigen erschrickt*].'[235]

A year later, in his 'Preface' to his *Theory of Colour* (*Zur Farbenlehre*) (1810), Goethe encourages us to listen to nature – but what we will hear, it is suggested, is something terrible:

Let us shut our eyes, let us open our ears and sharpen our sense of hearing. From the softest breath to the most savage noise, from the simplest tone to the most sublime harmony, from the fiercest cry of passion to the gentlest word of reason, it is nature alone that speaks, revealing its existence, energy, life, and circumstances, so that a blind man to whom the vast world of the visible is denied may seize hold of an infinite living realm through what he can hear.

Thus nature also speaks to other senses which lie even deeper, to known, misunderstood, and unknown senses. Thus it converses with itself and with us through a thousand phenomena. No one who is observant will ever find nature dead or silent.[236]

The task for us, it seems, is to translate the archaic sense of *das Schaudern* (awe, sacred terror) into *das Erstaunen* (marvel, a sense of wonder). When confronted with the archaic, indifference is not an option.

Whilst it would be easy to portray such thinkers as Heidegger, Klages, Jung, Grassi – or, for that matter, even Goethe and D. H. Lawrence – as intellectual recidivists or reactionary *Querdenker*, it is not sufficient (for either supporters or critics) simply to push them into the political Right, as if that were where they belonged (and as if, even if they did, there were no need to articulate a more sophisticated response to them). That there are real political dangers that lurk here should not blind us to the genuine opportunities, in the sense of opportunities for authenticity, that their thought also presents. After all, we have already noted the apodictic style of Foucault, and the tendency within postmodernism for the absence of principle itself to be elevated into a principle. In an age subservient to, as the combative Joseph Ratzinger once put it, the 'dictatorship of relativism',[237] the archaic is opposed to relativism, because it insists that something – some *thing* – actually *is*, that there is something *unconditional*, which is the *condition* for everything else. The archaic is not an argument, it is an apprehension: in all senses of the word. As such, the concept of the archaic does not merely bridge the divide between the religious and the secular, or the Christian and the pagan, it transcends all categorization. (Indeed, there is room for a humanist understanding of the archaic; humanist, that is, in the organic and vitalist sense espoused by Grassi and others.) Although the sacred, *das Heilige*, the numinous, the primal or the primordial are all worthy synonyms, in this overview of the history of the concept of the archaic – because the archaic *itself* is without or beyond or prior to all history – I have suggested that one route to understanding the archaic lies through the aesthetic, inasmuch as both reveal the precariousness of temporality.

For we can read the archaic as the aesthetic, and not simply as the sublime, in the specific sense that, in his third critique, Kant defined beauty as 'purposefulness without purpose'.[238] That is to say, our apprehension of the archaic relies to a considerable extent on an 'as-if'. Whether the archaic is 'nature within' or 'nature without' is irrelevant (it is both), since to speak of the archaic as a manifestation of autonomous forces – as spirits, as *genii loci*, as gods – is to acknowledge that it *appears as if they exist*. The archaic is autonomy without agency: it is the world as it brings itself into being. And for the French philosopher Bertrand Vergely, each of us has a sense of the world-as-a-world-brought-into-being, inasmuch as to mature and to grow old is 'to approach the great mystery of existence', or 'to get back in touch with its originary radicality, with the originary absolute', as the poets show us.[239] Nostalgia, says Vergely, is not a regression, but an advance into the mystery of life, in which beauty approaches us.[240]

Because aesthetic intuition exists outside time, it offers unique access to the archaic, which is timeless. In *Timaeus*, time is presented as 'a moving image of eternity',[241] as an Eternal Now, unchanging, unchangeable, forever inexpressible, for 'we say that it "was", or "is", or "will be", but the truth is that "is" alone is properly attributed to it, and that "was" and "will be" are only to be spoken of becoming in time.'[242] The archaic is the 'is', the extension of the past into the present which relates us to the past; thus establishing, or so Ludwig Klages believed, the ontological priority of the past over the future. Klages criticizes the overprivileging of the future as leading to a neglect of the present, regretting the 'devaluation' (*Entwertung*) and 'deprivation of its rights' (*Entrechtung*) of the present moment, the 'subjugation of life through purposes and the future', and the 'excessive emphasis on purposefulness and planning', which finds its counterpart in 'the "historical sense" that cannot get enough of ever more comprehensive records of data' but to its own cost never really understands the past.[243] The archaic is not irrelevant to everyday life in the modern world; its persistence can explain much about why the modern world is the way it is.

For there is, as Vergely reminds us, a deep relation between the ancient and the modern, between the archaic and modernity. 'Because modernity is modern, it loves what is new; and because it loves what is new, modernity loves what is old, because the old is nothing other than the appearance of the new.'[244] He explains elsewhere that 'modernity is the maturity of the past, indeed it is its old age', and 'the Ancients are among us, today, and not outside us, in the past'.[245] As Pascal observed, 'those we call old [*anciens*] were really new [*nouveaux*] in all respects [. . .] and in us one may discover this antiquity [*cette antiquité*] which we revere in others.''[246] Thus antiquity (*l'ancienneté*) – or, one might say, the archaic – is 'not so much old as close to the beginning' (*les origines*); and, being close to the beginning (*les origines*), it is 'close to the origin' (*l'originaire*) – to the origin (*l'originaire*) 'which, linked to what it is original' (*l'originel*), 'speaks to us of our destination'. For 'we are called to be' what we originally, or originarily (*originellement*), are: that is to say, 'to live with an authenticity older than any memory'.[247] 'Whoever draws close to what is original [*l'originel*], and to the authentic being behind it, is rejuvenated.'[248] And so we must listen to the call of the old, the call of the ancient, the call of the archaic.

For the archaic calls to us throughout the day but above all in the night, like the horns in the forest in Schubert's setting of 'Nachtgesang im Walde';[249] even as we fell vast tracts of the Amazon jungle and concrete over our own fields and woods, still the archaic urges itself upon us with ever great insistence from the forests of the mind. The archaic is not simply something dark, it is *the* dark – it is what the

Egyptians (according to Damascius, the last of the neo-Platonists) called 'a *thrice unknown darkness*', it is (in the words of an Orphic hymn) 'Night, parent Goddess, source of sweet repose, / From whom at first both Gods and men arose'.[250] It is no coincidence that so many of the thinkers discussed in this volume are German, or used the medium of the German language;[251] across music, painting, art, and philosophy, German culture has tracked the archaic with a subtlety and a sophistication, as well as a relentlessness and a near-obsession, that is virtually unparalleled.[252] Yet the archaic is not simply a 'German' problem. If, as Jonathan Meades has suggested, in Caspar David Friedrich's famous painting *Chalk Cliffs On Rügen* (1818–1820), the three figures depicted are not so much gazing out to sea as turning away from the immensity of the forest behind them,[253] maybe it is time for us all to turn and face what is standing behind us.

To put it another way: the archaic is the past in the present, the past *as* the present, the *nunc stans* or Eternal Now which reveals the Eternal Then. On Mount Horeb Yahweh tells Moses 'I AM'; by contrast, the caves and the forests, the planets and the sap in the plants all tell us: 'IT WAS ALWAYS THUS.' The archaic is not *post factum*, it *is* the *factum*, whether we like it or not; to argue for the archaic is not to argue *post hoc, ergo propter hoc*, but to affirm that it *is* the *hoc*. What matters is not so much that it is 'always already there' as that we are 'always already afterwards'. So whether we respond to our perpetual lateness with the despondency of gloomy mourning; whether we react to the incomprehensible facticity of the archaic as a source of terror beyond articulation; or whether we find in the archaic, as something that can never be exploited, expropriated, or instrumentalized, never dominated, subjugated or tamed, never even controlled, audited, or otherwise managed, a source of exuberant celebration – this remains for each of us to decide. For as Leucippus, the philosopher for whom the archaic was being *and* non-being, reminds us in one of the sole surviving fragments of his thought, 'authentic joy is the goal of the soul'.

Notes

1 To guide myself and the reader through this potentially, and actually, vast topic of the archaic, I have chosen the following articles as a starting point: H. R. Schweizer, 'Archaisch', in *Historisches Wörterbuch der Philosophie*, vol. 1, ed. J. Ritter, Basel and Stuttgart: Schwabe, 1971, cols. 495–97; T. Trappe, 'Ungrund; Urgrund', and H. Holzey and D. Schoeller Reich, 'Ursprung', in *Historisches Wörterbuch der Philosophie*, vol. 11, ed. J. Ritter, K. Gründer, and G. Gabriel, Basel: Schwabe, 2001, cols. 168–72 and 417–24; and R. McKirahan, 'Archē', in *Routledge Encyclopedia of Philosophy*, ed. E. Craig, vol. 1, London and New York: Routledge, 1998, pp. 359–61.

2 See the entries for *archaios* and *archē* in F. Astius, *Lexicon Platonicum sive vocum platonicarum index*, vol. 1, Leipzig: Weidmann, pp. 282–85; and R. Bombacigno, I. Ramelli and E. Vimercati (eds), *Lexicon: Plato*, ed., Milan: Biblia, 2003, pp. 156–57.

3 *Laws*, book 3, 680 d and 865 d, in Plato, *Laws*, tr. R. G. Bury, vol. 2, London: Heinemann; New York: Putnam, 1926, pp. 179 and 241; see Schweizer, 'Archaisch', col. 495.

4 *Theaetetus*, 180 d, in Plato, *Theaetetus; Sophist*, tr. H. N. Fowler, London: Heinemann; New York: Putnam, 1921, p. 143; see Schweizer, 'Archaisch', col. 495.

5 *Cratylus*, 418 c, in Plato, *Cratylus; Parmenides; Greater Hippias; Lesser Hippias*, tr. H. N. Fowler, London: Heinemann; Cambridge, MA: Harvard University Press, 1970, p. 119; see Schweizer, 'Archaisch', col. 495.

6 *Symposium*, 193 c, in Plato, *Lysis; Symposium; Gorgias*, tr. W. R. M. Lamb, London: Heinemann; New York: Putnam, 1932, p. 147; see Schweizer, 'Archaisch', col. 495.

7 *Politics*, book 2, chapter 8, 1268 b 39; Aristotle, *Basic Works*, ed. Richard McKeon, New York: Random House, 1941, p. 1163.

8 *Rhetoric*, 1387 a 16; *Basic Works*, p. 1399.

9 Diels-Kranz, 11 A 12 = Aristotle, *Metaphysics*, 983 b, in J. Barnes (ed.), *Early Greek Philosophy*, Harmondsworth: Penguin, 1987, p. 63.

10 Diels-Kranz, 12 A9, B1 = Simplicius, *Commentary on the Physics of Aristotle*, 24.13–25, in *Early Greek Philosophy*, p. 75; cf. Aristotle, *Physics*, 203 b, in *Early Greek Philosophy*, pp. 75–76. For further discussion of Anaximander's notion of *apeiron* or the universe as 'boundless', 'unlimited', or 'infinite', see J. Barnes, *The Presocratic Philosophers*, London: Routledge and Kegan Paul, 1979, chapter 2, 'Anaximander on Nature', pp. 19–37.

11 Diels-Kranz, fragment 6: 'The being of things, which is eternal, and nature in itself admit of divine and not human knowledge, except that it was impossible for any of the things that are and are known by us to have come to be, if the being of the things from which the world-order came together, both the limitings things and the unlimited things, did not preexist. But since these *beginnings* preexisted and were neither alike nor even related, it would have been impossible for them to be ordered, if a harmony had not come upon them, in whatever way it came to be'; fragment 8: '[the] one is the first *principle* [*starting-point*] of all things'; and fragment 13: 'the brain [contains] the *origin* of man, the heart the origin of animals, the navel the origin of plants, the genitals the origin of all (living things)' (Carl A. Huffman, *Philolaus of Croton: Pythagorean and Presocratic: A Commentary on the Fragments and Testimonia with Interpretive Essays*, Cambridge: Cambridge University Press, 1993, pp. 123–24, 345, 307; and see part 2, chapter 3, 'Philolaus' Use of ἀρχαί and the method of hypothesis', pp. 78–92); see R. McKirahan, 'Archē', p. 359.

12 Diels-Kranz, 12 A 9, in *Early Greek Philosophy*, p. 75.

13 Diels-Kranz, 13 A 7 = Hippolytus, *Refutation of all Heresies*, Book 1, Chapter 7, 1–9, in *Early Greek Philosophy*, pp. 77–78.

14 Diels-Kranz, 22 B 30 = Clement, *Stromata*, book 5, chapter 14, §104, in *Early Greek Philosophy*, p. 122.

15 Diels-Kranz, 22 A 1 = Diogenes Laertius, *Lives and Opinions of the Philosophers*, book 9, chapter 1, §8, in *Early Greek Philosophy*, p. 107.

16 Diels-Kranz, 28 B 8 = Simplicius, *Commentary on the Physics of Aristotle*, 144.25–146.27, in *Early Greek Philosophy*, pp. 134–35.

17 Diels-Kranz, 67 A 8 = Cicero, *Academica*, book 2, chapter 37, §118; in *Les Présocratiques*, ed. J.-P. Dumont, Paris: Gallimard, 1988, p. 734.

18 Diels-Kranz, 68 B 9 = Sextus Empiricus, *Against the Mathematicians*, book 7, 135–36; in *Early Greek Philosophy*, p. 253.

19 Arisotle, *On the Soul*, book 1, chapter 2, 404 a, in *Basic Works*, pp. 538–39. The comparison inspires a useful work on Democritus, see J. Salem, *Démocrite: Grains de poussière dans un rayon de soleil*, Paris: Vrin, 2002.

20 Diels-Kranz, 67 B 2 = Stobaeus, *Anthology*, book 1, chapter 4, 7c; in *Early Greek Philosophy*, p. 243.

21 Diels-Kranz, 67 A 37 = Clement of Alexandria, *Stromata/Miscellanies*, book 2, chapter 21, §129; in J.-P. Dumont (ed.), *Les Présocratiques*, Paris: Gallimard, 1988, p. 745.

22 *Phaedo*, 101 e: 'the principle' (Plato, *The Collected Dialogues*, ed. Edith Hamilton and Huntington Cairns, Princeton, NJ: Princeton University Press, 1989, p. 83); *Phaedrus*, 245 d: 'Now a first principle cannot come into being, for while anything that comes to

be must come to be from a first principle, the latter itself cannot come to be from anything whatsoever; if it did, it would cease any longer to be a first principle' (Plato, *Collected Dialogues*, p. 492); and *Republic*, book 6, 510 b: 'In one part of it a soul, using as images the things that were previously imitated, is compelled to investigate on the basis of hypotheses and makes its way not to a beginning but to an end; while in the other part it makes its way to a beginning that is free from hypotheses [. . .]', and book 7, 533 c: 'Only the dialectical way of inquiry proceeds in this direction, destroying the hypotheses, to the beginning itself in order to make it secure; and when the eye of the soul is really buried in a barbaric bog, dialectic gently draws it forth and leads it up above' (*The Republic of Plato*, ed. and tr. Allan Bloom, New York: Basic Books, 1968, pp. 190 and 212); cf. Bloom's note on p. 464: 'The word translated as "beginning" can also be rendered as "starting-point", "principle" or "cause" '; see R. McKirahan, 'Archē', p. 359.

23 Aristotle, *Metaphysics*, Book 5, Chapter 1, 1013 a, in *The Basic Works of Aristotle*, tr. R. McKeon, New York: Random House, 1941, p. 752; see R. McKirahan, 'Archē', p. 359.

24 Aristotle, *Metaphysics*, 1013 a, in *Basic Works*, p. 752.

25 Aristotle, *Metaphysics*, 1013 a, in *Basic Works*, p. 752.

26 Compare with his discussion of Melissus and Parmenides in *Physics*, book 1, chapter 2, 186 a, in *Basic Works*, p. 221–22, of Anaxagoras in *Physics*, book 1, chapter 4, 187 a, in *Basic Works*, pp. 224–25, and of the principles of Empedocles in *Physics*, book 1, chapter 6, 189 a, in *Basic Works*, p. 228; see R. McKirahan, 'Archē', p. 359.

27 *Physics*, book 1, chapters 6–7, 189 a–191 a, in *Basic Works*, pp. 228–33; see R. McKirahan, 'Archē', p. 359.

28 *Posterior Analytics*, book 1, chapter 10, 76 a, in *Basic Works*, p. 124; see R. McKirahan, 'Archē', p. 359.

29 McKirahan, 'Archē', p. 361.

30 Plotinus, Sixth Ennead, First Tractate, §30, in Plotinus, *The Enneads*, tr. S. MacKenna, rev. B. S. Page, London: Faber and Faber, 1969, p. 470; see Schweizer, 'Archaisch', col. 496.

31 'When the primal Being blends with an inferior principle, it is hampered in its relation to the highest, but without suffering the loss of its own nature since it can always recover its *earliest state* by turning its tendency back to its own' (Fourth Ennead, Seventh Tractate, §9, in *The Enneads*, p. 353); see Schweizer, 'Archaisch', col. 496.

32 Plotinus, Second Ennead, Third Tractate, §15, in *The Enneads*, p. 101; see Schweizer, 'Archaisch', col. 496.

33 Second Ennead, Third Tractate, §8, in *The Enneads*, p. 97; see Schweizer, 'Archaisch', col. 496.

34 Plato, *Parmenides*, 160 b, in *Collected Dialogues*, p. 950; see MacKenna's note in Plotinus, *The Enneads*, p. 361.

35 Fifth Ennead, Second Tractate, §1, ll. 1, 7ff. (*The Enneads*, p. 361).

36 J. Bussanich, 'Plotinus's metaphysics of the One', in Lloyd P. Gerson (ed.), *The Cambridge Companion to Plotinus* (Cambridge: Cambridge University Press, 1996), pp. 38–65 (p. 38).

37 Ovid, *Metamorphoses*, book 3, 154–64; Ovid, *Metamorphoses*, tr. A. D. Melville, Oxford and New York: Oxford University Press, 1987, pp. 55–56.

38 Seneca, *Epistulae ad Luciulium*; Seneca, *Letters to Lucilius*, tr. E. P. Barker, 2 vols, Oxford: Clarendon Press, 1932, vol. 2, pp. 127–28.

39 R. Otto, *Das Heilige: Über das Irrationale in der Idee des Göttlichen und sein Verhältnis zum Rationalen* [1917], Munich: Beck, 1987; R. Otto, *The Idea of the Holy: An Inquiry into the non-rational factor in the idea of the divine and its relation to the rational*, tr. J. W. Harvey, London, Oxford, New York: Oxford University Press, 1958.

40 *Faust II*, ll. 6271–74; in J. W. von Goethe, *Faust*, ed. Cyrus Hamlin, tr. Walter Arndt, 2nd edn, New York and London: Norton, 2001, p. 178.

41 P. Hadot, *La Philosophie comme manière de vivre: Entretiens avec Jeanne Carlier et Arnold I. Davidson*, Paris: Albin Michel, 2001, p. 172.

42 St. Irenaeus of Lyon, *Against the Heresies* [*Contra haereses*], book 1, chapter 1, §1; in Bentley Layton (ed.), *The Gnostic Scriptures*, London: SCM Press, 1987, p. 281. For a recent discussion of Valentinian Gnosticism, see Michel Onfray, *Le christianisme hédoniste* [*Contre-histoire de la philosophie*, vol. 2], Paris: Grasset, 2006, 'Valentin et «les semences de l'élection»', pp. 59–62.

43 Schopenhauer, 'On the Fourfold Root of the Principle of Sufficient Reason', chapter 2, §8, 'Spinoza', in *Werke in fünf Bänden*, ed. L. Lütkehaus, vol. 3, Zurich: Haffmans, 1988, p. 28; see Trappe, 'Ungrund; Urgrund', cols 168–69.

44 See Holzhey and Schoeller Reisch, 'Ursprung', col. 417.

45 See G. W. F. Hegel, *Vorlesungen über die Geschichte der Philosophie*, vol. 3, Berlin: deb, 1984, part 2, section 1, subsection B, §6 and part 3, section 1, subsection B, pp. 82–85 and 164–86; and R. Steiner, *Mysticism at the Dawn of the Modern Age* [based on lectures given in 1900], tr. K. E. Zimmer, Blauvelt, NY: Rudolf Steiner Publications, 1980.

46 Sermon 47, 'Haec est vita aeterna', in Meister Eckhart, *Sermons and Treatises*, ed. and tr. M. O'C. Walshe, vol. 2, Longmead: Element Books, 1987, pp. 27–30 (p. 27); *Die deutschen und lateinischen Werke: Die deutschen Werke*, vol. 2, *Predigten*, ed. J. Quint, Stuttgart: Kohlhammer, 1971, pp. 378–88 (here: p. 379); see Holzhey and Schoeller Reisch, 'Ursprung', col. 418.

47 Sermon 88, 'Mandatum novum do vobis ut diligatis invicem sicut dilexi vos', in Eckhart, *Sermons and Treatises*, vol. 2, pp. 279–83: 'Supposing my eye were a light and strong enough to absorb the full force of the sun's light and unite with it, then it would see not only by its own power, but it would see with the light of the sun in all its strength. So it is with my intellect. The intellect is a light, and if I turn it right away from all things and in the direction of God, then, since God is continually overflowing with grace, my intellect becomes illumined and united with love, and therein knows and loves God as He is in Himself. By this we are shown how God continually flows out into rational creatures with the light of grace and how we must approach this light of grace with our intellect to be drawn out of ourselves and ascend into that one light that is God Himself' (p. 281); see Holzhey and Schoeller Reisch, 'Ursprung', col. 418.

48 Cited in Adolf Lasson, *Meister Eckhart, der Mystiker: Zur Geschichte der religiösen Speculation in Deutschland*, Berlin: Wilhelm Hertz, 1868, chapter 2, §2, 'Die Gottheit', pp. 109–12; see Trappe, 'Ungrund; Urgrund', col. 169.

49 Lasson, *Meister Eckhart*, p. 112.

50 The Book of Wisdom, 18: 14–15 (DRV).

51 Sermon 1, 'Dum medium silentium omnia et nox in suo cursu medium inter haberet', in Eckhart, *Sermons and Treatises*, tr. M. O'C. Walshe, Longmead: Element Books, 1979, vol. 1, pp. 1–13 (here: p. 3). For further discussion, see E. Grassi, *Kunst und Mythos*, Frankfurt am Main: Suhrkamp, 1990, pp. 130–37; and *Die unerhörte Metapher*, ed. E. Hidalgo-Serna (Frankfurt am Main: Hain, 1992).

52 Sermon 6, 'Intravit Jesus in templum dei et ejiciebat omnes verdentes et ementes', in *Sermons and Treatises*, vol. 1, pp. 55–62 (here: p. 61).

53 Sermon 36, 'Adolescens, tibi dico: surge', in *Sermons and Treatises*, vol. 1, pp. 263–66 (pp. 265–66).

54 Johannes Tauler, Sermon 26 ('Repleti sunt omnes Spiritu Sancto'), in *Die Predigten Taulers: Aus der Engelberger und der Freiburger Handschrift sowie aus Schmidts Abschriften der ehemaligen Straßburger Handschriften*, ed. Ferdinand Vetter, Berlin: Weidmann, 1910, p. 107; see Holzhey and Schoeller Reisch, 'Ursprung', col. 418.

55 Tauler, Sermon 83 ('Audi, Israel'), in *Die Predigten Taulers*, p. 277.
56 Heinrich Suso, *Das Büchlein von der ewigen Weisheit*, Neu-Ruppin: Alfred Oehmigke, 1861, part 1, §1 and §2, pp. 4 and 8.
57 For an overview of Boehme and his teaching, see R. M. Jones, *Spiritual Reformers in the 16th and 17th Centuries*, London: Macmillan, 1914, chapters 9, 10 and 11, pp. 151–234; and J. Boehme, *Essential Readings*, ed. R. Waterfield, Wellingborough: Crucible, 1989.
58 J. Boehme, 'De incarnatione verbi, oder Von der Menschwerdung Jesu Christi', in *Sämtliche Schriften*, ed. W.-E. Peuckert, vol. 4, Stuttgart: Frommanns-Holzboog, 1957, part 2, chapter 1, §8, pp. 120–21.
59 See A. Schopenhauer, *The World as Will and Representation*, tr. E. F. J. Payne, 2 vols, New York: Dover, 1969, vol. 1, §36, where he describes genius as 'the ability to leave entirely out of sight our own interest, our willing, and our aims, and consequently to discard entirely our own personality for a time, in order to remain *pure knowing subject*, the clear eye of the world [*klares Weltauge*]' (p. 186); cf. §38 (p. 198), §54 (p. 282) and vol. 2, chapter 30, p. 371. In *The World as Will and Representation* Schopenhauer describes the self-knowledge of the will as 'the objectivity, the revelation, the mirror of the will' (§29, p. 165), the subject purified of will as 'the clear mirror of the inner nature of the world' (§36, p. 186).
60 Boehme, 'De incarnatione verbi', p. 121. Here, 'magia' means the transition from a possible to a real existence in a material sense (see 'Glossar', in J. Boehme, *Von der Gnadenwahl*, ed. R. Pietsch, Stuttgart: Reclam, 1988, p. 244). See also *The Signature of All Things* (*De Signatura Rerum*), chapter 3, §2: 'Without nature God is a mystery, understand in the nothing, for without nature is the nothing, which is an eye of eternity, an abyssal eye, that stands or sees in the nothing, for it is the abyss; and this same eye is a will, understand a longing after manifestation, to find the nothing; but now there is nothing before the will, where it might find something, where it might have a place to rest, therefore it enters into itself, and finds itself through nature' (J. Boehme, *"The Signature of All Things" with Other Writings*, tr. C. Bax, London and Toronto: Dent; New York: Dutton, 1912, p. 22).
61 J. Boehme, 'Mysterium Pansophicum, Oder Gründlicher Bericht von dem Irdischen und Himmlischen Mysterio' [1620], in *Sämtliche Schriften*, vol. 4, text 1, p. 97.
62 H. L. Martensen, *Jacob Boehme: His Life and Teaching, or Studies in Theosophy*, tr. T. Rhys Evans, London: Hodder and Stoughton, 1885, p. 7; cf. C. Wilson, *Religion and the Rebel*, Bath: Ashgrove Press, 1984, pp. 153–54, who sees in Boehme's experience a confirmation of Paracelsus' principle that 'we may look into Nature in the same way that the sun shines through a glass'.
63 Schopenhauer, 'On the Fourfold Root', in *Werke*, vol. 3, p. 28.
64 Spinoza, *Ethics*, Part 1, definition 1; Part 1, proposition 16, corollary 1; in Spinoza, *Selections*, ed. J. Wild, London: Scribner, 1928, pp. 94 and 113.
65 Schopenhauer, 'On the Fourfold Root', in *Werke*, vol. 3, pp. 27, 29. For an account of Schelling's thought, see X. Tillette, *Schelling: Une philosophie en devenir*, 2 vols, Paris: Vrin, 1970; A. Gulyga, *Schelling: Leben und Werk*, tr. from the Russian E. Kirsten, Stuttgart: Deutsche Verlags-Anstalt, 1989; and, for an overview of the recent debate over his thought, J. Norman and A. Welchman (eds), *The New Schelling*, London and New York: Continuum, 2004.
66 F. W. J. Schelling, *Philosophical Inquiries into the Nature of Human Freedom*, tr. James Gutmann, La Salle, IL: Open Court, 1936, p. 33; Schelling, *Werke*, ed. M. Schröter, Munich: Beck, 1946–1962, vol. I.7, p. 359.
67 Schelling, *Philosophical Inquiries*, p. 87; *Werke*, vol. I.7, p. 406; see Schopenhauer, *On the Fourfold Root*, §8, in *Werke*, vol. 3, p. 28.
68 Schelling, *Werke*, vol. I.10, p. 189; see J. Gutmann, 'Introduction' to *Philosophical Inquiries*, p. xlvi.

69 Schelling, *Werke*, II.3, 123; see J. Gutmann, 'Introduction' to *Philosophical Inquiries*, p. xlvii.

70 Indeed, in his essay entitled 'Freud's Position in the History of Modern Thought' (1929), Thomas Mann (1875–1955) wondered about the relation between 'the great return to the noctural, the sacred and the origin [*Heilig-Ursprüngliche*], the vital and pre-conscious, to the mythical-historical-Romantic womb' (in the work of Ernst Moritz Arndt, Johann Joseph von Görres, and the Brothers Grimm), the fascination with 'the chthonic, night, death, the daimonic' and 'a pre-Olympian primal and earth religiosity' (in Carl Gustav Carus and Schopenhauer, as well as Jörgen Zoega, Friedrich Creuzer, Karl Otfried Müller, and Bachofen), and 'the powerlessness of mind and reason in comparison to the powers of the depths of the soul, the dynamics of passion, the irrational, and the unconscious' (in Klages and Oswald Spengler) (Thomas Mann, 'Die Stellung Freud in der modernen Geistegeschichte', in *Schriften und Reden zur Literatur, Kunst und Philosophie*, vol. 1, Frankfurt am Main: Fischer, 1968, pp. 367–85 [p. 371]); cf. his comments in 'Pariser Rechenschaft' (in Thomas Mann, *Über mich selbst*, Frankfurt am Main: Fischer, 1994, pp. 271–357, esp. p. 309).

71 Schelling, *Philosophical Inquiries*, pp. 87, 88–89; *Werke*, vol. I.7, pp. 406, 407–08; see Trappe, 'Ungrund; Urgrund', col. 169. For further discussion of the *Ungrund*, see M. Gabriel, 'The Mythological Being of Reflection – An Essay on Hegel, Schelling, and the Contingency of Necessity', in M. Gabriel and S. Žižek, *Mythology, Madness and Laughter: Subjectivity in German Idealism*, London and New York: Continuum, 2009, pp. 15–94 (esp. p. 59).

72 Schelling, *Philosophical Inquiries*, pp. 87, 89; *Werke*, vol. I.7, 406, 408.

73 Schopenhauer, 'On the Fourfold Root', in *Werke*, vol. 3, p. 28; or literally, *hoc quidem vere palmarium est*.

74 Schelling, 'Andere Deduktion der Principien der positiven Philosophie', in *Werke*, vol. II.4, pp. 335–56 (here: p. 337). See Gabriel, 'The Mythological Being of Reflection', pp. 57–58.

75 Schelling, 'Andere Deduktion der Principien', in *Werke*, vol. II.4, p. 337.

76 Schelling, 'Andere Deduktion der Principien', in *Werke*, vol. II.4, p. 341.

77 Kant, *Critique of Pure Reason*, ed. and tr. Paul Guyer and Allen W. Wood, Cambridge: Cambridge University Press, 1997, A 11–12 (B 26), p. 133.

78 See *Religion with the Boundaries of Mere Reason*, Part One, chapter 4, 'Concerning the Origin of Evil in Human Nature', where Kant defines *Ursprung* – or 'origin (the first origin)' – as 'the descent of an effect from its first cause, i.e., from that cause which is not in turn the effect of another cause of the same kind', distinguishing between '*origin according to reason*' and '*origin according to time*', where in the case of the former 'the effect's *being* is considered' and in the latter 'its *occurrence*, and hence, as an event, it is referred to its *cause in time*' (I. Kant, *Religion and Rational Theology*, ed. and tr. A. W. Wood and G. di Giovanni [Cambridge Edition of the Works of Immanuel Kant], Cambridge: Cambridge University Press, 1996, pp. 39–215 (here: p. 85); see Holzhey and Schoeller Reisch, 'Ursprung', cols 418–19.

79 Kant, *Metaphysik L1: Kosmologie, Psychologie, Theologie nach Pölitz*, in *Vorlesungen über die Metaphysik* (1821), in *Gesammelte Schriften*, ed. Königlich Preußische Akademie der Wissenschaften [Akadamie-Ausgabe], 29 vols, Berlin: Riemer; Walter de Gruyter, 1902–1980, vol. 28/1, pp. 193–350 (here: p. 324).

80 Kant, *Critique of Practical Reason*, in *Practical Philosophy*, ed. and tr. M. J. Gregor, Cambridge: Cambridge University Press, 1986, pp. 133–271 (here: p. 254).

81 Kant, *Critique of Pure Reason*, A 587/B 615 (p. 561); see Trappe, 'Ungrund; Urgrund', col. 169.

82 Kant, *Critique of Judgement*, §65 and §74, in *Critique of the Power of Judgment*, ed. P. Guyer and E. Matthews [Cambridge Edition], Cambridge: Cambridge University Press, 2000, pp. 247 and 268; see Trappe, 'Ungrund; Urgrund', col. 169.

83 Kant, *Critique of Judgement*, §79, p. 285; see Trappe, 'Ungrund; Urgrund', col. 169.

84 See Holzhey and Schoeller Reisch, 'Ursprung', col. 422.

85 See Trappe, 'Ungrund; Urgrund', col. 169.

86 For further discussion of Obereit, see C. Behle, ' "Allharmonie von Allkraft zum All-Wohl": Jacob Hermann Obereit zwischen Aufklärung, Hermetismus und Idealismus', in T. Lange and H. Neumeyer (eds), *Kunst und Wissenschaft um 1800*, Würzburg: Königshausen & Neumann, 2000, pp. 151–73; and M.-G. Dehrmann, *Produktive Einsamkeit: Studien zu Gottfried Arnold, Shaftesbury, Johann Georg Zimmermann, Jacob Hermann Obereit und Christoph Martin Wieland*, Hanover: Wehrhahn, 2002, chapter 3.2, 'Die einsamen Heldengeister – Johann Hermann Obereit', pp. 92–113.

87 Romans 11:36; cf. the conclusion to the Canon of the Mass.

88 See J. H. Obereit, 'Obereits Widerruf für Kant: Ein psychologischer Kreislauf', in ΓΝΩΘΙΣΑΥΤΟΝ, vol. 9, no. 2 (1792), 107–143 (reprinted in K. P. Moritz, *Die Schriften in dreissig Bänden*, ed. P. and U. Nettelbeck, vol. 9, ΓΝΩΘΙΣΑΥΤΟΝ *oder Magazin zur Erfahrungsseelenkunde als ein Lesebuch für Gelehrte und Ungelehrte*, Nördlingen: Greno, 1986, pp. 172–98 (here: pp. 182–83); see Trappe, 'Ungrund; Urgrund', cols 169 and 172.

89 'Obereits Widderuf für Kant', pp. 196–97.

90 For a helpful account of Fichte's thought and its premises, see F. Neuhouser, *Fichte's Theory of Subjectivity*, Cambridge: Cambridge University Press, 1990.

91 Fichte, *Darstellung der Wissenschaftslehre: Aus dem Jahre 1801*, in J. G. Fichte, *Gesammelte Werke*, ed. I. H. Fichte, vol. 2, Berlin: Veit, 1845, pp. 3–163 (here: p. 63); see Holzhey and Schoeller Reisch, 'Ursprung', col. 420.

92 Fichte, *Darstellung der Wissenschaftslehre* [1801], in *Gesammelte Werke*, vol. 2, p. 67; see Holzhey and Schoeller Reisch, 'Ursprung', col. 420.

93 Fichte's thought can appear to be, to say the least, highly obscure, but a sense of what Fichte is aiming at can be gained from his 'Annalen des philosophichen Tons' (1797), a satirical attack on styles of reviewing, but which contains the following critique of the Kantian *a priori*:

>   What does that mean: concepts are *a priori* there?
>   'Well, it means they are present prior to all experience in the mind.'
>   Good: what whatever it is in which something else is present, must itself be present?
>   'Indeed.'
>   Thus prior to all experience a mind must be present. – At this point they fall silent

(Fichte, *Gesammelte Werke*, vol. 2, pp. 459–89 [here: pp. 476–77]). For further discussion of the *Wissenschaftslehre*, see G. J. Seidel, *Fichte's 'Wissenschaftslehre' of 1794: A Commentary on Part 1*, West Lafayette, ID: Purdue University Press, 1993.

94 Fichte, 'Second Introduction to the *Wissenschaftslehre*, For Readers Who Already Have A Philosophical System of Their Own' [1797], §5, in J. G. Fichte, *Introductions to the 'Wissenschaftslehre' and Other Writings (1797–1800)*, ed. and tr. D. Breazeale, Indianapolis, ID: Hackett, 1994, pp. 36–105 (here: p. 46); *Werke*, ed. I. H. Fichte, vol. 1, Berlin: Veit, 1845, pp. 453–518 (here: p. 463). For further discussion of intellectual intuition, see D. Simpson (ed.), *German Aesthetic and Literary Criticism: Kant, Fichte, Schelling, Schopenhauer, Hegel*, Cambridge: Cambridge University Press, 1984, p. 21; and X. Tilliette, *Recherches sur l'intuition intellectuelle de Kant à Hegel*, Paris: Vrin, 1995.

95 Fichte, 'Second Introduction to the *Wissenschaftslehre*', §5, in Fichte, *Introductions to the 'Wissenschaftslehre' and Other Writings*, pp. 46–47; *Werke*, vol. 2, p. 463.

96 Schelling, 'Allgemeine Übersicht der neusten philosophischen Literatur' (1797/1798), in F. W. J. Schelling, *Werke* [Akademie-Ausgabe], vol. I/4, ed. W. G. Jacobs, W. Schieche, and H. Buchner, Stuttgart: Frommann-Holzboog, 1988, pp. 57–190 (here: p. 147); see Holzhey and Schoeller Reisch, 'Ursprung', col. 420.

97 Schelling, 'Allgemeine Übersicht der neusten philosophischen Literatur', in *Werke* [Akademie-Ausgabe], vol. I/4, p. 147.

98 For a discussion of Spinozism in Germany, see F. Beiser, *The Fate of Reason: German Philosophy from Kant to Fichte*, Cambridge, MA, and London: Harvard University Press, 1987, pp. 48–61 and 75–83.

99 Schelling, 'Philosophische Briefe über Dogmatismus und Kriticismus' (1795), Letter 8, in *Sämmtliche Werke*, ed. M. Schröter, vol. I.1, *Sämmtliche Werke 1792–1797*, Stuttgart and Ausgburg: Cotta, 1856, pp. 281–341 (here: p. 318).

100 Spinoza, *Ethics*, part 5, proposition 23, scholium; in *Selections*, p. 385.

101 Schelling, *Einleitung in die Philosophie der Offenbarung*, Lecture 12, in *Werke*, vol. II.3, p. 245.

102 Or under the influence of Plotinus: see W. Beierwaltes, 'The Legacy of Neoplatonism in F. W. J. Schelling's Thought', *International Journal of Philosophical Studies* 10, 2002, 393–428.

103 Schelling, *Philosophical Investigations*, p. 34; *Werke*, vol I.7, p. 359; see Holzhey and Schoeller Reisch, col. 420.

104 Schelling, *Philosophical Inquiries into the Nature of Human Freedom*, p. 34; *Werke*, vol. I.7, pp. 359–60. For a discussion of Schelling drawing on a phrase from this passage, see S. Žižek, *The Indivisible Remainder: An Essay on Schelling and Related Matters*, London: Verso, 1996.

105 Schelling, *System of Transcendental Idealism*, part 6, §3, in Simpson (ed.), *German Aesthetic and Literary Criticism*, p. 128; *Werke*, I.3, p. 625. On this phrase, see Schelling's alternate version of this passage, reprinted in *German Aesthetic and Literary Criticism*, p. 268, n. 21; *Werke*, vol. I.3, p. 625, fn. 1.

106 Fichte, *Grundlage des Naturrechts nach Prinzipien der Wissenschaftslehre*, in *Werke*, ed. I. H. Fichte, vol. 3, Berlin: Veit, 1845, pp. 83–84.

107 *The Will to Power*, §796; in F. Nietzsche, *The Will to Power*, ed. W. Kaufmann, tr. W. Kaufmann and R. J. Hollingdale, New York: Vintage, 1968, p. 419; *Sämtliche Werke: Kritische Studienausgabe*, ed. G. Colli and M. Montinari, 15 vols, Berlin and New York: Munich: Walter de Gruyter; Deutscher Taschenbuch Verlag, 1967–1977; 1988, vol. 12, 2[114], p. 119. Henceforth referred to as KSA, followed by volume, section or fragment, and page number.

108 See Nietzsche, *The Birth of Tragedy*, §5, §24, and 'Attempt at a Self-Criticism', §5; in *Basic Writings*, ed. and tr. W. Kaufmann, New York: Modern Library, 1968, pp. 52, 141, 22; KSA vol. 1, pp. 47, 152, 17.

109 For further discussion of this term and the works it covers, see R. Abbey, *Nietzsche's Middle Period*, Oxford: Oxford University Press, 2000.

110 Nietzsche, *Human, All Too Human*, vol. 1, §1, in F. Nietzsche, *Human, All Too Human*, tr. R. J. Hollingdale, Cambridge: Cambridge University Press, 1986, p. 12; KSA, vol. 2, p. 23. See Holzhey and Schoeller Reisch, col. 421.

111 Nietzsche, *Daybreak*, §44; in F. Nietzsche, *Daybreak*, tr. R. J. Hollingdale, Cambridge: Cambridge University Press, 1982, pp. 43–44; KSA, vol. 3, pp. 51–52. See Holzhey and Schoeller Reisch, col. 421.

112 Nietzsche, *Human, All Too Human*, vol. 1, §1, p. 12; KSA, vol. 2, p. 24.

113 Nietzsche, *On the Genealogy of Morals*, Preface, §3, §2; *Basic Writings*, pp. 452–53, 451; KSA, vol . 5, pp. 249, 248. See Holzhey and Schoeller Reisch, col. 421. For a discussion of the different uses of the word *Ursprung* and their relation to such terms as *Entstehung* or *Herkunft*, see Michel Foucault, 'Nietzsche, Genealogy, History'

[1971], in Paul Rabinow (ed.), *The Foucault Reader*, Harmondsworth: Penguin, 1984, pp. 76–97; and see below.

114 *The Birth of Tragedy*, §16; *Basic Writings*, p. 104; KSA, vol. 1, p. 108.

115 *The Birth of Tragedy*, §22; *Basic Writings*, pp. 131–32; KSA, vol. 1, p. 141.

116 Nietzsche, fragment of a continuation of *The Birth of Tragedy*, written in 1871; KSA 7, 10[1], 334: *Die Menschheit* [. . .] *darf* [. . .] *als die fortgesetzte Geburt des Genius bezeichnet werden: von jenem ungeheuren allgegenwärtigen Gesichtspunkte des Ur-Einen aus ist in jedem Moment der Genius erreicht, die ganze Pyramide des Scheins bis zu ihrer Spitze vollkommen.* See Trappe, 'Ungrund; Urgrund', col. 168.

117 Ibid., KSA 7, 10[1], 335: *des in völliger Selbstvergessenheit mit dem Urgrunde der Welt eins gewordenen Menschen, der jetzt aus dem Urschmerze heraus den Wiederschein desselben zu seiner Erlösung schafft.*

118 *The Birth of Tragedy*, 'Attempt at a Self-Criticism', §2, in *Basic Works*, p. 18; KSA, vol. 1, p. 13.

119 Nietzsche, KSA, vol 11, 41[4], p. 679.

120 See K. Gründer (ed.), *Der Streit um Nietzsches 'Geburt der Tragödie': Die Schriften von E. Rohde, R. Wagner, U. v. Wilamowitz-Möllendorff*, Hildesheim: Olms, 1969; 1989.

121 Walter Benjamin, *The Origin of German Tragic Drama*, tr. John Osborne (London: NLB, 1977), p. 62.

122 See M. W. Jennings, *Dialectical Images: Walter Benjamin's Theory of Literary Criticism*, Ithaca and London: Cornell University Press, 1987, pp. 166–69 (on the symbol) and 170–78 (on allegory); and M. Pensky, *Melancholy Dialectics: Walter Benjamin and the Play of Mourning*, Amherst: University of Massachusetts Press, 1993, pp. 112–17.

123 Benjamin contrasts Friedrich Creuzer's view of the symbol in *Symbolik und Mythologie der alten Völker, besonders der Griechen*, which regards the symbol as being characterized by 'the momentary, the total, the inscrutability of its origin, the necessary' and by its 'clarity [. . .] brevity [. . .] grace and beauty' (*Symbolik und Mythologie der alten Völker, besonders der Griechen*, 1. Teil, 2 edn, Leipzig and Darmstadt, 1819, p. 118, 66–67, 70–71) with Görres' dismissal of the view that the symbol is 'being' and allegory is 'a sign' (cited in *Symbolik und Mythologie*, p. 199), and even Creuzer's argument that, because the symbol is a 'momentary totality', it is allegory, not the symbol, that 'embraces myth' (*Symbolik und Mythologie*, pp. 70–71) – a view shared by Johann Heinrich Voss in his *Antisymbolik*, vol. II (Stuttgart 1826, p. 223). Furthermore, Creuzer distinguishes between 'mystic' (or 'formal') symbols, 'plastic' symbols, and 'real' symbols (or 'emblems', *Sinnbilder*), as Görres points out (Creuzer, *Symbolik und Mythologie*, pp. 63–64; cf. Görres, cited in *Symbolik und Mythologie*, p. 147).

124 Benjamin, *The Origin of German Tragic Drama*, p. 166. Whereas Greek tragedy (in Nietzsche's view) represents a totality, operates by means of the symbol, and offers an ecstatic affirmation of life through its Dionysian pessimism, baroque drama (in Benjamin's view) is characterized by postponement and deferral, operates by means of allegory, and offers a melancholy view of life as requiring religious salvation. Thus whereas classical tragedy hinges on the symbol and its presentation, however imperfect, of a timeless truth in beautiful appearance, allegory represents the very temporariness of truth in the ruination of appearance.

125 Benjamin, *The Origin of German Tragic Drama*, p. 102.

126 See Hermann Cohen, *Logik der reinen Erkenntnis*, Berlin: Cassirer, 1902; second edn, 1914.

127 Benjamin, *The Origin of German Tragic Drama*, p. 45; *Der Ursprung des deutschen Trauerspiels*, p. 28.

128 M. Foucault, *L'archéologie du savoir*, Paris: Gallimard, 1969, pp. 12 (see Holzhey and Schoeller Reisch, col. 421) and 17.

129 Foucault, *L'archéologie du savoir*, p. 12.

130 L. Ferry and A. Renaut, *La Pensee 68*, Paris: Gallimard, 1985, p. 105.

131 'Nietzsche, Genealogy, History', in P. Rabinow (ed.), *The Foucault Reader*, Harmondsworth: Penguin, 1984, pp. 76–100; 'Nietzsche, la généalogie, l'histoire', in *Philosophie: Anthologie*, ed. A. I. Davidson and F. Gros, Paris: Gallimard, 2004, 393–423.

132 'Nietzsche, Genealogy, History', in *The Foucault Reader*, pp. 76, 78–79.

133 'Nietzsche, Genealogy, History', in *The Foucault Reader*, p. 78.

134 'Nietzsche, Genealogy, History', in *The Foucault Reader*, p. 78; 'Nietzsche, la généalogie, l'histoire', in *Philosophie*, p. 396.

135 R. Enthoven, *Les nouveaux chemins de la connaissance*, 'La folie 1/5: Michel Foucault', broadcast on France Culture, 11 January 2010: 'Il y a toujours chez Foucault ce sentiment étrange que Foucault permet de penser des choses extravagantes, il ouvre la possibilité [. . .] de penser autrement, et il a une façon de le dire, j'allais dire jusqu'à même dans sa voix, qui semble contredire cette ambition, c'est-à-dire qu'il a une présentation – c'est la plus classique, des penseurs vertigineux.'

136 T. W. Adorno, 'Der Begriff der Unbewußten in der transzendentalen Seelenlehre', in *Philosophische Frühschriften* [*Gesammelte Schriften*, vol. 1], Frankfurt am Main: Suhrkamp, 1997, pp. 79–322.

137 See Axel Honneth, 'L'esprit et son objet: Parentés anthroplogiques entre la dialectique de la raison et la critique de la civilisation dans la philosophie de la vie', in Gérard Raulet, *Weimar, ou, l'explosion de la modernité: Actes du colloque 'Weimar et la modernité'*, Paris: Anthropos, 1984, pp. 97–112; G. Stauth and B. S. Turner, 'Ludwig Klages (1872–1956) and the Origins of Critical Theory', *Theory, Culture and Society*, 9/3 (1992), 45–63; G. Stauth, 'Critical Theory and Pre-Fascist Social Thought', *History of European Ideas*, 18 (1994), 711–27; and M. Großheim, ' "Die namenlose Dummheit, die das Resultat des Fortschritts ist" – Lebensphilosophische und dialektische Kritik der Moderne', *Logos: Zeitschrift für systematische Philosophie*, 3 (1996), 97–133.

138 Richard Block, 'Selective Affinities: Walter Benjamin and Ludwig Klages', *Arcadia*, 35 (2000), 117–136; Werner Fuld, 'Die Aura: Zur Geschichte eines Begriffes bei Benjamin', *Akzente* 26 (1979), 352–70; Werner Fuld, 'Walter Benjamins Beziehung zu Ludwig Klages', *Akzente* 28 (1981), 274–87; M. Pauen, *Dithyrambiker des Untergangs: Gnostizismus in Ästhetik und Philosophie der Moderne*, Berlin: Akademie Verlag, 1994, pp. 139 and 187–89 (on Benjamin), pp. 139–40 and pp. 189–90 (on Adorno); and Michael Großheim, 'Archaisches oder dialektisches Bild? Zum Kontext einer Debatte zwischen Adorno and Benjamin, *Deutsches Vierteljahrsschrift für Literaturwissenschaft und Geistesgeschichte*, 71 (1997), 494–517.

139 T. W. Adorno and W. Benjamin, *The Complete Correspondence 1928–1940*, ed. H. Lonitz, tr. N. Walker, Cambridge: Polity, 1999, p. 38.

140 Adorno, letter to Benjamin of 5 December 1954; *Complete Correspondence*, p. 61.

141 See Adorno's letters to Benjamin of 17 December 1934, 2–4 August 1935, and 6 September 1936 (pp. 68–9, 106–07, 145–46).

142 Adorno, *Zur Metakritik der Erkenntnisstheorie; Drei Studien zu Hegel* [*Gesammelte Schriften*, ed. Rolf Tiedemann, vol. 5], Frankfurt am Main: Suhrkamp, 1971, 'Einleitung', pp. 12–47 (here: p. 28).

143 Adorno, *Negative Dialectics*, tr. E. B. Ashton, London: Routledge and Kegan Paul, 1973, p. 155; *Negative Dialektik: Jargon der Eigentlichkeit* [*Gesammelte Schriften*, vol. 6], Frankfurt am Main: Suhrkamp, 1975, p. 7–412 (here: p. 158).

144 Lessing, *Minna von Barnhelm*, Act 2, scene 9: 'But let me tell you how reasonable this reason, how necessary this necessity is' (*wie vernünftig diese Vernunft, wie notwendig diese Notwendigkeit ist*).

145 In conversation with Eckermann on 13 February 1829, Goethe had declared that 'the understanding will not reach [nature]; man must be capable of elevating himself to the highest reason, to come into contact with the divinity, which manifests itself in the primitive phenomena (*Urphänomenen*), which dwells behind them, and from which they proceed' (*Conversations of Goethe*, pp. 294–95)

146 Cf. *Human, All Too Human*, vol. 1, 'Among Friends, An Epilogue', §2; *KSA*, Bd. 2, S. 366.

147 Nietzsche, KSA, vol. 8, 3[75], 36. This 'highest reason' of Goethe and Nietzsche lies above and beyond the 'logic' championed by Robert A. Segal in his 'Reply' to my paper.

148 See Carl Kerényi's discussion of Malinowski's *Myth in Primitive Society* [1926] in his 'Prolegomena', in C. G. Jung and C. Kerényi, *Essays on a Science of Mythology: The Myth of the Divine Child and the Mysteries of Eleusis* [1941], Princeton, NJ: Princeton University Press, 1963, p. 6. For an application of this view of myth to a literary text (in this case, Thomas Mann's novella *Death in Venice*), see Heidi M. and Robert J. R. Rockwood, 'The Psychological Reality of Myth in *Der Tod in Venedig*', *The Germanic Review*, 59 (1984), 137–41.

149 J. Derrida, 'Structure, sign and play', in *Writing and Difference*, tr. A. Bass, London and New York: Routledge, 2001, pp. 351–70; 'La structure, le signe et le jeu', in *L'écriture et la différence*, Paris: Seuil, 1967, pp. 409–28: 'The history of metaphysics, like the history of the West, is [. . .] the determination of Being as *presence* [*présence*] in all senses of this word. It could be shown that all the names related to fundamentals, to principles, or to the centre have always designated an invariable presence – *eidos*, *archē*, *telos*, *energeia*, *ousia* (essence, existence, substance, subject), *alētheia*, transcendentality, consciousness, God, man, and so on' (p. 353; pp. 410–11); see Holzhey and Schoeller Reisch, col. 421.

150 J. Derrida, 'Différance', in *Margins of Philosophy*, tr. A. Bass, Brighton: Harvester Press, 1982, pp. 1–27 (pp. 7, 15, and 22); 'La différance', in *Marges de la philosophie*, Paris: Minuit, 1972, pp. 1–29 (pp. 7, 16 and 23).

151 'Différance', p. 11; 'La différance', p. 12. Cf. ' "Older" than Being itself, such a *différance* has no name in our language. But we "already know" that if it is **unnameable**, it is not provisionally so, not because our language has not yet found or received this *name*, or because we would have to seek it in another language, outside the finite system of our own. It is rather because there is no *name* for it at all [. . .]' (*Plus "vieille" que l'être lui-même, une telle différance n'a aucun nom dans notre language. Mais nous "savons déjà" que, si elle est **innommable**, ce n'est pas par provision, parce que notre langue n'a pas encore trouvé ou reçu ce* nom, *ou parce qu'il faudrait le chercher dans une autre langue, hors du système fini de la nôtre. C'est parce qu'il n'y a pas de* nom *pour cela* [. . .]) (p. 26; p. 28).

152 See G. Deleuze and F. Guattari, *Milles Plateaux*, Paris: Minuit, 1980; *A Thousand Plateaus*, tr. Brian Massumi, New York and London: Continuum, 2004.

153 See C. Kerslake, 'Rebirth through incest: On Deleuze's Early Jungianism', *Angelaki* 9(1), 2004, 135–57.

154 See C. Hauke, *Jung and the Postmodern: The Interpretation of Realities*, London and Philadelphia: Routledge, 2000.

155 Derrida, 'Sign, Structure, and Play', in *Writing and Difference*, pp. 369–70; 'La structure, le signe et le jeu', in *L'écriture et la différence*, p. 427.

156 Derrida, 'Sign, Structure, and Play', in *Writing and Difference*, p. 370; 'La structure, le signe et le jeu', in *L'écriture et la différence*, p. 428.

157 See Bart Van der Straeten, 'The Uncanny and the Architecture of Deconstruction', in *Image and Narrative: Online Magazine of the Visual Narrative*, issue 5, 'The Uncanny'. Retrieved from http://www.imageandnarrative.be/uncanny/bartvander-straeten.htm

158 *Timaeus*, 49 a–b (*Collected Dialogues*, p. 1176).

159 *Timaeus*, 50 b; *Collected Dialogues*, p. 1177.

160 *Timaeus*, 51 a–b; *Collected Dialogues*, p. 1178.

161 See J. Derrida and P. Eisenman, *Chora L Works*, ed. J. Kipnis and T. Leeser, introd. B. Tschumi, NY: Monacelli Press, 1997.

162 See the papers issuing from a conference held on the island of Capri in 1994 in J. Derrida and G. Vattimo (eds), *Religion*, tr. D. Webb, Cambridge: Polity Press, 1998, especially Derrida's 'Faith and Knowledge: The Two Sources of "Religion" at the Limits of Reason Alone', pp. 1–78.

163 See J. D. Caputo, *The Prayers and Tears of Jacques Derrida: Religion without Religion*, Bloomington, IN: Indiana University Press, 1997; J. H. Otlhuis (ed.), *Religion With/Out Religion: The Prayers and Tears of John D. Caputo*, London: Routledge, 2002; and Y. Sherwood and K. Hart (eds), *Derrida and Religion: Other Testaments*, London: Routledge, 2005.

164 See D. Tacey, 'The role of the numinous in the reception of Jung', and S. Rowland, 'Jung and Derrida: The Numinous, deconstruction, and myth', in D. Tacey and A. Casement (eds), *The Idea of the Numinous: Contemporary Jungian and Psychoanalytic Perspectives*, London and New York: Routledge, 2006, pp. 213–28 and 98–116.

165 Robert J. Kozljanič speaks of 'the entire Western-philosophical, vitalist undercurrent, the entire cryptic tradition of *Lebensphilosophie*' as 'a suppressed and subterranean tradition of thought in Western philosophy that is true-to-life' (*Ernesto Grassi: Leben und Denken*, Munich: Fink, 2003, p. 105).

166 K. Kailo, 'C. G. Jung's Transcendent Function and Temenos: Valid Concepts for EcoFeminist Literary Criticism?', in G. Schmitt (ed.), *Die Welt im Symbol: C. G. Jung Symposion 7.-8.9.2000*, Oulu: University of Oulu, 2001, pp. 102–26 (p. 102).

167 R. J. Kozljanič, *Lebensphilosophie: Eine Einführung*, Stuttgart: Kohlhammer, 2004, pp. 188–89.

168 See the sections and references in 'Der Anfang und das anfängliche Denken', 'Das anfängliche Denken (Entwurf)', 'Das anfängliche Denken', and 'Das anfängliche Denken: Warum das Denken aus dem Anfang?', in *Beiträge zur Philosophie (Vom Ereignis)* [*Gesamtausgabe*, vol. III.65], ed. F.-W. von Herrmann, Frankfurt am Main: Klostermann, 1989, pp. 55–60; and *Über den Anfang* [*Gesamtausgabe*, vol. III.70], ed. P.-L. Coriardo, Frankfurt am Main: Klostermann, 2005.

169 M. Heidegger, *Mindfulness*, tr. P. Emad and T. Kalary, New York and London: Continuum, 2006, pp. 364–69; *Besinnung* [*Gesamtausgabe*, vol. III.66], ed. F.-W. von Herrmann, Frankfurt am Main: Klostermann, 1997, pp. 411–17.

170 M. Heidegger, 'Zur Bestimmung der Philosophie', §8, in *Zur Bestimmung der Philosophie: 1. Die Idee der Philosophie und das Weltanschauungsproblem (Kriegsnotsemester 1919); 2. Phänomenologie und transzendentale Wertphilosophie (Sommersemester 1919)* [Gesamtausgabe, vol. 56/57], ed. Bernd Heimbüchel, Frankfurt am Main: Klostermann, 1987, p. 96; see Holzhey and Schoeller Reisch, col. 421.

171 M. Heidegger, *Being and Time*, tr. J. Macquarrie and E. Robinson, Oxford: Basil Blackwell, 1978, §15, pp. 276–77; *Sein und Zeit*, Tübingen: Niemeyer, 1986, pp. 233–34.

172 Heidegger, *Being and Time*, §45, p. 277; *Sein und Zeit*, p. 234.

173 Heidegger, 'Vom Wesen des Grundes' [1929], in *Wegmarken* [1967] [*Gesamtausgabe*, vol. I.9], Frankfurt am Main: Klostermann, 1976, pp. 123–75 (here: pp. 172, 174).

174 Heidegger, *An Introduction to Metaphysics*, tr. R. Manheim, New Haven: Yale University Press; London: Oxford University Press, 1959, p. 2 (tr. modified); *Einführung in die Metaphysik* [1953], Tübingen: Niemeyer, 1958, p. 2.

175 Heidegger, *An Introduction to Metaphysics*, p. 3; *Einführung in die Metaphysik*, p. 2; see Holzhey and Schoeller Reisch, col. 421.

176 Heidegger, *An Introduction to Metaphysics*, p. 6; *Einführung in die Metaphysik* [1953], Tübingen: Neimeyer, 1958, pp. 4–5.

177 M. Heidegger, 'The Origin of the Work of Art', in *Poetry, Language, Thought*, tr. A. Hofstadter, NY: Harper & Row, 1971, pp. 17–87; ' Der Ursprung des Kunstwerkes', in *Holzwege*, Frankfurt am Main: Klostermann, 1980, pp. 1–72.

178 'The Origin of the Work of Art', p. 77; 'Der Ursprung des Kunstwerkes', p. 63.

179 'The Origin of the Work of Art', p. 76, tr. modified; 'Der Ursprung des Kunstwerkes', p. 62.

180 'The Origin of the Work of Art', pp. 77–78, tr. modified; 'Der Ursprung des Kunstwerkes', p. 64.

181 'The Origin of the Work of Art', p. 78, tr. modified; 'Der Ursprung des Kunstwerkes', p. 65. See Hölderlin, 'The Journey' (*Die Wanderung*), in *Poems and Fragments*, tr. M. Hamburger, London: Anvil, 1994, pp. 414–21 (here: p. 415): 'For whatever dwells / Close to its origin is loath to leave the place.'

182 Heidegger, 'Der Spruch des Anaximander', in *Holzwege*, pp. 317–68.

183 *An Introduction to Metaphysics*, p. 5; *Einführung in die Metaphysik*, p. 4.

184 M. Heidegger and E. Fink, *Heraclitus Seminar*, tr. C. H. Seibert, Evanston, IL: Northwestern University Press, 1993, p. 115. See also Heidegger's accounts of his visits to Greece of 1962 and 1967, in Heidegger, *Zu Hölderlin; Griechenlandreisen* [*Gesamtausgabe*, vol. III.75], ed. C. Ochwadt, Frankfurt am Main: Klostermann, 2000, pp. 211–73.

185 'The gods exist' (*Die Götter sind da*), in U. von Wilamowitz-Moellendorff, *Die Glaube der Hellen*, 2 vols, Darmstadt: Wissenschaftliche Buchgesellschaft, 1973, vol. 1, p. 17. For further discussion, see E. Flaig, 'Towards "*Rassenhygiene*": Wilamowitz and the German New Right', in I. Gildenhard and M. Ruehl (eds), *Out of Arcadia: Classics and Politics in Germany in the Age of Burckhardt, Nietzsche and Wilamowitz*, London: Institute of Classical Studies, 2003, pp. 105–27 (esp. pp. 113 and 115–16).

186 W. F. Otto, *Die Wirklichkeit der Götter: Von der Unzerstörbarkeit griechischer Weltsicht*, Reinbek bei Hamburg: Rowohlt, 1963, pp. 66 and 71.

187 For the persistence of the questions about the fundamentals of Being (*Seyn*) and the *Ab-grund*, of 'inceptual thinking' (*das anfängliche Denken*) and the gods, see Heidegger's *Mindfulness* (*Besinnung*) – described by its translators as his 'second being-historical treatise' (p. xiii) – especially the section entitled 'Gods: Project-opening What is to be Thought Beforehand in Every Inquiring Naming of the Godhood of Gods' ('Entwurf des zuvor zu Denkenden in jeglicher fragenden Nennung der Gottschaft der Götter') (*Mindfulness*, pp. 201–25; *Besinnung*, pp. 227–56).

188 See 'The "Face to Face" Interview', conducted by John Freeman for the BBC in 1959, W. McGuire and R. F. C. Hull, *C. G. Jung Speaking: Interviews and Encounters*, Princeton, NJ: Princeton University Press, 1977, pp. 424–39 (here: p. 428).

189 Nor was Jung alone in his interest in the persistence of the archaic: so was, to an extent that is often unappreciated, Freud. See his comments in 'Thoughts for the Times on War and Death' (1915) on the relation to death of 'primaeval, historic man' and the one 'which persists in every one of us, but which conceals itself, invisible to conscious-ness, in the deeper strata of our mental life', and on how 'the man of prehistoric times survives unchanged in our unconscious' (*Standard Edition of the Complete Psychological Works*, ed. J. Strachey and A. Freud, 24 vols, London: Hogarth Press, 1953–1974, vol. 14, pp. 274–302 [pp. 292 and 296]).

190 C. G. Jung, *Collected Works*, ed. Sir H. Read, M. Fordham, G. Adler, and W. McGuire, 20 vols, London: Routledge and Kegan Paul, 1953–1983, vol. 6, *Psychological Types*, para. 684.

191 Jung, *Psychological Types*, CW 6, para. 781. Compare with his remarks on 'primitive human perception of objects' (*ursprüngliche Objekterfassung*) in 'Psychological Aspects of the Mother Archetype' (1938/1954), paras. 148–98 (here: para. 187).

192 Jung, *Psychological Types*, CW 6, para. 684.

193 See F. Schiller, *On the Aesthetic Education of Man* [1795], ed. and tr. E. M. Wilkinson and L. A. Willoughby, Oxford: Clarendon Press, 1982, Letter 13, pp. 84–93.

194 Jung, 'Introduction to the Religious and Psychological Problems of Alchemy', in *Psychology and Alchemy*, CW 12, paras. 1–43 (here: para. 15; cf. para. 20).

195 For Jung's argument that archetypes derive from recurring situations, see 'The Concept of the Collective Unconscious' (1936), CW 9/i, paras. 87–110 (here: para. 99); 'The Significance of the Father in the Destiny of the Individual' (third edn, 1949), CW 4, paras. 693–744 (here: para. 728); and *Nietzsche's 'Zarathustra': Notes of the Seminar given in 1934–1939*, ed. J. L. Jarrett, London: Routledge, 1989, vol. 1, pp. 22–23 and 240–41.

196 J. Campbell, *The Hero With a Thousand Faces* [1949], London: Fontana, 1993, p. 4.

197 M.-L. von Franz, *C. G. Jung: His Myth in our Time* [1972], tr. W. H. Kennedy, New York: Putnam, 1975, pp. 13–14.

198 Compare with the title of Roderick Main's recent study, *Revelations of Chance: Synchronicity as Spiritual Experience*, Albany, NY: State University of New York Press, 2007.

199 For an introduction to the thought of Klages, see (in German) R. J. Kozljanič, 'Ludwig Klages: biozentrisches Denken, Lebensausdruck und geschautes Bild', in *Lebensphilosophie*, pp. 149–90; and (in English) R. Furness, 'Ludwig Klages', in *Zarathustra's Children: A Study of a Lost Generation of Writers*, Rochester, NY: Camden House, 2000, pp. 99–122; and P. Bishop, 'The Battle between Spirit and Soul: Messianism, Redemption and Apocalypse in Klages', in W. Cristaudo and W. Baker (eds), *Messianism, Apocalypse and Redemption in 20th Century German Thought*, Adelaide: ATF Press, 2006, pp.181–94.

200 L. Klages, *Der Geist als Widersacher der Seele*, Bonn: Bouvier Verlag Herbert Grundmann, 1981, pp. 1125–26.

201 M. Pauen, *Dithyrambiker des Untergangs: Gnostizismus in Ästhetik und Philosophie der Moderne*, Berlin: Akademie Verlag, 1994, p. 145.

202 In his 'biocentric metaphysics', Klages equates the following 'primal images' (*Urbilder*): 'earth = seed = egg = womb = mother' (*Der Geist als Widersacher der Seele*, p. 1329). According to Klages, the true 'eternal recurrence' is the 'circular time' of the ever-renewed, ceaselessly flowing images (p. 1349), or what he terms 'the secret of the *rebirth of the soul*' (p. 1350).

203 Klages, *Der Geist als Widersacher der Seele*, p. 1338.

204 Klages, *GWS*, p. 1424.

205 L. Klages, *Rhythmen und Runen: Nachlass herausgegeben von ihm selbst*, Leipzig: Barth, 1944, 'Ausgeführter Anfang', pp. 414–24 (here: p. 417).

206 Klages, *Rhythmen und Runen*, p. 244.

207 *The Peloponnesian War*, book I, §23 (Thucydides, *The Peloponnesian War: A New Translation; Backgrounds; Interpretations*, ed. Walter Blanco and Jennifer Tolbert Roberts, tr. Walter Blanco, New York and London: Norton, 1998, p. 11).

208 For an introduction in German to Grassi, see E. Bons, *Der Philosoph Ernesto Grassi: Integratives Denken, Antirationalismus, Vico-Interpretation*, Munich: Fink, 1990; and R. J. Kozljanič, *Ernesto Grassi: Leben und Denken* (see above). For a study of the immediate political context of Grassi's thought, see W. Büttemayer, *Ernesto Grassi – Humanismus zwischen Faschismus und Nationalsozialismus*, Freiburg: Alber, 2009.

209 E. Grassi, *Kunst und Mythos*, Frankfurt am Main: Suhrkamp, 1990, p. 7. For further discussion, see R. J. Kozljanič, *Kunst und Mythos: Lebensphilosophische Untersuchungen zu Ernest Grassis Begriff der Urwirklichkeit*, Oldenburg: Igel, 2001.
210 Grassi, *Kunst und Mythos*, p. 62.
211 Grassi, *Kunst und Mythos*, pp. 42–44 and 84–87.
212 Grassi, Kunst und Mythos, pp. 87–88; cf. *The New Science of Giambattista Vico: Translated from the third edition (1744)*, tr. T. G. Bergin and M. H. Fisch, Ithaca, NY: Cornell University Press, 1948, §17, p. 10. For further discussion, see the essays collected in E. Grassi, *Vico and Humanism: Essays on Vico, Heidegger and Rhetoric* [*Emery Vico Studies*, vol. 3], New York: Lang, 1990.
213 Grassi, *Kunst und Mythos*, pp. 93–94.
214 See Grassi, *Kunst und Mythos*, pp. 124–28.
215 See Grassi, *Kunst und Mythos*, p. 125, citing M. Heidegger, *Hölderlins Hymne 'Der Ister': Freiburger Vorlesung Sommersemester 1942* [*Gesamtausgabe*, vol. III.53], ed. W. Biemel, Frankfurt am Main: Klostermann, 1984, p. 18; *Hölderlins Hymn 'The Ister'*, tr. W. McNeill and J. Davis, Bloomington: Indiana University Press, 1996, p. 17. According to Heidegger, 'the symbolic image' has its 'ground' in the 'framework' of 'the distinction that is made between a sensuous and a nonsenuous realm', and 'in every employment of symbolic images we presuppose that this distinction has been made' (p. 17). For further discussion, see E. Grassi, *Heidegger and the Question of Renaissance Humanism*, Binghamton, NY: Center for Medieval and Early Renaissance Studies, 1983.
216 E. Grassi, *Kunst und Mythos*, p. 128.
217 E. Grassi, *Renaissance Humanism: Studies in Philosophy and Poetics*, tr. W. F. Veit, Binghamton, NY: Center for Medieval and Early Renaissance Studies, 1988, p. 39.
218 Grassi, *Renaissance Humanism*, p. 40.
219 Grassi, *Renaissance Humanism*, p. 88.
220 Grassi, *Renaissance Humanism*, p. 113.
221 M. Eliade, *The Sacred and the Profane: The Nature of Religion*, tr. W. R. Trask, New York and London: Harcourt Brace Jovanovich, 1959, p. 23.
222 Eliade, *The Sacred and the Profane*, p. 23–24.
223 S. M. Wasserstrom, *Religion after Religion: Gershom Scholem, Mircea Eliade and Henry Corbin at Eranos*, Princeton, NJ: Princeton University Press, 1999.
224 Eliade, *The Sacred and the Profane*, p. 63. See also Eliade's comments on sacred time, pp. 68–69.
225 Grassi, *Kunst und Mythos*, pp. 40–42.
226 P. Gauguin, *Noa noa*, tr. O. F. Theis, New York: Brown, 1919, p. 44 and 49–51.
227 D. H. Lawrence, *Fantasia of the Unconscious*, London: Secker, 1928, pp. 37–38.
228 S. Maitland, *A Book of Silence*, London: Granta, 2008, p. 63.
229 Pascal, *Pensées*, tr. A. J. Krailsheimer, Harmondsworth: Penguin, 1966, no. 206, p. 95; *Pensées*, ed. L. Brunschvicg, Paris: Hachette, [n.d.], no. 206, p. 428.
230 Maitland, *A Book of Silence*, p. 81.
231 Maitland, *A Book of Silence*, pp. 81–82.
232 Cited in Maitland, *A Book of Silence*, p. 53.
233 *Maxims and Reflections*, no. 433; in *Goethe on Science: An Anthology of Goethe's Scientific Writings*, ed. and tr. J. Naydler, Edinburgh: Floris Books, 1996, p. 108 (tr. modified).
234 *Maxims and Reflections*, no. 412; *Goethe on Science*, p. 108 (tr. modified).
235 *Goethes Gespräche*, ed. Flodoard von Biedermann, vol. 2, Leipzig: F. W. v. Biedermann, 1909, p. 40; translated in P. Hadot, *The Veil of Isis: An Essay on the History of the Idea of Nature*, tr. M. Chase, Cambridge, MA and London: Belknap Press of Harvard University Press, 2006, p. 258.

236 Goethe, *Theory of Colour*, 'Preface,' in *Scientific Writings* [Goethe Edition, vol. 12], ed. and tr. D. Miller, New York: Suhrkamp Publishers, 1988, p. 158.

237 Joseph Ratzinger, Homily at the Votive Mass for the Election of a New Pope, 18 April 2005. For further discussion, see the symposium 'A "Dictatorship" of Relativism? Symposium in Response to Cardinal Ratzinger's Last Homily', in *Common Knowledge*, vol. 13, nos 2–3 (Spring-Fall 2007), 214–455. To be sure, Ratzinger's version of the archaic is different from the ones presented elsewhere in this volume, not least because, for him, the true *archē* is *logos*.

238 Kant, *Critique of the Power of Judgment*, §15, p. 111.

239 M. de Hennezel and B. Vergely, *Une vie pour se mettre au monde,* Paris: Carnets Nord, 2010, p. 118.

240 De Hennezel and Vergely, *Une vie pour se mettre au monde*, p. 119. Vergely sustains his argument with allusions to Chopin's *Nocturns* and the famous poem 'The Lake' (*Le Lac*) by Alphonse de Lamartine.

241 Plato, *Timaeus*, 37 d; *Collected Dialogues*, p. 1167.

242 Plato, *Timaeus*, 37 d; *Collected Dialogues*, p. 1167.

243 See Klages, 'Bachofen als Erneuerer des symbolischen Denkens' [1937], cited in H. E. Schröder, 'Gedenkblatt für ein hundertjähriges Buch' [1961], in *Schiller – Nietzsche – Klages: Abhandlungen und Essays zur Geistesgeschichte der Gegenwart*, Bonn: Bouvier Verlag Herbert Grundmann, 1974, pp. 438–43 [here: p. 450]); (*Sämtliche Werke*, ed. Ernst Frauchiger [*et al.*], 9 vols, Bonn: Bouvier Verlag Hermann Grundmann, 1964–1999, vol. 3, pp. 409 and 421 and vol. 4, p. 707); and *Der Geist als Widersacher der Seele*, p. 635. For a discussion contrasting Klages' attitude to the future with that of Ernst Bloch, see M. Großheim, 'Das Prinzip Hoffnung und das Prinzip Gegenwart', in M. Großheim and H.-J. Waschkies (eds), *Rehabiliterung des Subjektiven: Festschrift für Hermann Schmitz*, Bonn: Bouvier, 1993, pp. 143–178; and for a discussion of Klages's 'negation of the future' and a comparison with the Homeric conception of time, see B. Müller, *Kosmik: Prozeßontologie und temporale Poetik bei Ludwig Klages und Alfred Schuler: Zur Philosophie und Dichtung der Schwabinger Kosmischen Runde*, Munich: Telesma, 2007, pp. 146 and 30–33.

244 De Hennezel and Vergely, *Une vie pour se mettre au monde*, p. 164–65.

245 B. Vergely, *Boulevard des philosophes: De la Renaissance à aujourd'hui*, Toulouse: Milan, 2005, p. 121.

246 Pascal, 'Fragment of a Treatise on Emptiness', in *Pensées et Opuscules*, ed. L. Brunschvicg, Paris: Hachette, n.d., 74–84 (p. 81).

247 De Hennezel and Vergely, *Une vie pour se mettre au monde*, pp. 162–63.

248 De Hennezel and Vergely, *Une vie pour se mettre au monde*, pp. 163. Cf. 'The very old and the very new communicate with each other. Life comes from far away and this distant side is behind us, far in the past. But this distance of life which comes from far away is for all that not old. Expressing the novelty of life, it is young. The further one goes into the antiquity [*l'ancienneté*] of things, the further one goes into their novelty. To bring back the past which makes us go into the age of the world [*la vieillesse du monde*] acquaints us with the novelty of the world' (p. 164).

249 Franz Schubert, 'Nachtgesang im Walde' (op. Posth. 139, composed April 1827), a setting for male voice choir and four horns of a text by Johann Gabriel Seidl. I am grateful to Alan Cardew for drawing this short, yet intensely atmospheric, piece of music to my attention.

250 Damascius, *Peri Archon* (*On First Things*), cited in Thomas Taylor, *The Mystical Hymns of Orpheus*, London: Dobell; Reeves and Turner, 1896, 'Introduction', p. xxiv; and 'To Night', p. 10.

251 For historically and culturally conditioned, yet nevertheless insightful, overviews of German culture, before and after the catastrophes of the Thirties and Forties, see F. Johnson, *The German Mind, as reflected in their literature from 1870 to 1914: together with two supplementary chapters on New Movements which have arisen between 1914 and 1921*, London and Sydney: Chapman and Dodd, 1922; and H. Kohn, *The Mind of Germany: The Education of a Nation*, New York: Scribner, 1960.

252 There is an important link between, on the one hand, the German search for the archaic and the primordial, and, on the other, modernism; see August K. Wiedmann, *The German Quest for Primal Origins in Art, Culture, and Politics 1900–1933: Die Flucht in 'Urzustände'*, Lewiston, Queenston, Lampeter: Mellen, 1996.

253 J. Meades (presenter), *Magnetic North*, Episode 1, broadcast BBC 4, 21 February 2008, 21:00.

Chapter 2

'Archaic theories of history': Thucydides, Hesiod, Pindar

The originary character of language

Ernesto Grassi

1. Preliminary remark

A preliminary remark: in the title of my contribution I use the term *archaic* as an adjective, and thus the expression contained with it – *arche* – in a dual sense.[1] Considered from a chronological, i.e., historical, perspective: *When* in the Western tradition are the *first* 'archai' of a theory of history worked out? Considered *theoretically*: *how* were these 'archai' defined at the beginning of Western thought and why are they of importance for us today?

The brief observations of my paper are dedicated to Gajo Petrović, the founder of the periodical *Praxis* and, in my view, the most important southern Slavic commentator on the Stalinist-Marxist view of history.[2] My remarks are a response to his essay on 'the concept of *Gelassenheit*', which he published in the form of an open letter to me in the third volume of *Zurich Conversations* (1991).[3]

The original theoretical problematic from which I begin is the *humanist rejection* of any priority to a rational, abstract starting point of philosophical activity.

2. The need for a demythologization of the theory of history: Thucydides

In the discussion of my theme with reference to three figures – Thucydides, Hesiod, and Pindar (with particular emphasis on the latter) – I shall attempt to show [that] the relation between them, precisely within the framework of the definition of the origin and the nature of history and within the framework of the problem of the originary character of language, has not, in my view, been sufficiently emphasized. Thus it may cause some surprise to the reader to see the names of these three authors associated with the problem of the 'arche' of history.

Note: 'Archaische Theorien der Geschichte: Thukydides, Hesiod, Pindar – Die Ursprünglichkeit der Sprache' was given as a lecture to the Faculty of Philosophy at the University of Zagreb on 6 November 1990, and subsequently published in *Synthesis philosophica*, no. 13, vol. 7/1 (1992), 85–94. It is translated and published here by the kind permission of Emilio Hidalgo-Serna. All footnotes have been added by the translator, with a view to adding factual explanatory material.

Thucydides starts from the assumption that the 'archai' of human history are defined by a 'fundamental necessity', or *anankē*,[4] with which the human being must come to terms in order to rescue his existence. From this there arise passions and basic attitudes, which are decisive for human actions.

Then again, for *Hesiod* the roots of history are the works, *erga*, in which the human being experiences a compulsion to overcome an originary Eris,[5] or an *inner discord*, which is experienced when he confronts reality.[6]

For *Pindar* it is the emergence of power and the originary lack of derivation of the *agonistic metaphorical word*, and not rational language, or any kind of undefined, work-related ergon,[7] that is the actual 'arche' of human actions.

In the following remarks we want just to touch on Thucydides and Hesiod, in order to understand better Pindar's central thesis.

Let us begin with Thucydides: he regards as the *megala*,[8] as the highest principle, as the starting point of his historical challenge, the *Peloponnesian War* and not, for example, the Homeric epic poems about the Trojan War.[9] He gives the following reason for this: 'What is now called Hellas does not seem in the past to have yet been stably settled' (I, 2);[10] so historical research into a Hellas that did not yet exist would be senseless. Thucydides specifies that 'there were continual migrations, and individual tribes learned to abandon their land with little ado as a result of the *coercion* of the *superior power* of others [...] and because they believed they could satisfy their basic need for daily bread [*anagkaion trophēs*] anywhere, they were permanently ready to *migrate*' (I, 2).[11]

Thus the 'ananke trophes', the need for daily bread – the need of concrete existence – is the root of emigration and hence of human history. Correspondingly our author underlines the folllowing essential point: 'Ownership of the richest land [*aretēn gēs*] led to some becoming more *powerful* than others, which led to civil strife, in which they died, and encouraged other tribes to attack them [...] so from the whole of Hellas the oppressed and the outcast, always the most powerful, migrated to Athens as a place of *security*' (I, 2).[12]

Hence all the great passionate events in human history – love and hate; destruction and reconstruction; war and peace – have their historical 'arche' in the *ananke of nourishment*, that is, in the necessity of survival, and not in any abstract, philosophical, or mythological considerations.

It has been maintained that the work of Thucydides should be read like a Greek tragedy: that is certainly true, but with one essential difference that, *instead of* a mythology or a set of myths developed by dramatists, it is the problem and the experience of the demands made by nature on human beings by nature that must be placed in the foreground.

Thus the task of history is the demythologization of what happens in Greece and the move away from any kind of abstract metaphysics.

From this basic thesis concerning the 'arche' of human history result Thucydides' reflections on tyranny, on democracy, on aristocracy, on the passion for power, for violence, for desperate attempts to defend oneself, to find *security* or to attain a place of peace, even in the awareness that history shows they always fail. Let us recall the speech that Thucydides ascribes to Pericles.[13] Such an outlook is generally recognized as fundamental for the West's theoretical definition of writing history.

This is Thucydides' famous and, as it were, conclusive declaration: 'Some listeners will perhaps be displeased by my *unpoetic account* [in contrast to the Homeric account of the Trojan War, cf. Book 1, §11]:[14] those, however, who want to see clearly things as they were and thus, given human nature, as they will one day be again, may find this book *useful* as a basis for judgement, and this is enough for me' (22, 4).[15] History should not, as in the case of Homer, be carried out on the basis of mythological thought or theology. In the 'ananke trophes' lies the archaic moment, where the writer of history must begin.

3. Hesiod's theory of 'Eris' – as strife and conflict – as the suffered and pathic 'archē' of history

Whilst Thucydides, in his reflections, defines quite generally the theory of the 'ananke trophēs' as the 'archē' of human history as it comes into being, in his text *Works and Days* Hesiod deepens and defines the fundamental claim that nature makes on the human being *as 'Eris'*, as discord, and the experience and *overcoming* of this originary 'conflict', this necessity, – through work, through human works or *erga* – is the 'arche' of history.[16] But what kind of *works*? The conquest of nature, the development of social institutions, legal entanglements, the arts, and religious demands that are always changing.

The interpreters of Hesiod's work have highlighted as the novelty of this text the way its theoretical discussion begins *with a concrete example* – his argument, his strife with his brother, Perses – which 'in his poem is only an artistic form in order to lend it greater power'.[17] But there is a more important principle at stake.

By beginning his discussion with the concrete example of his argument with his brother, Hesiod points to a fundamental demand ('arche') experienced by the human being: that of *Eris*, of *Strife*, whose structure, experience, and function is not single but multiple: *ouk ara ēn Erisôn ... eisi duô* [not one kind of Strife ... but two] (11–12).[18]

Eris, 'Strife', 'Dissonance' – which is not susceptible to derivation because of its archaic character and, as such, is unfathomable – is experienced 'pathically' by the human being in two ways: on the one hand, as something *negative* and, as such, *blameworthy*, and on the other, as something positive and as such *praiseworthy* (12–13).[19]

Is the fundamental *contradictoriness* of all that appears the 'arche' of history? Not rationality, but *contradiction* and *passion* – is this the source of human actions? The basic form of the answer to this question on Hesiod's part is *not rational*, not logical, but one that takes the form of a eulogy, or in other words begins with praise of what is pathically suffered and blame for what is passionately suffered: praise for desire, blame for anguish: *the need to overcome the contradiction*. 'The two aspects of Eris', our author maintains, 'have a *different* [thumos] effect, a *pathic* effect' (13) on the human being.[20] The first is negative: 'No one loves " Eris", or Strife', *ou tis tēn ge philet protos* (i.e., labour, anguish, what we experience as evil).

Through this Eris, through this Strife, through the anguish to which it gives rise, we experience an archaic, negative necessity, *ananke*. Then again, in the suffering caused by this 'Eris' we experience at the same time, surprisingly enough, a positive *ananke*, namely the necessity, the urge, to overcome Strife, Eris – in whose realm we originally find ourselves – and to do so through work, through human 'erga'. Is *contradiction*, the non-rational, the non-logical identity, the 'arche' of history?

Hence this is Hesiod's problem: which passion has priority in the human being? The persistence in strife, in the disorder of pain, or its *overcoming* through 'work'? Or do the two necessities as they are experienced point to a single, archaic, undifferentiated *passion*?

This is the first Hesodian thesis: night, darkness, op-position, strife, anguish – what the human being originally suffers and which constitutes the first archaic aspect of Eris as it is experienced – lies originally *in the earth*. Or as it is written in Hesiod's text: 'Cronus [. . .] set it in the roots of the earth' (l. 18).[21]

On the other hand, Hesiod's second fundamental thesis is: the experience of Eris has a positive function, inasmuch as it instigates the fight against a lack of working, against laziness, and the necessity to recognize *ergon*.

For Hesiod emphasizes: 'This is why the lazy potter looks at the other potters, the handworker at the other handworkers' (24–25).[22] And hence the fundamental thesis: 'The gods keep the secrets of life hidden in the darkness of the earth: if this were not the case [. . .] human beings would become lazy, would persist in a state of not working, *aergan conta*' (42).[23]

The original function of work – not as a curse, as labour – and the history corresponding to it are far removed from later Socratic, Platonic, metaphysical ideas and doctrines. We are in the fifth century BC. The fact that Hesiod in his text then refers to Pandora – in order to explain the existence of labour, pain, and confusion among human beings – is in fact an expression of a falling back into a *mythological* realm of theology and a renunciation of the doctrine of Eris.

4. Pindar: the 'archē' of the agonistic word

Do we we find in the Greek tradition, instead of the Thucydidean thesis of the necessity of nourishment (*ananke trophes*), instead of the Hesiodian reference to Eris, to strife (which is to be overcome through the concept of the work), an indication of another *archaic, eristic* function, one that is specific to the human being alone?

Pindar defines it as *agon*, as contest, struggle, but through the experience of the original, metaphorical *word*, which is experienced in the realm of a festive, 'elevated situation' – the Olympic Games.[24] For the time being it remains a complete mystery: the *how* and *why* in the blink of an eye [*Augen-blick*], in an *in-stant* [*in-stans*], the *hic-stare*, in the announcement of a power that descends upon us, which archaically reveals itself in the struggle for the metaphorical word (language as the expression of a *metis*, an acumen) and emerges, not in the national, but in the 'metic', the ingenious character. In '*Metis*', which reveals an *ability* and at the same time a *fundamental structure* of the ever changing and different announcement of what comes into appearance. The metic, ingenious word defines the struggle, in whose realm everything can suddenly appear different, *swiftly* in accordance with temporal and geographical circumstance, it is intimately bound up with *kairos*, it is the constant attempt to capture something which assumes a different shape, of something secretly metaphorical.

Let us follow Pindar's instructions. The first verses of Olympian Ode II confirm and demonstrate the lack of derivation of the fantastic word. 'Leading hymns [*anaxiphormigge himnoi*], which god, which human shall we let resound?' (II, 1–2).[25] The realm of 'letting resound', of the 'echo', of 'ringing out' what is original [*des Ursprünglichen*] is the realm of the word, of language, which Apollo and Heros announce in the 'agon', in the Eris of contest, of metaphorical language. The word, language, then, as the original, actual, human 'ergon'. What does this mean?

The first verses of Olympian Ode I confirm and refer to the revelatory function of the struggle in the sphere of language: 'In first place there is water, gold gleams like fire lit in the night (*dia purei nukti*), but when you, my heart, lay claim to win prizes through song, do not search for them in an external, warm sun of the sky, *for this is empty*'.[26]

Only in the joy of the victory of the word can suffering blossom and perish (Olympian II, 18).[27] Pindar claims that 'one after another moments of joy and suffering come over the human individual', but that 'only in the victory of metaphorical language does the despair of care dissolve' (Olympian II, 33 and 51).[28]

So, only through the searching and finding in agonistic, linguistic, metic, ingenious contest, and not in some kind of casual, technical 'ergon', is what is original [*das Ursprüngliche*] announced. Only in Eris, in agonistic language, in its 'Metis', in its acumen, in its metaphorical function do we attain the indicative historical sign of what happens to us, on the basis of which we can realize ourselves. 'Whether it is right or whether it is not right, even time itself cannot destroy this work' (Olympian II, 15).[29]

To search for oneself in the struggle for the right word, to identify oneself, to find one's own personality, to reveal the meaning of the action of a moment: the original [*ursprünglich*], agonistic, eristic event as the human being's experience of an ever present urge of the *eurgon* of the word. The pleasure at the blow of fate in the festive contest of language permits the meaning of the real to shine forth in the night of complete misunderstanding in which we live.

In the contest, in the relation to the Other, in the moving forward [*Fort-Schreiten*] to a deeper sense of language, is realized a desire for a goal, the attainment of which only *false praise* can destroy: 'With praise comes a danger that has nothing to do with justice, but is the expression of cowardly and longing individuals, who want through their words of praise to bury beautiful deeds in darkness' (Olympian II, 95).[30]

All this is rooted in the claim that is announced to the passions of our existential, historical hopes and illusions, inclinations and rejections, in the agonistic word: the 'arche' of our original history [*unserer ursprünglichen Geschichte*].

'Great danger turns the coward back. He is condemned to death, so why squander growing old without name, without reputation? It is this dispute I wish to intervene. And you, my friend, *give me the event* [*tu de praxin philan didei*]. Thus he spoke and had recourse to powerful words' (Olympian I, 84).[31]

Images of the fantasy – images that show, but don't prove [*weisende und nicht be-weisende*] – are sought for and suffered, in order to illuminate the various historical situations, to escape the greedy fingers of death; the invocation of the deeds of life through the hope of overcoming anxiety in the *ergon* of language, so that the leaves of the laurel wreath, which we urge ourselves towards, do not fade too soon. Hence the search in the *Agon* of contest for the unity that corresponds to the moment, the *kairos*. 'The greatest good is given to each by the day' (Olympian I, 99).[32]

The original Eris counts as the experience of overcoming the tensions among the various historical tribes in the agonistic Olympian festival of Hellas to forge a unity – without the destruction of war. Alexander the Great gave the order to destroy Thebes, but to save the house of Pindar.[33]

Eris demands of the working individual that he meet her demands as they arise from the dictates of the seasons, from the vicissitudes of various geological situations. Every invocation of a 'sacred' task is, without the saving metaphorical word, senseless. 'Sometimes for humans it is the wind, other times it is the rain that is the son of the clouds. When effort triumphs, songs resound, the beginnings of later victorious words' (Olympian XI, 1–6).[34]

Pindar identifies the *kairos* of metaphorical language as the measure of everything. 'There is a measure for all things and the *kairos* is the best way to understand it' (Olympian XIII, 47).[35]

To pray to *kairos* or to to beseech it is, because of its lack of rational derivation – and moreover its essence is secret – impossible. And it is in this realm that we live and compete with one other. In this state of distress, our contribution within the context of place and time becomes increasingly in demand. Then are words the waves of a river that sets the stones, over which we daily stumble, in motion.

In measuring oneself, in and through the 'agon' of the word, it is *not a question of scholarliness*, but on the contrary, in the contest of conversation, within the context of friendship, one rejects all abstract knowledge, and has recourse only to our original [*ursprünglichen*] passions [which] constitute the real 'ousia', the substance of our language.

One further invocation by Pindar: 'O Olympia, Mother of the Games with a golden crown, Governess of Truth, Location of the Games in which divining prophets burn as sacrifices, and as they invoke – in flashes of blinding lightning – and search for a favourable sign from Zeus for humankind, they are concerned with attaining success and overcoming suffering' (Olympian VIII, 1–7).[36]

The totality, what is 'sacred', the composure [*Gelassenheit*] in the contest of language to which Pindar refers, proves in fact to be the uncovering of a *unity* in the harmony, in the framework of language. It is in the agonistic festive battle that 'I try to sing what becomes yet more attractive through the racing contest of the chariot, *that shows me the way of the words [eurun hodon logôn]*' (Olympian I, 111).[37]

In the preceding thoughts on the 'archaic' theories of history [have we] reached an insight into the original understanding of *Gelassenheit*? Can we enter into a dialogue with Petrović? Has he not always intervened on behalf of the original character [*die Ursprünglichkeit*] of the word, the struggle for the freedom of the word, its *archaic* character?

The freedom, the inscrutability of the word that Pindar declares is the affirmation of the Metis, of 'acumen', of difference, that is announced in the *diversity* of the *meaning of the words* – hence in their metaphorical structure. Because the word constantly announces something different, it is time and again the experience of a transposition, hence of unity in difference.

Witness the testimony of Hölderlin in the tradition of German philosophy: 'The great word, the "*en diapherôn heautô*" (the One that is different in itself) of Heraclitus, this could *only* have been found by a Greek, for it is the essence of beauty, and before it was found, there was no philosophy' (Hölderlin, *Hyperion*, vol. 1, book 2).[38]

Is it only pre-Socratic Greek philosophy that stays away from all abstract thought? Here I must break off the preceding considerations and deal with the relation between Western ontology and the onset of the humanistic problem of the priority of the question about the word *rather than the question about what is*.

5. Ontology as the traditional starting point of language problems

The traditional rational philosophy of the West places emphasis on the rational inquiry into what is and believes it [can] identify this goal through the *concipere mente simplici primo ens* ['to conceive being by the first simple mind'], through the *concept* and its *kinds* on the basis of *categories*, or claims, and thus the *proprietates rei* ['properties of the object'].[39]

According to such a traditional way of understanding, language has its source and validity in the *ontological* definition of what is and, as such, language does not just have to be a-historical, but it must also exclude – because of its non-rational character – all metaphorical, poetic language from the realm of philosophy.

For the traditional way of understanding of philosophy the laurel wreath, the highest distinction for a poet, is *only* recognized as the highest symbol *if* poetry announces the goal of an eternal, unhistorical truth, i.e., something that was ontologically determined.

For Scholasticism, for example, the fundamental criterion for the translation of a term is given by the '*res*', *not* by the *verb*, for the latter is only a '*flatus vocis*' ['breath of the voice'], and knowledge, *scientia*, should not and cannot be bound to variants that are associated with time and place. Only the rational definition of the *res* through the clarification of its species can claim to uncover the universal, what is of necessity, which is abstracted from all temporal and spatial 'accidentals'.

In the course of his endeavours to translate classical Greek texts and thereby secure the continuity of tradition, the humanist Leonardi Bruni (1369–1444)[40] was forced in his work – more precisely, in his *experience* that he thereby acquired – to recognize that the logical fixity of an expression is untenable (L. Bruni, *Humanistische philosophische Schriften*, L. Bruni Teubner, 1928, p. 128, rr. 21–22).[41]

Within various systems of coordinates, from the experience of the '*circumstantiae*' as the imposition of demands which we experience, is revealed the significance of what appears, and hence a philosophy that is not abstract: 'philosophiam . . . cuius est fontibus haec omnis nostra derivatur humanitas' ['philosophy . . . from whose sources all this humanity of ours is derived'](L. Bruni, *Dialogus ad P. P. Histrum*, in: *Prosatori latini del Quattrocento*, a cura di E. Garin, Milano 1952, p. 149).[42]

The transition from traditional ontology to specifically humanistic philosophical thought takes place via the fundamental thesis of the priority of the problem of the word as opposed to the priority of the rational definition of the real. (Cf. E. Grassi, *Einführung in philosophische Probleme des Humanismus*, Wissenschaftliche Buchgesellschaft, Darmstadt 1986).[43]

6. The humanist tradition

According to the definition of G. Pontano (1426–1503),[44] the word is the essence of human fate, 'itaque ut *fatum* a fando, id est a dicendo deductum est' ['thus "fate" is derived from *fari*, that is, from "speaking"] (G. Pontano, *I dialoghi*, a cura di G. Previtera, Sansoni Firenze 1943, Aegidius, pp. 270, rr. 17–18).[45]

The sought-for and desired 'Latin philosophy' of humanism is discovered and seen in a thought that is *not* based on abstract, universal rational principles, but springs from times, places, and personal experiences: 'Ut ratio habeatur rerum, temporum, personarum, locorum' ['So thought comes from objects, times, persons, and places'] (ibid., p. 284, rr. 8–9).[46]

History becomes the true 'text', in relation to which the various significances of phenomena that appear and assail us first become susceptible to interpretation. 'Historia . . . descriptio est, quae nostra *vidit* aut *videre* potuit aetas' ['History . . . is the representation which our age saw or was able to see'].[47] And Guarino Veronese[48] refers, as he emphasizes this essential point, to the etymology of the expression *historia*: 'historein videre Graeci dicunt et *historia* spectaculum' ['the Greeks say *historein* to mean "to see" and *historia* to mean "a public show" '] (Guarino Veronese, Epist. 296, vol. 2, p. 460, rr. 73–74).[49]

Thus we recognize that the interpretation of what appears in the concrete demands which we experience historically mean[s] that we have to refer to the testimonies which have in each case led to an insight into the real.

In this way humankind is given, corresponding to a remark made by Guarino, a stage, a *palaestra*, a place of exercise (*tibi parata quaedam est palaestra*) – the historical stage – where he carries out his experiments (*in qua . . . praestabis experimentum*) (Epist. 785, vol. 3, p. 438, rr. 45–46).[50]

The problem of philosophical thought is thus not one of an abstract ontology, but one of the experience of the real *in its historicity*.

Vives (1492–1540)[51] is well aware of this: his relation to the humanists is proven by, among other things, his remark, found in his introduction to *De disciplinis*, in which he refers to Leonardo Bruni (the Aretino, as he calls him in accordance with the customs of the fifteenth century): '*Memoria patrum et avorum coeptum est in Italia revocari studium linguarum per discipulos . . . inter quos maximi nominis fuere* Leonardus Aretinus Angelus Politianus' ['Within the lifetime of our fathers and grandfathers, a study of languages began to be recalled among the students {of Peter of Ravenna for Latin and of Manuel Crisolara for Greek} . . . Among those of greatest renown were Leonardo d'Arezzo . . . and Angelo Polazzo'] (I. L. Vives, *De causis corruptarum artium*, in: *Opera Omnia*, Valentiae, vol. 4, book 4, chapter 4, p. 171).[52]

Let us remember the basic instruction that G. B. Vico gives us in his *Scienza Nuova*:[53] 'Chi fa cose stesso, ess lo narri' (369). 'Whoever makes things happen, should also narrate them'.[54]

This is the task for Petrović in the new situation that arises for our *praxis* and which we – who are not southern Slavs – expect from him.

Zagreb, 4 November 1990
Translated by Paul Bishop

Notes

1 In Greek philosophy *arche* translates as 'origin', 'beginning', or 'first principle', and it forms a major philosophical question in the writings of the early pre-Socratics, such as Thales, Anaximander, and Anaximenes, as well as such later thinkers as Pythagoras and Empedocles.

2 Born in Karlovac, Croatia, in 1927, Gajo Petrović was a critic of the philosophical views expressed by Stalin and an exponent of Marxist humanism. Following his doctoral dissertation on the materialist philosopher Georgi Plekhanov, he taught philosophy at the University of Zagreb, and was founding editor of the journal *Praxis*. He and Ernest Grassi corresponded over many years. He died in Zagreb in 1993.

3 G. Petrović, 'Marx, Arbeit und Gelassenheit: Ein Brief an Ernesto Grassi', in *Synthesis philosophica*, no. 13, vol. 7/1 (1992), 103–24; and in E. Grassi and H. Schmale (eds), *Arbeit und Gelassenheit: Zwei Grundformen des Umgangs mit der Natur* [*Zürcher Gespräche*, vol. 3], Munich: Fink, 1994, pp. 235–63. The 'Zurich conversations' were founded in 1976 by Ernesto Grassi and Hugo Schmale, professor for psychology at the University of Hamburg, as a series of international, interdisciplinary seminars, held twice a year; the earlier volumes of the *Zürcher Gespräche* are entitled *Ein semiotisches Problem* (1982) and *Anspruch und Widerspruch* (1987).

4 In Greek mythology Ananke is the goddess of necessity (destiny or fate); according to one tradition, she is also the mother of the Moirae, the three fates whose father is Zeus.

5   In Greek mythology Eris is the goddess of strife or discord. In *Works and Days* Hesiod distinguishes two kinds of strife, one bad and one good – war and competition.
6   See also Grassi's discussion of Hesiod's genealogy of the goddess Iris in E. Grassi and M. Lorch, *Folly and Insanity in Renaissance Literature*, Binghamton, NY: Center for Medieval and Early Renaissance Studies, 1986, pp. 22–24.
7   In Greek *ergon* means 'deed' or 'performance'.
8   Compare with the following commentary: 'The Peloponnesian War is worthy of study because it exceeded in magnitude any previous conflict. In the opening chapter, Thucydides tells us that he expected the war would be "great" (1.1.1: *megas*). It proved to be the "greatest shock" (1.1.2: *kinêsis megistê*) ever to befall the Greek world and a good portion of the barbarians. Later, Thucydides concludes his survey of early Greek history with the claim that the accumulated resources available at its start were "greater" (*meizôn*) than any sudden alliance could muster. In between, Thucydides argues his thesis that previous events were "not great" (1.1.3: *ou megala*). The term *megas* recurs fifteen times as Thucydides calculates the magnitude of one entity or another. Sparta can summon considerable military force, but Athens is clearly portrayed as the expanding power, and this expansion is linked [. . .] to the accumulation of money' (Gregory Crane, *Thucydides and the Ancient Simplicity: The Limits of Political Realism*, Berkeley, Los Angeles, London: University of California Press, 1998, p. 170).
9   The Peloponnesian War, 431–404 BCE, was 'in its more prominent features a struggle between Athens, a democratic State and a sea-power, which had converted the Delian Confederacy (designed to resist the Persians) into an empire under her own rule, and most of the States of the Peloponnese together with Boeotia and headed by Sparta, an oligarchical and conservative power, whose land army was the most efficient military force of the day' (Sir P. Harvey, *The Oxford Companion to Classical Literature*, Oxford: Clarendon Press, 1937, p. 310). Whilst recording that more modern theories see the reason for the war in the commerical rivalry of Athens and Corinth, Harvey notes that the truest explanation of the conflict, according to Thucydides, was 'the rise of Athens to greatness, which caused the Spartans to become afraid of them'; the third edition of *The Oxford Classical Dictionary* remarks that most of the war was recorded by Thucydides (c. 460–c. 400 BCE), 'and that is the most interesting thing about it' (S. Hornblower and A. Spawforth (eds), *The Oxford Classical Dictionary*, Oxford and New York: Oxford University Press, 1996, p. 1134). In translating the passages from Thucydides cited by Grassi, I have consulted *The Peloponnesian War: A New Translation; Background; Interpretations*, ed. W. Blanco and J. T. Roberts, tr. W. Blanco, New York and London: Norton, 1998. For further discussion and detailed commentary on Thucydides' text, see R. B. Strassler (ed.), *The Landmark Thucydides*, New York, London, Toronto, Sydney: Free Press, 1996.
10 Cf. Thucydides, *The Peloponnesian War*, Book 1, §2.
11 Cf. Thucydides, *The Peloponnesian War*, Book 1, §2. For further discussion of Thucydides' view of the earliest human societies, see Crane, *Thucydides and the Ancient Simplicity*, chap. 5, pp. 125–47.
12 Cf. Thucydides, *The Peloponnesian War*, Book 1, §2.
13 See Pericles' funeral oration as contained in Thucydides' *History of the Peloponnesian War*, book 2, §35 to §46; cf. his speeches giving advice to the Athenians in response to Sparta (book 1, §140 to §143) and trying to restore the confidence of the Athenians (book 2, §60 to §64).
14 Cf. 'If we can trust the evidence of Homer – which, considering that he was a poet, was probably exaggerated [. . .]' (Book 1, §10).
15 Cf. Thucydides, *The Peloponnesian War*, Book 1, §22.
16 The early Greek poet Hesiod lived during the eighth century BCE, and is known for two main surviving works; his *Theogony*, a hexameter poem recounting the genealogy of

the gods, and *Works and Days*, also in hexameters, which seeks to resolve a quarrel with his brother, Perses, whilst speaking in praise of honest work. In *The Oxford Companion to Classical Literature*, Sir Paul Harvey comments that *Works and Days* 'represents the life-experience of a single close-fisted peasant, schooled in adversity, circumspect, grumbling but courageous, and is marked by simplicity and a sense of human misery' (p. 451). For further discussion, see Hesiod, *Works and Days: A Translation and Commentary for the Social Sciences*, tr. D. W. Tandy and W. C. Neale, Berkeley: University of California Press, 1996. See also Hesiod, *Theogony; Works and Days; Testimonia*, ed. and tr. G. W. Most, Cambridge, MA, and London: Harvard University Press, 2006.

17 Unidentified quotation. In *Works and Days* Hesiod repeatedly addresses his brother, Perses (see *Works and Days*, ll. 10, 213, 274, 286, 299, 397, 611, 633 and 641).

18 See Hesiod, *Works and Days*, ll. 11–12.

19 See Hesiod, *Works and Days*, ll. 12–13.

20 Hesiod, *Works and Days*, l. 13.

21 Hesiod, *Works and Days*, l. 18 (p. 89).

22 See Hesiod, *Works and Days*, ll. 23–26: 'For a man who is not working but one who looks at some other man, a rich one who is hastening to plough and plant and set his house in order, he envies him, one neighbour envying his neighbour who is hastening towards wealth: and this Strife is good for mortals. And potter is angry with potter, and builder with builder, and beggar begrudges beggar, and poet poet' (p. 89).

23 Hesiod, *Works and Days*, l. 42: 'For the gods keep the means of life concealed from human beings. Otherwise you would easily be able to work in just one day so as to have enough for a whole year even without working, and quickly you would store the rudder above the smoke, and the work of the cattle and of the hard-working mules would be ended' (p. 91).

24 The ancient Greek lyric poet Pindar (c. 518–c. 443 BCE) is known for his victory odes, or *epinicia*, written in honour of winners at games: the Olympian, the Pythian (held at Delphi), the Isthmian (held near Corinth), and the Nemean. Sir Paul Harvey notes that his second Olympian ode reflects a doctrine of life after death and resurrection that suggests Pindar's acquaintance with ancient Orphic teachings (*The Oxford Companion to Classical Literature*, p. 328). For further discussion of Pindar with reference to his reception in Germany, see J. T. Hamilton, *Soliciting Darkness: Pindar, Obscurity, and the Classical Tradition*, Cambridge, MA, and London: Harvard University Department of Comparative Literature, 2003; and the 'Introduction' in *Pindar's Victory Songs*, tr. F. J. Nisetich, Baltimore and London: The Johns Hopkins Press, 1980, pp. 1–77. References in the footnotes are to Frank J. Nisetich's translations of Pindar.

25 Olympian Ode II, ll. 1–2: 'Songs, lords of the lyre, / what god, what hero, what man / shall we celebrate?' (p. 88).

26 Olympian Ode I, ll. 1–6: 'Water is preeminent and gold, like a fire / burning in the night, outshines / all possessions that magnify men's pride. / But if, my soul, you yearn / to celebrate great games, / look no further / for another star / shining through the deserted ether [. . .]' (p. 82).

27 Olympian Ode II, l. 18: '[. . .] malignant pain / perishes in noble joy [. . .]' (p. 88).

28 Olympian Ode, II, ll. 33 and 51: 'The shifting tides of good and evil / beat incessantly upon mankind. [. . .] A man forgets the strain of contending / when he triumphs' (p. 89).

29 Olympian Ode II, l. 15: 'What has been done / with justice or without / not even time the father of all can undo' (p. 88).

30 Olympian Ode II, ll. 95–98: 'But praise falls in with surfeit / and is muted, not in justice / but because of boisterous men, whose noise / would obscure beauty' (p. 91); cf. 'But praise is attacked by envy – envy, not mated with justice, but prompted by besotted

minds, envy that is ever eager to babble, and to blot the fair deeds of noble men' (Pindar, *The Odes*, tr. John Sandys, London; New York: Heinemann; Putnam, 1927, p. 29).

31 Olympian Ode I, ll. 84–87: ' "[. . .] Great danger / does not come upon / the spineless man, and yet, if we must die, / why squat in the shadows, coddling a bland / old age, with no nobility, for nothing? / As for me, I will undertake this exploit. / And you – I beseech you: let me achieve it." / He spoke, and his words found fulfilment' (p. 84).

32 Olympian Ode I, ll. 98–99: 'A single's day blessing / is the highest good a mortal knows' (p. 85).

33 It is recorded that, during the sack of Thebes in 335 BCE, Alexander the Great ordered that the only buildings to be spared were the citadel, the temples, and the house where Pindar had lived. See Milton's sonnet, 'When the assault was intended to the City', in J. Milton, *Complete Shorter Poems*, ed. J. Carey, Harlow: Longman, 1997, p. 289.

34 Olympian Ode XI, ll. 1–6: 'Sometimes men need the winds most, / at other times / waters from the sky, / rainy descendants of the cloud. / And when a man has triumphed / and put his toil behind, / it is time for melodious song / to arise, laying / the foundation of future glory, / a sworn pledge securing proud success' (p. 123).

35 Olympian Ode XIII, l. 47: 'In every matter / measure is the thing – / to know it / is all tact' (p. 145).

36 Olympian Ode VIII, ll. 1–7: 'Mother of contests for the golden crown, / queen of truth, Olympia, / where men of prophecy, / consulting Zeus' sacrificial fire, / probe his will! / God of the white-flashing bolt, / what has he to say / of the contenders, struggling / for glory, breathless until they hold it?' (p. 119).

37 Olympian Ode I, ll. 109–11: '[. . .] still sweeter the triumph I hope / will fall to your speeding chariot, / and may I be one to praise it [. . .]' (p. 85).

38 'Das große Wort, das *en diapherôn heautô* (das Eine in sich selber unterschiedne) des Heraklit, das konnte nur ein Grieche finden, denn es ist das Wesen der Schönheit, und ehe das gefunden war, gabs keine Philosophie' (Hölderlin, *Sämtliche Werke* [Stuttgarter Ausgabe], vol. 3, Stuttgart: Kohlhammer, 1957, p. 81; cf. Hölderlin, *Sämtliche Werke* [Frankfurter Ausgabe], vol. 11, Frankfurt am Main: Roter Stern, 1982, p. 682. Hölderlin's reference here is to the following fragment from Heraclitus: 'They do not comprehend how, in differing, it [i.e., the universe] agrees with itself – a backward-turning connection, like that of a bow and a lyre' (Diels-Kranz 22 B 51; J. Barnes, ed. and tr., *Early Greek Philosophy*, Harmondsworth: Penguin, 1987, p. 102). In the *Symposium*, Erixymachus refers explicitly to this image, when he remarks that 'medicine is under the sole direction of the god of love, as are also the gymnastic and the agronomic arts', as well as music, 'which is, perhaps, what Heraclitus meant us to understand by that rather cryptic pronouncement, "The one in conflict with itself is held together, like the harmony of the bow and the lyre" ' (*Symposium*, 187a; Plato, *The Collected Dialogues*, ed. E. Hamilton and H. Cairns, Princeton, NJ: Princeton University Press, 1989, p. 540).

39 Cf. E. Grassi, *Renaissance Humanism: Studies in Philosophy and Poetics*, tr. W. F. Veit, Binghamton, NY: Center for Medieval and Early Renaissance Studies, 1988, p. 7.

40 The Italian humanist and chancellor of Florence, Leonardo Bruni (c. 1369–1444), also known as Leonardo Aretino, is considered to be the first modern historian. Cf. Grassi's discussion of Bruni in *Heidegger and the Question of Renaissance Humanism: Four Studies*, Binghamton, NY: Center for Medieval and Early Renaissance Studies, 1983, pp. 19–21. For further discussion, see the 'Introduction' in G. Griffths, J. Hankins, and D. Thompson (eds), *Leonardi Bruno: Selected Texts*, Binghamton, NY: Center for Early Medieval and Early Renaissance Studies, 1987, pp. 3–50; and for a general discussion of rhetoric in Renaissance humanism in relation to modern concerns, see

G. Remer, 'Genres of political speech: Oratory and conversation, today and in antiquity', *Language and Communication* 28 (2008), 182–96.

41 Leonardi Bruni, *Humanistisch-philosophische Schriften mit einer Chronologie seiner Werke und Briefe*, ed. Hans Baron, Leipzig: B. G. Teubner, 1928, p. 128. Cf. passage cited in *Renaissance Humanism*, p. 21: '*Verba autem ipsa inter se quam molliter componenda et coagmentanda sunt eorumque continuatio et quasi textura omnis aurium sensu* [. . .] *moderanda*' ['For the very words and what follows them and all the fabric, as it were, in the sense of hearing are to be constructed and arranged as subtly as possible'] ('Prefatio in Orationes Demosthenis', *Schriften*, p. 128).

42 Bruni, 'A Petrum Paulum Histrum Dialogus, in E. Garin (ed.), *Prosatori latini del Quattrocentro*, Milan: R. Ricciardi, 1952, pp. 44–99 (p. 54). Cf. Grassi, *Renaissance Humanism*, p. 22.

43 E. Grassi, *Einführung in philosophische Probleme des Humanismus*, Darmstadt: Wissenschaftliche Buchgesellschaft, 1986; translated by Walter F. Veit as *Renaissance Humanism*. As the foreword explains, this volume constitutes the third of Grassi's triptych of studies on Renaissance humanism, the preceding volumes being *Heidegger and the Question of Renaissance Humanism: Four Studies* (1983) and *Folly and Insanity in Renaissance Literature* (1986).

44 Giovanni Gioviano Pontano (1426–1503), or Iovanius Pontanus, was an Italian humanist prose writer and poet. Cf. Grassi's discussion of Pontano in *Heidegger and the Question of Renaissance Humanism*, pp. 51–54. For further discussion, see E. Grassi, 'Humanistic Rhetorical Philosophizing: Giovanni Pontano's Theory of the Unity of Poetry, Rhetoric, and History', *Philosophy and Rhetoric*, vol. 17, no. 3, 1984, 135–55; and R. Weiss, 'The Humanist Discovery of Rhetoric as Philosophy: Giovanni Giovano Pontano's "Aegidius" ', *Philosophy and Rhetoric*, vol. 13, Winter, 1980, 25–42.

45 Pontano, *I dialoghi*, ed. Carmelo Previtera, Florence: Sansone, 1943, p. 269; cf. Grassi, *Renaissance Humanism*, p. 37, citing 'Aegidius', in *Dialoge*, tr. H. Kiefer, Munich: Wilhelm Fink, p. 573: *Itaque ut fatum a fando, idest a dicendo deductum est*.

46 Pontano, *I dialoghi*; cf. Grassi, *Renaissance Humanism*, p. 41: 'a concrete contemplation of things, times, personalities and places' (cited from 'Aegidius', in *Dialoge*, p. 602).

47 Pontano, *I dialoghi*.

48 Guarino da Verona (1370–1460), or Guarino Veronese, was an Italianist humanist translator and commentator. For further discussion of his relation to the humanist outlook, see A. T. Grafton and L. Jardine, 'Humanism and the School of Guarino: A problem of evaluation', *Past and Present* 96 (August, 1982), 51–80.

49 *Epistolario di Guarino Veronese*, ed. R. Sabbadini, 3 vols (Turin: Bottega d'Erasmo, 1915–1919; 1967), vol. 2, p. 460; cf. Grassi, *Renaissance Humanism*, p. 54.

50 *Epistolario di Guarino Veronese*, vol. 1, p. 108; cf. Grassi, *Renaissance Humanism*, p. 56.

51 Juan Luis Vives (1492–1540), also known by his Catalan name as Joan Lluís Vives i March, was a Spanish humanist scholar, born in Valencia. See Grassi's discussion of Vives's *Fabula de homine* in *Rhetoric as Philosophy: The Humanist Tradition*, University Park and London: The Pennsylvania State University Press, 1980, pp. 10–13. For further discussion of Vives, see C. G. Noreña, *Juan Luis Vives*, The Hague: Nijhoff, 1970; on his view of language, see J. M. Navarro, 'Sprachbewußtsein und Sprachtheorie in Juan Luis Vives, Vorläufer der Pragmalinguistik', in K.-H. Wagner and W. Wildgen (eds), *Studien zur Grammatik und Sprachtheorie*, Bremen: Institut für Allgemeine und Angewandte Sprachwissenschaft, Universität Bremen, 1990, 37–46; and of his view of history, see I. Bejczy, ' "Historia praestat omnibus disciplines": Juan Luis Vives on history and historical study', *Renaissance Studies* 17/1 (2003), 69–83.

52 Vives, *De disciplinis*, part 1, 'De causis corruptarum artium', in *Joannis Ludovici Vivis Valentini Opera omnia*, ed. G. M. y Siscar and F. F. y Feuro, 7 vols in 8, Valencia: In officina Benedecti Montfort, 1782–1790, vol. 6, p. 171; cf. Grassi, *Renaissance Humanism*, p. 66. For further discussion of 'De causis corruptarum artium', see the 'Introduction' in J. L. Vives, *Against the Pseudodialecticians: A Humanist Attack on Medieval Logic*, tr. R. Guerlac, Dordrecht, Holland; Boston; London: Reidel, 1979, pp. 1–43.

53 The famous Italian philosopher, historian, and rhetorician Giovanni Battista (or Giambattista) Vico (1668–1744) was born in Naples; his major work, *Scienza nuova*, was published in 1725, and reissued with revisions in 1730. See Grassi's discussion of Vico in *Heidegger and the Question of Renaissance Humanism*, pp. 26–28; and in *Rhetoric and Philosophy*, pp. 4–8 and 35–47. For Grassi, Vico, 'in taking up again the humanistic tradition, affirms that the ground of human historicity and human society is not the rational process of thinking but the imaginative act' (*Rhetoric as Philosophy*, p. 65).

54 See *Scienza nuova*, Book 1, Section 4, 'Method', §349: 'History cannot be more certain than when he who creates the things also narrates them' (*ove avvenga che chi fa le cose esso stesso le narri, ivi non può essere più certa l'istoria*) (*The New Science of Giambattista Vico: Unabridged Translation of the Third Edition (1744) with the addition of 'Practice of the New Science'*, tr. T. G. Bergin and M. H. Fisch, Ithaca and London: Cornell University Press, 1968, p. 104.

Chapter 3

Genius loci and the numen of a place

A mytho-phenomenological approach to the archaic

Robert Josef Kozljanič

The archaic as enchanting beginning

The English word 'archaic' can be etymologically traced back to the Greek word 'arché'. The main meanings of 'arché' include 'beginning', 'origin', 'cause', 'principle', 'main', 'command', 'reign', and 'regime'. Words combined with 'arch-' – such as 'arch-enemy', 'arch-rogue', and 'archbishop' – are based on a similar meaning. An 'arch-enemy' is an enemy from the outset, one's main enemy; an 'arch-rogue' is a rogue *ab initio*, right from the start; and the 'archbishop' is the chief bishop, a central episcopal figure.

In what follows I shall focus on one of these aspects of the archaic. It is only one aspect, but a very important one: namely, that the archaic is a beginning. In his poem 'Steps' (*Stufen*), Hermann Hesse emphasizes this aspect very well: 'For every beginning has a special magic / That nurtures life and bestows protection' (*Und jedem Anfang wohnt ein Zauber inne, / Der uns beschützt und der uns hilft, zu leben*).[1] Hesse shows that the archaic is always something primary: it is an initiation, an opening, a genesis, a creation. And, as his poem also points out, the archaic is a particular beginning. A beginning that brings a special magic: a magic that nurtures, fosters life, and helps us to live. A magic that protects, shelters, and guards us on our way through life. That this magic has the ability to nurture and protect shows that it possesses an inherent force or energy. This magic contains an intrinsic power – the power to change, to configure, to create. This power works primarily not with physical causes, but rather with psychic forces. So it is the fascination, the spell, the allure of the archaic beginning that motivates, moves us forward, and brings forth something new. Which means that the magic of the beginning always shows two sides: a passive and an active side, one that allures and hypnotizes, and one that empowers and stimulates. In summary: one could say that the archaic is, in an important way, a beginning that simultaneously

Note: Sincere thanks to Paul Bishop for his help with the revision and translation of this essay.

enchants and activates, something with the potential or the power to create and to form, to reform and to transform something or somebody.

Now, when Hesse speaks of the beginning, he means the beginning of a new period of life, a new step on the stairway of personal development. It is evident, however, that the meaning of the archaic as a beginning extends much further than this. In other words, every beginning that brings forth a new creature (a plant, an animal, or a human being), a new period (be it biographical, social, cultural, or historical), even every beginning that brings forth a new idea (a concept, an opinion, an ideal, a virtue, a point of view) is – as long as the enchantment and the empowerment lasts – in this sense archaic. Moreover, the changing, configuring, creative power of the beginning is power that endures; it is much more than a simple kick-start. For as long as the creature, or the period, or the idea lasts, so long, too, does the power of the beginning. To put it another way: the archaic power of beginning lasts as long as the magic of the creature, or the period, or the idea does. It is in some sense the 'soul' of the creature, of the period, or of the idea.

Ever since the time of the ancient Greeks, many philosophical treatises have been written on the concept and the problem of the arché and the archaic. The task of rethinking this interesting subject has a certain charm, but in what follows I shall choose another path. In this essay I shall take a look at a pre-philosophical – i.e., mythological – concept of the archaic-as-beginning: the old Roman concept of the 'genius', the tutelary or attendant spirit of a person. Then I shall turn to the modern philosophical concept of the 'genius loci' – the presiding spirit or god of a place – as the 'atmosphere' of a site, which brings with it a number of fundamental phenomenological implications. In this way, via what may initially appear as a detour, I hope to provide an example of the phenomenon of the archaic as an enchanting beginning, and to enrich our understanding of it.

The personal genius as spirit of genesis

According to the traditional mythical belief of the Romans, every human being has his personal tutelary or protecting spirit; the Romans called this tutelary spirit the 'genius'.[2] This genius accompanies the individual throughout his life, and so Horace calls it 'comes', a companion or an escort (*Epistles*, 2, 2, 187).[3] It was also called one's 'tutela', or patron. The genius is sometimes related to what the Romans called 'anima' or 'animus', in other words, the soul.[4] And yet it is much more than just the individual's soul. It is also his character, or fate, in a positive as well as a negative sense. The genius is primarily responsible for how a human individual develops, for what he becomes, and for when he dies. Hence the genius could also be called the spirit of one's destiny; or, as the classical philologist Theodor Birt (1852–1933) called it, one's 'spirit of becoming' (*Werdegeist*).[5] This expression also suggests the etymology of the term. Indeed, 'the origin of the name', according to another classical philologist, Walter F. Otto (1874–1958), 'from the root *gen* found in *gignere* is evident, nor was it ignored by the ancients'.[6] *Gignere* means 'to produce' or 'to beget', 'to bring forth', 'to give birth to'. Thus the genius is a productive spirit that gives birth to something or brings something into being.

But the genius is not only a spirit that brings things into being, it is also a spirit that demands and (at least if one is fortunate enough) actually promotes joy. This fact can be found in the earliest accounts in Roman literature, above all in the comedies of Titus Maccius Plautus (c. 250–184 BCE). Here we find on repeated occasions the advice that we should look after our genius – particularly in a culinary sense ('genio suo bona facere', 'genium suum meliorem facere') (*Persa*, 263; *Stichus*, 622).[7] The genius can be encouraged or pleased,[8] or on the contrary – as frequently happens in the Plautine comedies with misers – neglected: 'genium suum defraudare', to deceive one's genius, to do harm to one's genius, as Plautus puts it (*Aulularia*, 278).[9]

At this point it should be noted that all the designations for the genius – the genius as soul, as character, as the spirit of fate, or becoming, or joy – are in some way vague, even misleading. For these designations are understood today in a way that is far too abstract and psychological. They suggest that the genius is something purely interior to the soul, something invisible, something limited to the human psyche. But such a view, although it is not entirely false, falls short. For the genius is more than this, it is also something much more concrete. For example, the genius is, among other things, thought of as existing outside the individual human being. It is an independent and (sometimes) autonomous *divine being*, that can act out its own intentions. It is to this divine being that the Roman would pray and make offerings (for instance, on birthdays or on the occasion of important family celebrations). Moreover, this divine being can, principally in the form of a snake or serpent, appear outside the individual human being and act independently of him. These non-venomous serpents, which many Romans kept as if pets or tolerated in their houses as 'co-habitees',[10] were often seen as the embodiments of the geniuses of the people living there or as geniuses of the place (as the 'good spirits of the house'). They were treated with respect, offerings were made to them, plates of food were set out for them. Of such serpents Seneca wrote that they 'crawl between goblets and the folds of robes in a harmless glide' (*De Ira*, 2, 31, 6).[11] An interesting story has come down to us which sheds a clearer light on this aspect. This story is about the mother of the elder Scipio, Publius Scipio. She 'found one evening, when her husband was absent, a large serpent lying next to her in bed; a few days later she realized she was pregnant') (Aulus Gellius, 6, 1, 3; Livy, 26, 19, 7).[12] As the Romans saw it, the serpent functioned here as the embodiment of the genius of the husband (or as the embodiment of an ancestral spirit). And precisely because the genius is the creative principle (or, so to speak, the 'arché' that brings forth), it or its embodiment, the serpent, can produce offspring – even in the husband's absence.

The main feast-day of the genius was considered to be the individual's birthday. On this day the particular genius was hailed by the Romans as Natalis, and by the Greeks as *daímon genéthlios* – in other words, in both cases as the spirit or god of birth. Offerings of flowers and wine were made (Horace, *Epistles*, 2, 1, 144),[13] as well as offerings of incense. But offerings involving blood, though rare, were also possible: for instance, of lambs and pigs (Horace, *Carmina*, 4, 11, 8 and 3, 17, 14–15).[14] The idea behind such offerings or sacrifices was that, at the moment of birth, the genius joined the individual and bound itself to him or her for the individual's entire lifespan, and for this reason (even after the death of the individual) would be celebrated on the individual's birthday and treated to food and drink.[15] Similarly, on anniversaries of the family or the household, offerings were made to the genius – specifically, the genius of the father of the family and the head of the household ('pater familias') – at the fireplace or in the private shrine or chapel (the *lararium*). A wall painting at Pompeii gives us a good impression of such a sacrificial offering.[16]

The genius and its cult are essentially bound up with the life of the individual human being, and with the significant events in his life: such events as birth, celebrations of birthdays, as well as death and the continuing presence of the genius after the individual's death. It was also the custom to swear by one's genius, especially regarding important matters. One did not simply swear – as was the custom since the days of Augustus – by the genius of the emperor (cf. Horace, *Epistles*, 2, 1, 16),[17] but also by the genius of one's friend, one's master or one's patron (cf. Horace, *Epistles*, 1, 7, 91–95; Seneca, *Epistles to Lucilius*, 1, 12, 2).[18]

The central significance attributed to the genius for the individual's life can only be understood in its fullest extent if one remembers that the genius came to the fore not only in matters of birth, the twists and turns of fate, and death, but also in matters of love. After all, the genius was closely related to such key events in life as marriage and procreation. The marriage bed was named after it as the 'lectus genialis' (as, for example, in Horace, *Epistles*, 1, 1, 87; Servius, *Commentary on Virgil's Aeneid*, 6, 603; Arnobius, 2, 67);[19] the expression 'pulvinar geniale' can be found in Catullus (64, 47).[20] An adulterer was even regarded as having shown contempt for the 'genius sacri fulcri' (Juvenal, 6, 22).[21] 'This much is to be considered as ancient and accurately handed down', according to W. F. Otto, 'that the marital bed took its name from the genius, and was consecrated to it, and perhaps he was called to it', for 'whatever has life, has its genius; if it does not show itself, there is no life', and 'so it is quite natural to call it to the place in which new life is supposed to be created, and to honour it there'.[22]

Thus the genius, as a personal tutelary spirit, exercises a protective function and is closely related to the significant, fateful moments and stages of life – procreation, birth, marriage, family, oaths, love, and death. Although it is similar to the human soul, the genius is not identical with it, but as a divine spirit leads its own kind of life, which does not keep it bound to the individual's body and soul, nor the beginning and end of the individual. That the spirit is primarily a 'spirit of becoming' is demonstrated by its close relation to the individual's birth, to the anniversary of his birthday, as well as to other important initiatory situations, such as procreation and marriage. The genius is perhaps the most beautiful mythological representation of what, earlier, I called the archaic as an enchanting beginning. At the same time further aspects and functions of the archaic are revealed and illustrated by the genius. As I have said, the genius can also be described as a 'spirit of fate', for it determines in an important way how an individual human being develops, what he becomes, and when he dies. As long as the individual is alive, the genius remains active in him as an originary principle. Or, in Aristotelian terms, the genius as the originary principle is the *télos* of the human being (*télos*, that is, in the sense of 'goal' as well as of 'limit'), the individual's 'entelechy'. Expressed in more general terms: just as, as a 'spirit of becoming', the genius is not only concerned with procreation and birth, so, too, the archaic as the originary principle is not solely concerned with the individual's starting point in time. As already mentioned, the archaic possesses a power that does not just initiate, but also creates, forms, shapes, re-shapes, and transforms – its effectiveness persists throughout the individual's life. In the genius as the tutelary spirit that determines the individual's fate and (trans)forms his life, the persistent effect of the archaic becomes clear.

But the genius is not just a 'spirit of fate', of 'becoming', but is also a 'spirit of joy'. It wants the individual human to lead a good life, indeed to lead *the* good life: in the culinary, sexual, and general sensory sense. A life of individual and social sensuous pleasure promotes the genius and its productive power; an ascetic and puritanical life diminishes it. The demands of one's genius mean one is guided by the reins of sensual pleasure. Which means that sensual pleasure displays what the genius wants, or to be even more explicit: the genius has an influence on the individual human being by means of sensual pleasure. When we said above that the fascination, the spell, the allure of the archaic beginning motivates, moves us forward, and brings forth something new, we can now state more specifically: the genius is the spell cast or the attraction wrought by a pleasurable sight or a pleasurable desire, by means of which the genius has a motivating and productive effect on the individual. The archaic joy of the genius is and remains a joy that brings something forth: that is, which is productive and creative. Wherever it leads only to a shallow satisfaction of desire, or where the satisfaction of desire is exhausted in the frenzy of pure pleasure, the desire for pleasure loses its originary and archaic character and, as a result, its connection to the genius.

Thus pleasure has to exercise a dual magic on the human being: on the one hand, it casts a spell, it thrills and excites, but on the other, it activates and stimulates. Wherever the desire for pleasure merely casts a spell and causes excitement, wherever it does not awaken the formative and individuating powers of the individual, it remains uncreative. It is not only the puritanical curtailment of pleasure and desire, but precisely an unproductive, crude, crass hedonism that severs the link with the genius. So the genius illustrates how the originary archaic is not simply something to whose spell the individual should succumb, allowing it to work its magical power, but also something that one should take hold of and shape, in order to form, nurture, and develop.

The provisional definition given in the opening section can thus be reformulated more precisely in the case of the genius as tutelary spirit. Here the archaic reveals itself as a beginning that is bound up with the individual, that stands in a relationship of dialogue with the individual, that casts a spell and gets things moving: it is a beginning that has the capacity and the power to initiate life and to shape it, as well as to re-shape and transform it – but only when the individual human being does not lose his or her pleasurable and sensuous connection with the 'spirit of becoming', and when he or she consciously takes hold of and creatively develops this spirit.

The 'genius loci' as the numen of a place

According to the ancient Romans, every individual human being had his *own* genius; similarly, certain places also had their own genius: in other words, a 'local spirit'. This spirit was equally regarded as a protective or tutelary one, and thus sometimes called and honoured as the 'tutela' or 'tutela loci' (see Petronius, 57, 2),[23] or as the 'genius loci'.

In Herculaneum a Roman wall painting has survived which shows in a very graphic way how the genius of a place, a 'genius loci', was conceived and what it signified. In the painting we can see a small circular altar, around which a serpent winds its way upwards. Its tail touches the earth, its head stretches over the top of the altar. The serpent – unambiguously identified by the inscription to its right '*GENIUS HUIUS LOCI MONTIS*' as a manifestation of 'the genius of this part of the mountain' –[24] is eating the offering of food that has been placed on the altar. To the left of the picture a young boy is approaching the altar. In his right hand he is holding a branch. On his head he is wearing a garland. He could be a shepherd boy who, ritually dressed, is coming to this part of the mountain to make a sacrifice or offering to the genius for the welfare of his flock. This conjecture is supported by another passage found elsewhere: 'The shepherd in spring does not drive out his flock until he, together with Pales, Faunus, and the Lares, has made an offering to this genius [*salso farre*]', or in other words to the 'genius loci' (Calpurnius Siculus, 5, 26).[25]

Another witness account provides further insight into the cultic practices that were associated with places over which the 'genius loci' presided. There is an instructive passage in the text *De agri cultura* by Marcus Porcius Cato (234–149 BCE), which classical philologists have overlooked in this context but to which the human geographer Yi-Fu Tuan has explicitly drawn attention.[26] Now, although the actual expression 'genius loci' does not occur in this text, it nevertheless demonstrates very clearly how the set of cultic expectations surrounding what later (in imperial times) was explicitly referred to by the concept of the 'genius loci' already existed very early (in the age of the Republic). In this text Cato tells us 'the Roman formula to be observed in thinning a grove':

A pig is to be sacrificed, and the following prayer uttered: "Whether thou be god or goddess [*Si deus, si dea es*] to whom this grove is dedicated [*quoium illud sacrum est*], as it is thy right to receive a sacrifice of a pig for the thinning of this sacred grove, and to this intent, whether I or one at my bidding do it, may it be rightly done. To this end, in offering this pig to thee I humbly beg that thou wilt be gracious and merciful to me, to my house and household, and to my children. Wilt thou deign to receive this pig which I offer thee to this end.[27]

From this source one may conclude that, according to ancient Roman ideas, certain divine local spirits inhabited particular numinous places. These local deities or spirits are respected and honoured. If human interventions are needed in such places, then expiatory offerings are made: a sign that one has to behave with reverence and awe on these sites. Offerings are made to the spirits where they dwell, so that the human beings living there come to no harm, and so that life can proceed along sacred and fruitful paths. From this passage, as from the wall painting in Herculaneum, it becomes evident that the concept of the 'genius loci' is related to clearly demarcated sites, or to (so to speak) 'micro-locations': to this particular grove, next to an estate; to this particular spot of the mountain, far away from the village, amid the summer meadows. Elsewhere I have provided further examples of sites relevant to the concept of the 'genius loci', most of which show how it refers to a specific, limited and thus clearly and vividly characterisable appearance of a particular location and its numen.[28]

So far the examples I have mentioned are related to predominantly rural sites over which a 'genius loci' presided. But a 'genius loci' can also preside over domestic and urban sites, in houses and homes, villages, and cities, in which many numinous places can be found where a local spirit or divinity was honoured. Numerous imperial testimonials provide direct evidence of the domestic significance of the 'genius loci', or to the genius of the home.[29] Many inscriptions bear witness to a village or civic cult: to the genius of a village, a district, the place for the assembly, the food stores or the granary, the customs office, the watch-station, the theatre, the threshing floor, the meat market, the school, or the baths.[30] Even in a military context the genius is not without its significance – as the 'genius exercitius'. This genius of the army is honoured in the troops' barracks, and its statue stands at the centre of the camp. In Speyer there is a good example: at the foot of the statue of the genius of the army there is a dedicatory inscription dated 181 CE, dedicated to the genius of the military unit, the local genius, and to the harmony of the various divisions.[31]

Let us examine a typical example of a location which is both artificial and numinous over which a 'genius loci' presides: the spring of the nymph Juturna. The spring can be found in the main square in ancient Rome, the Roman Forum, which was for centuries the centre of public life in the city. Juturna is the daughter of the river god Volturnus, wife of Janus, and mother of Fontus (or Fons) (Arnobius, 3, 29).[32] In an inscription found near her spring,[33] Juturna is called *Genius Stationis Aquarum* – the genius of the watering place.[34]

The spring shrine of Juturna can be found on the south side of the Roman Forum, next to the temple of Castor and Pollux. Originally a marsh-like valley lay where the Forum came to be built. The first prehistoric settlements were constructed on the surrounding hills. From the 10th century until the 8th century BCE the swampy marsh was simply used as a burial ground. Not until the (mythological) founding of the city in 753 BCE were the dead buried on the Esquiline Hill. When the Tarquins built the Cloaca Maxima, the main drainage canal in Rome, and drained the marsh, the Roman Forum was constructed bit by bit along with its meeting places, shrines, basilicas, porticos, commemorative statues, and triumphal arches. The Forum thus became the religious, economic, legal, and political centre of the city.[35]

Until the completion of the first aqueduct in 312 BCE the spring of Juturna, whose waters were considered to possess healing properties (Varro, *De lingua Latina*, 5, 71),[36] and which were later still used for official offerings (Servius, *Commentary*, 12, 139),[37] supplied the Romans with drinking water. Already in the time of the Republic (2nd century BCE) the spring water was held in a large basin ornamented with a group of figures. This basin, called the Lacus Iuturnae, whose appearance was altered for the last time by Trajan, as well as (ten metres southeast of it) a small chapel with a sacrifical altar, can still be seen today. The reconstructed aedicule goes back to the second century BCE. In front of it stands a small circular altar with the inscription 'M BARBATIVS POLLIO / A ED CVR / IVTURNAI SACRVM / TEAL' and a square altar pedestal; on it, a relief depicts Juturna and her brother, Turnus. Turnus (betrothed to Lavinia, until her father gives her in marriage to Aeneas) was to fall in battle with Aeneas, despite the support of his sister (see Virgil, *Aeneid*, book 12). The gesture, simultaneously of protection and farewell, which on the relief Juturna makes to her brother who is armed and dressed in armour, reminds us of this mythico-historical event. In the middle of the Juturna basin today, one finds the base of an altar decorated with a relief of the Dioscuri. This relief, like the Temple of the Dioscuri that stands nearby, is a reminder of the fateful combat of Turnus and Aeneas, the Trojan ancestor of the Romans. After the battle at Lake Regillus against the Latins, the Dioscuri (Castor and Pollux) are supposed to have given their horses water to drink at the 'Lacus Iuturnae', at this very spring basin (Ovid, *Fasti* 1, 706; Dionysios of Halicarnassus, 6, 13).[38]

So much, then, for examples of ancient sites over which a 'genius loci' presides. As has become clear, the local spirit functions as a tutelary or protective spirit (of mountain meadows and farmland, of domestic, village, and urban spaces), one that is called upon to protect the well-being of humans and animals, from whose spring healing waters are drawn, on the site of which a general protective influence may be detected. But in these cases we can see that it is not just a question of a protective spirit, but also of a 'spirit of becoming'. Hence, among other things, its supportive and life-promoting influence: for the protective spirit supports all life that is developing and undergoing the process of becoming, insofar as it lies within its sphere of influence. Not for nothing does Cato say: 'To this end, in offering this pig to thee I humbly beg that thou wilt be gracious and merciful to me, to my house and household, and to my children.' For the entire house, with all its human and animal inhabitants, is supposed to be encouraged to flourish: in its fecundity, its propensity to procreate, as well as in its health and its well-being. The example of Juturna shows yet a further aspect of the 'spirit of becoming', an aspect which can most obviously be observed in artificial, city-based sites: that the life of the city should be supported and promoted in a 'socio-historical' sense. Thus it is not only a question of supporting 'biological' development, but rather of supporting the creation of identity and the work of memory in a social space. The spring nymph, Juturna, protects and nurtures (with her healing spring waters) precisely not just the 'biological' well-being of the citizens, but also (by means of the mythical stories with which she and her site are implicitly associated) their social well-being. She preserves cultural memories from the time of the mythical foundation of Rome, and creates (socio-historical) identity.

A 'spirit of becoming' can be present in the site of a 'genius loci' in not only a 'biological' and 'socio-historical', but also in a 'geological sense', however. This is demonstrated by a charming passage in Ovid's *Metamorphoses*, which tells of a numinous grotto created by a natural spirit:

> In its most sacred nook there was a well-shaded grotto, wrought by no artist's hand [*arte laboratum nulla*]. But Nature by her own cunning had imitated art [*simulaverat artem / ingenio natura sua*]; for she had shaped a native arch of the living rock and soft tufa [*nam pumice vivo / et levibus tofis nativum duxerat arcum*]. A sparkling spring with its slender stream babbled on one side and widened into a pool girt with grassy banks.[39]

Here the grotto is not artificially constructed through art or artistry ('arte laboratum nulla') in order, for instance, to fulfil an aesthetic or cultic function; rather, it has been made by the spirit of nature ('ingenio natura sua') in an artistic fashion ('simulaverat artem'). The spirit ('ingenium') of this site has made, out of living ('vivo') stone, a 'natural' ('nativem') arch. And this is why this grotto is sacred (to Diana and the nymphs): in it the divine-numinous power of self-creation and self-formation of this site is made manifest, in it we perceive the activity of a divine 'spirit of becoming'.

Of course, the concepts of the 'genius' and the 'ingenium' are admittedly not completely identical. The concept of the 'ingenium' covers a different spectrum of significance: unlike the 'genius', it has a different, more 'psychological' or 'philosophical' and less 'mythological' or 'religious' meaning. Nevertheless, there is an obvious similarity to which the expression 'ingenium loci' in Ovid points (Ovid, *Tristia*, 5, 10, 18; *Ex Ponto* 2, 1, 52 and 4, 7, 22).[40] By 'ingenium loci' Ovid means something 'natural', its character or, more accurately, the natural disposition of a site, meaning the composition of the terrain as well as its vegetation.[41]

In a purely mythico-religious context, one speaks less of the 'ingenium' and more about the 'numen' of a place. In the 'numen' of a place is revealed the most profound aspect of what it means to talk about a 'genius loci' as a 'spirit of becoming'. The word 'numen' originally refers (in the pre-Augustan age) to the action and the power of a divinity. Frequently it is a question of the numen of the gods, without the precise identity of the gods being specified, as for example in Cicero (*In Catilinam*, 3, 19), Catullus (64, 134; 76, 4), and Gaius Julius Caesar (*De bello Gallico*, 6, 16, 3).[42] In other cases the numen is clearly attributed to a particular divinity, for example in a reference to 'Iovis numen', the numen of Jupiter (Nonius Marcellus, *De Conpendiosa Doctrina*, 173, 27).[43] In all these cases, numen is usually synonymous with divine power (in Latin: 'vis'), divine action, divine will. Numen and 'vis' (power) are thus synonyms, as Cicero's *De divinatione* (2, 124) shows particularly clearly.[44]

The characteristic combination of the word 'numen' with a name of a god in the genitive case can still be found in the Augustan and post-Augustan ages. Such as, for example, when Virgil in the *Aeneid* speaks of a 'numen Iononis, a 'numen Phoebi' (*Aeneid*, 3, 437; 9, 661).[45] At the same time, one also finds here for the first time the use of the concept of numen as synonymous with the concept of god or goddess. This is a further, and novel, aspect of the concept. 'Numen' now no longer just means divine power or energy, the workings of this god or that goddess or of the gods in general, but numen means above all divinity itself; 'in the pre-Augustan era "numen" means the quality of a subject, in the Augustan and post-Augustan era it can also dennote the subject itself'.[46] This becomes clear in Virgil's didactic poem on agriculture, in which he says: 'And you Fauns, the rustics' ever present gods [*agrestum praesentia numina*] / (come trip it, Fauns and Dryad maids withal!), / 'tis of your bounties I sing' (*Georgica*, 1, 10–12).[47] Even more clearly in the *Aeneid* (6, 68), numen and divinity are placed as equals next to each other, when Virgil speaks of 'the wandering gods and storm-tossed powers of Troy' (*errantisque deos agitataque numina Troiae*).[48] Along with this extension of meaning there is a further development: the concept of numen can also be understood in a more personal, more concrete, more sharply profiled and more site-specific way. Thus Horace (*Carmina*, 4, 5, 35) calls the personal tutelary spirit of Augustus his 'numen'.[49] Ovid speaks of poets in whom a numen dwells ('numen inest illis', *Ars amatoria*, 3, 548), as well as of the numina of the field and the woods (*Metamorphoses*, 6, 392), of mount Parnassus (*Metamorphoses*, 1, 320), and of a wood in which likewise a numen dwells ('numen inesse loco', *Amores* 3, 1, 1–2).[50] In Virgil (*Aeneid* 12, 181), the poet invokes the springs and rivers, the aether and the numina of the sea.[51] Sites where a numen dwells were considered sacred, as Ovid (*Fasti*, 3, 264) says of a lake: 'est lacus, antiqua religione sacer'.[52]

Because after the Augustan age the concept of the numen became increasingly personal, concrete, and specific to a particular site, its proximity to the concept of the 'genius loci' became correspondingly close. This is why Friedrich Pfister is right to speak about the 'numina locorum': 'The *numina locorum* can sometimes be more precisely determined, on other occasions however they are completely unknown, as is the case in the ritual of the Priests of Arval (Henzen, *Acta fratr. Arval*, 146): "May it be that you are a god or goddess, under whose patronage this grove or site is lying" (*sive deo sive deae, in cuius tutela hic lucus locusve est*).[53] To such a divinity a solemn offering can be made, when they are known as *numini*, which does not refer to gender, or as *genio numinis*, as in CIL, 6, 151 (*genio numini fontis*) [. . .] or *genio loci* [. . .]) or also as *Tutelae loci* (Petronius, 57, 2; CIL, 6, 216 and 777; 13, 440 [. . .]).[54] Here one can see clearly how closely 'numen' and 'genius loci' are related. In fact, one could even say that it is primarily sites that have (in an essential, not an accidental sense) a numen that also have a genius. The 'genius loci' refers to the spirit of a numinous site. The numen that pertains to a site in an essential sense is the *conditio sine qua non* of the 'genius loci'; it is its 'spirit of becoming'.

Similarly, this is also true, not just for natural, but also for artificial sites, irrespective of whether it is a tutelary spirit of a house, a village, a district, a city, or a military camp. This does not mean that the artificially created, historically developed site has its numen in the same way that a site that has naturally come into being has one. As we saw above, this was not the case: the natural grotto that Ovid described had been formed by the ingenium of nature or, more precisely, by the 'spirit of becoming' of the site. The same cannot be said of a house and its spirit, a city and its spirit, of the theatre or the curia, of a military camp and its genius of the army or the place. Here, one could say, the motivating force is exclusively human skilfulness, art, technique ('ars', 'techné'). Yet nevertheless: whatever gives the human, artificially shaped site its numen does not lie in the sphere of human skilfulness, but in the sphere of biographical and/or historical life events or destiny. It is only inasmuch as significant or 'existential' meanings accrue to a site, or inasmuch as certain fundamental, reality-shaping, life-changing events that determine the fate of an individual, a family, a tribe, or a linguistic community are bound up with a location, that it has a numen at all. A numen, that is, which is essentially associated with a place, not because of its nature, but because of its historical fate (in a socio-historical sense). This is an aspect of the 'spirit of fate' which rightly comes to the fore and displaces the aspect of the 'spirit of becoming'. Or, to put it another way: in the naturally numinous site the 'spirit of becoming' reveals itself as a natural spirit, and in the artificially numinous site it reveals itself as a 'spirit of fate'. In both cases, however, the same concept of the numen shines through the general background significance of the naturally numinous and the artifically numinous site alike. For behind both lies the 'numinosum', something beyond human making and control, something prior: an agency of power and action that owes allegiance to older and more unconscious strata.

Thus the 'genius loci' is not only a protective or tutelary spirit, but also, as we have seen, a 'spirit of becoming' and a 'spirit of fate'. Grown up in a natural, historical, or life-historical way with a location, it represents the site's potential and power, which has not only contributed to the formation of the site's external appearance, but continues to operate on the site, exerting its influence to protect, to nurture, and to shape.

Now, the question arises: how and to what extent can one also call the 'genius loci' a spirit of joy or pleasure? Certainly not in the same sense as was the case with the personal tutelary spirit, as something that fosters and nurtures human (sensual) pleasure. For that is not what is at stake here. True, among the various sites of a 'genius loci' there are some that correspond to what literary and art history know as the 'locus amœnus': the bucolic, idyllic place of pleasure.[55] Of course, such places have an erotic, sensual aura, inciting pleasure in such a way that it is possible to ascribe to them a 'spirit of pleasure'. But what matters most – and what makes them sites of a 'genius loci' in the strict sense of the term – is in the end not so sweet and light and harmless, like a 'locus amœnus': what is revealed is the site's numen. And this numen draws attention to itself as a numinous aura, a numinous atmosphere. Such an atmosphere can certainly – as in the 'locus amœnus' – be homely, cosy, charming and beautiful; but it does not have to be.

That the atmosphere that greeted the humans of antiquity in a numinous location could be something entirely different – something sublime, uncanny, oppressive and overpowering – is shown by a description in Seneca. This passage can be found in the forty-first letter to Lucilius:

> If ever you have come upon a grove that is full of ancient trees which have grown to an unusual height, shutting out a view of the sky by a viel of pleached and intertwining branches, then the loftiness of the forest, the seclusion of the spot [*secretum loci*], and your marvel at the thick unbroken shade in the midst of the open spaces, will prove to you the presence of deity [i.e., the numinous: *numen*]. Or if a cave, made by the deep crumbling of the rocks, holds up a mountain on its arch, a place not built with hands but hollowed out into such spaciousness by natural causes, your soul will be deeply moved by a certain intimation of the existence of god [*religionis suspicione percutiet*]. We worship the sources of mighty rivers; we erect altars at places where great streams burst suddenly from hidden sources; we adore springs of hot water as divine, and consecrate certain pools because of their dark waters or their immeasurable depths.[56]

Here Seneca addresses the most important aspects of such a site: the sublimity of the location, surrounded by old, mighty trees; the sense of secrecy, the wonder and amazement aroused by the dark, shady grove; the fear stirred by a deep cave; the unfathomability and marvel of special wells and lakes. All these phenomena awaken in the psyche an intuition of reverence and numinosity, they make the psyche shudder, they give us the feeling that something divine hangs 'in the air', that a deity dwells and is at work in this place.

But it is always the same feeling-toned sensations, such as sublimity, secrecy, inscrutability, dread, and sacrality of a site, that awaken in the psyche an involuntary sense of wonder and amazement, making us shudder in intuitive reverence. The numinous atmosphere of a place rubs off on the psyche of a person who visits it. Someone who is present on such a site participates in – more accurately, is made part of – the place itself: the visitor's mood becomes attuned to the atmosphere of the site. Or, as Rudolf Otto (1869–1937) puts it, the 'objective' site whose numinous atmosphere is 'felt as external' to the perceiver (or a sacred site, building, or ritual) triggers in the individual a certain feeling, the feeling of the *mysterium tremendum*, a sense of the 'sacred' or the 'holy' (*das Heilige*):

> The feeling of it may at times come sweeping like a gentle tide, pervading the mind with a tranquil mood of deepest worship. It may pass over into a more

set and lasting attitude of the soul, continuing, as it were, thrillingly vibrant and resonant, until at last it dies away and the soul resumes its 'profane', non-religious mood of everyday experience. It may burst in sudden eruption up from the depths of the soul with spasms and convulsions, or lead to the strangest excitements, to intoxicated frenzy, to transport, and to ecstasy. It has its wild and demonic forms and can sink to an almost grisly horror and shuddering.[57]

At this juncture it is important to make clear, in order to avoid a possible misunderstanding, that a place's 'spirit of becoming' does not only operate on the individual human being by means of a numinous atmosphere, but also *in persona*: as a god that assumes form, as a sensorily perceptible, personal epiphany (and, as we saw, also in the form of an animal, above all in the shape of a serpent). There are innumerable ancient literary testimonies and accounts of experiences which relate how a 'genius loci' appeared *in person* to human beings, how it spoke to them and touched them: in other words, how a 'genius loci' can exert an influence on human beings, not just as a vague or indeterminate atmosphere, but also as a personal being. The 'genius loci', the spirit of a place, manifests itself not just as a numinous atmosphere, but also a daimonic, personal epiphany.[58] And there is (from a mytho-phenomenological perspective) no real reason to dismiss this mode of appearance as opposed to a numinous atmosphere as something derivative or somehow 'projected'. What Ulrich von Wilamowitz-Moellendorff (1848–1931) said about the Greeks and their gods also applies to the Romans and their gods and 'genii loci': 'Humankind knows that the gods are there, above all from sensory perception; some have seen them, and occasionally they show themselves'; for the Greek gods 'are everywhere, they belong, to use modern terminology, just as human beings do, to nature, in which all of them live [. . .] On earth the gods intervene in everything that concerns human beings. They attend the sacrifice, the common table with those who offer sacrifice, they join with the daughters of human beings and give them children'.[59]

But back to the aspect of the 'spirit of pleasure': as we have seen, the aspect of the 'spirit of pleasure' relating to the personal tutelary spirit was, in the case of the 'genius loci', something most accurately described as a numinous atmosphere. If we said above that the support of the personal genius guided one by the reins of sensual pleasure, in this case the following is true: the patronage or the veneration of the 'genius loci' guides one by the reins of the numinous atmosphere. The very fact that the numinous atmosphere of a site concerns the individual human being, and how it does this, shows him or her what the 'genius loci', so to speak, *wants*. More precisely: the numinous atmosphere conditions how the individual reacts to a location, how he or she interacts with it. For it is the numinous atmosphere that casts its spell – in part enticing, in part threatening – on the individual, changes his or her state of consciousness and hence makes possible experiences and events far removed from everyday life. Yet the numinous atmosphere does not simply enchant and hold the individual human being spellbound, it also – a typical characteristic of everything archaic and originary – motivates and stimulates him or her: to shape, care for, and remember the sacred place and the events associated with it; to tell and re-tell stories and myths; to educate and pass on cultic and ritual forms of interaction. What Hesse says – 'For every beginning has a special magic / That nurtures life and bestows protection' – is true not only for what is archaic and originary about new stages in the life of the individual, it also applies to primitive sites of a 'genius loci'. Such sites are inhabited by a numinous magic, which protects and nurtures the locations themselves, as well as the living mythico-religious communities associated with them.

In conclusion, we could modify our original definition of the archaic and the originary in relation to the 'genius loci' as follows: the archaic reveals itself to be an atmospheric beginning in a specific location that stands in a relationship of dialogue with us, that casts a spell and gets things moving, and which moreover has the numinous capacity and power to form a location and its inhabitants, to reshape and transform them; but it can only do this when the inhabitants do not lose their atmospheric connection to the numinous spirit of becoming, when they are moved by this spirit and its history, and when they preserve and protect the main things and meanings that have grown up around the site, as well as adapting them for their age and guiding them towards the future.

Towards a phenomenology of the 'genius loci'

The number of phenomenologists who have seriously concerned themselves with what the Romans called the 'numen' of a site, or with its numinous atmosphere, is decidedly modest. Thus far I have mentioned one, the phenomenologist of religion Rudolf Otto. Another phenomenologist, a far more radical one with close links to vitalist philosophy or *Lebensphilosophie*, was Ludwig Klages (1872–1956). According to Klages, 'the ancients knew the "genius loci", the nimbus, the aura, and we too speak of the "atmosphere" surrounding a person, a house, a district. Now, this "atmosphere", which can be picked up by so-called sensitive natures and felt by those alert to nuance, but remains unknown to more robust minds, is a reality that realizes itself [*eine wirkende Wirklichkeit*], which gives and enriches or soaks up and weakens, which surrounds and gives warmth or hollows out and makes cold, which accelerates and stimulates or checks and subdues, which broadens out or narrows in, which inspires or inhibits, and its effect is essentially different from being touched by another *body*' (or, one might add, by anything else objectively perceived).[60]

Someone else who has engaged in a truly differentiated way with the phenomenon of 'atmospheres', including the atmosphere of place, is the phenomenologist of the body, Hermann Schmitz, who was born in Leipzig in 1928. Schmitz defines 'atmospheres' as 'spatially extended, bodily perceptible feelings, constellated without being contained' (*räumlich ausgedehnte, jeweils randlos ergossene, leiblich spürbare Gefühle*).[61] The moods associated with certain landscapes are, for Schmitz, examples of such 'atmospheres'. The atmospheres of particular climates, times of the day, and seasons of the year, such as the tension-laden atmosphere of a storm, the crystalline atmosphere of a star-clear, crisply cold winter night, or the oppressive and sultry atmosphere of midday in summer: these are all atmospheres – or, more precisely, 'suprapersonal atmospheres'. As well as the suprapersonal atmospheres that are not related to a particular individual, there are also 'personal atmospheres', such as joy, mourning, bliss: these 'personal atmospheres' can move one person or a group of people, and be reflected by them. According to Schmitz, there are not just profane suprapersonal and profane personal, but also 'divine suprapersonal atmospheres'. The atmosphere of 'Christian love' or of the 'Holy Spirit', for example, or the ecstatic Dionysian atmosphere, the reflective and restrained Athenian atmosphere: these are all suprapersonal 'divine' atmospheres, which Schmitz analyses and explains phenomenologically.[62]

Into the category of 'divine atmosphere' falls what Schmitz calls 'local divine atmospheres'. A purely atmospherically experienced, numinous site of a 'genius loci': this is the Schmitzian paradigm of such a local divine atmosphere. It is not for nothing that Schmitz refers in this context to the same passage from Seneca that we mentioned above (*Epistulae ad Lucilium*, 4, 41, 3),[63] for this passage is particularly well suited to illustrate the atmospheric and numinous mode of experiencing place in the ancient world. Even more interesting in this context is a quotation taken by Schmitz from Pomponius Mela's *De Chorographia* (1, 13, 72–75):

> Above there is a grotto, known as the Corycian cave, which has a unique atmosphere [*Supra specus est, nomine Corycius, singulari ingenio*], so extraordinary it cannot be easily described. Right at the top of a mountain, standing in the landscape with a steep drop of 10 stadia, the cave opens the mount with a massive chasm. Near to the deep grotto, it becomes ever greener, thanks to the shrubs covering the sides of the ravine, and it is surrounded in its entirety by a circle of trees: it is so extraordinary, so beautiful, that at first glance it strikes fear into those that enter in, but once one has begun to contemplate the sight, one cannot look away. [. . .] Its interior is too frightening for anyone to penetrate, and thus it remains unexplored. As a whole the cave is majestic and truly sacred [*sacer*], worthy of being inhabited by gods and people believe it is, there is nothing in it that does not inspire reverence or reveal a kind of divine presence [*numen*].[64]

This passage is particularly interesting because it shows how Schmitz is trying to restrict the full mode of manifestation of the 'genius loci' – namely, as I showed above, as a numinous atmosphere *and* a daimonic, personal epiphany – to its (corporeally) phenomenologically attainable and atmospheric part. For in the Latin text of Pomponius Mela it does not say 'above there is the Corycian cave with a unique atmosphere' or *Stimmung*, but 'above there is the Corycian cave with its unique *ingenium*'. In this context (as in Ovid's *Metamorphoses*, 3, 157–62; see above) the word 'ingenium' refers to an autonomously operating 'spirit of becoming' that has assumed a shape and is in no way reducible to something purely atmospheric: in other words, it refers to a 'genius loci'. And this important aspect is overlooked by Schmitz. The reason why he does so is related to the specific hierarchies and valuations of his phenomenology of the body. Because, for his corporeally phenomenological approach, what is related to mood, atmosphere, and feeling is (because it can be sensed directly by the body) considered to be originary; what relates to a person or a character is considered by him to be something 'only' secondary, because there is no direct path through bodily perceptions to these phenomena.

This privileging and hierarchization can also be seen in Schmitz's concept of 'divine atmospheres', and perhaps nowhere more clearly than in a tenet of his phenomenology of the body which says that 'every god is a feeling' (daß jeder Gott ein Gefühl ist);[65] every god, which also means every 'genius loci'. Against the background of this hypothesis it is not surprising that Schmitz comes to the conclusion that 'the local divine atmospheres belong to the immense realm of suprapersonal and objective feelings, that are partly found without any location 'in the air', as it were, or more precisely in extended space [. . .] like the weather, partly also condensed in particular locations and around particular objects, often only as echoes left floating'.[66] Not only the profane atmospheres of landscape and place, but also the divine local atmospheres and, along with them, all other divine atmospheres are 'nothing but' derivative or condensed and embodied climatic and extensive feelings. Against this background Schmitz can then state as a matter of principle that 'the authority of the divine is thus at its root always the authority of feelings', adding: 'True, it can appear as the authority of persons (gods), insofar as divine atmospheres, which are feelings [. . .], are as if condensed in these persons [. . .], but their divine authority then derives from it being granted through these atmospheres.'[67]

If this is so, then the question arises: how do gods and 'genii loci' appear in shapes and persons? According to Schmitz, they do so through acts of personificatory 'interpretation' (Erdeutung) and 'concretion' (Konkretion). What does he mean by this? What he is saying is that originally the divine becomes apparent only as vague or climate-like, extended spatial atmospheres, and therefore it can only be vaguely perceived and designated by human beings. This vagueness is, however, not sufficient for human beings: rather they want to obtain understanding and reach agreement about the divine, they want to understand it and talk about it with clarity. In short: they want to make it concrete, to picture it, personify it, and anthropomorphize it. In the case of some divinely atmospheric experiences 'a feeling of pressure to make things concrete becomes noticeably perceptible, without it being possible to satisfy it; [. . .] the agonizing question remains: just what was that exactly? But it sometimes happens that making things concrete is achieved at the price of personifying them, and these instances are of prime importance for the transition from the divine' – i.e., as a divine atmosphere, hanging purely 'in the air', as it were – 'to the' – concrete, because personified – 'god'.[68]

Through this process of 'interpretation' (Erdeutung) that makes concrete and personifies, it is said, one ends up not just with the formation of clearly defined gods, but with the formation of 'genii loci'. In this respect, too, Schmitz detects a specifically human urge, a human drive to make divine atmospheres concrete:

> Similarly, the atmosphere of twilight anxiety would not be concrete without such figures as the Loreley and her Indian sister Aranyani, nor the enchanting and seductive force of the silent midday sultriness without the sirens as the incarnations and 'interpretations' [*Erdeutungen*] in personified form of these highly feeling-charged atmospheres [. . .] It is only through such personifying incarnations and in them that the discourses on such powers of nature gain the force of referring to something concrete, by means of which their obscurity and [. . .] their uncertain lack of precision falls away and an unmistakable face of reality comes to light, which takes its suggestive energy from the nimbus of a powerful atmosphere.[69]

However much Schmitz tries here to illuminate from the standpoint of the phenomenology of the body the phenomenon of the 'genius loci', he only ever succeeds in capturing its atmospheric aspect; its other aspect – that 'genii loci' are, for mythical-minded human beings, not 'interpreted' (*erdeuteten*) or similarly derivative phenomena, but rather very concrete, reality-based, clearly formed and personal appearances – remains, despite attempts to make them plausible as 'concretion' and 'personification', phenomenologically elusive. Schmitz knows this, and he has recourse to the stopgap argument that he adduces before every attempt at an explanation. 'How does it happen,' he writes,

> that in all times and with all peoples such a personification takes place, since atmospheres are, after all, in themselves not exactly persons? [. . .] This brings us to the decisive question of truth about the meaning of divine persons and religion as a whole. It cannot be finally settled by phenomenology on its own; for the belief in divine persons, understood as a metaphysical conviction, cannot be conclusively proved or refuted purely phenomenologically, just as little as the belief in fairies, witches, dwarves, Goethe's Mothers or Kant's *homo noumenon* or anything else one cares to imagine.[70]

Of course, what Schmitz calls the 'question of truth' of religion is never conclusively answered, proved or refuted by phenomenology (nor by any other philosophy or discipline). And of course one could dispute for ever 'metaphysical convictions' and 'anything else one cares to imagine'. Nevertheless, this is precisely what phenomenology is *not* about. And Schmitz knows this as well. In phenomenology one is dealing with the recognition and the acceptance, the investigation and the explanation of phenomenal reality in its complex, multilayered, unreduced and undisguised fullness; it is about the reality of unfiltered experience. And on this experiential level the gods and the 'genii loci' always reveal themselves (to those for whom they are *actual* experiences, hence to primordially mythical cultures) to be originary phenomena that are more than merely atmospheric. It would have arguably been more modest and more appropriate if Schmitz had not turned for help to a sweeping and ultimately polemical stopgap argument, but had admitted the explanatory limitations of his methodology; if he admitted, for example, that, by means of his method of corporeal phenomenology, he can show, differentiate, and explain with particular clarity the atmospheric dimension of the numinous, whilst necessarily leaving other dimensions of the same phenomena less clearly defined or even not defined at all. Schmitz's grandiose and pioneering corporeal-phenomenological insight into the 'world of atmospheres', like his perspicacious, shrewd, and conceptually highly differentiated description of atmospheric phenomena as they are constelled in extended space, condensed in certain directions, and anchored in particular locations, would thereby have lost nothing of its rich phenomenological content; quite the opposite.

The 'genius loci' as the archetype of a place?

The only phenomenologist who has done justice to the daimonic, personal dimension of the 'genius loci' that Schmitz (and, by the same token, Rudolf Otto) overlooked is Ludwig Klages. For Klages succeeded in throwing light on such phenomena as the 'genius loci' not only in their atmospheric and numinous, but also in their daimonic and essential aspects. Without being able to go here into detail about his doctrine – which, in some respects, is close to Jung's notion of archetypes, if in a less Platonic and more Heraclitean way – of the 'reality of the (primordial) images' (*Wirklichkeit der [Ur]Bilder*),[71] I should like to refer to it using a rather long, but highly suggestive, passage, in which Klages explains how 'the *dimension of the daimonic* is essentially identical with the temporal-spatial uniqueness of an image':

> For us [i.e., for modern, "intellect-oriented" humans] there exists [three-dimensional] space and bits of space, [linear] time and bits of time; for the life-bound [i.e., biocentric] spirit [of mythically minded humans] there are as many spaces and times as there are images to be spatially and temporally experienced: there is a nocturnal space and a diurnal space, a domestic space and a cosmic space, a forest space and a temple space, an eastern space, a southern space, western space, northern space, but in certain circumstances there are as many house spaces as there are different houses, and even as many different spaces in one and the same house as there are moments of dynamic internalization of spatial appearances that constitute the house [*Augenblicke wirksamer Innerung der hausgestaltigen Raumerscheinung*]. The earliest form and the root of knowledge of the reality of images [*die Wirklichkeit der Bilder*] is polydaimonism, and the multiplicity it announces multiplies every content of meaning [*die mit ihm sich bekundende Vielfachheit vervielfacht jeden Bedeutungsinhalt*].[72]

Klages goes on to explain what, in the context of his doctrine of the 'reality of the images', he means by talking about daimons:

The true daimon is the daimon of a place, a landscape, an element, which changes with the way their own appearance changes; which receives the offerings of those who acknowledge it in temporal rhythms, whose daily, monthly, or annual patterns are based on the forms in which it is revealed; and which is at work in those events or tasks that the [mythically minded] spirit was compelled to accept as powerfully at work through their images, of which pathic [and visionary] perception is aware.[73]

If Klages is right in his assessment, and there is much to suggest he is, then one could discern in the 'spirit of becoming' of a place as something archaic and originary one last aspect, and perhaps its most profound: namely, that at its heart lie powerful and evolving (primordial) images. The 'genius loci' is thus not only a tutelary or protective spirit and a 'spirit of becoming', not only a 'spirit of fate' and a 'spirit of development' (*Bildegeist*), but precisely also a 'spirit of images' (*Bilder-Geist*). Or, following C. G. Jung's formulation, originary sites of a 'genius loci' are inhabited by archetypes, such sites possessing an innate archetypal potency. And the numinous atmosphere of a site is, in the final analysis, the aura or the charisma, the herald or the harbinger of a local archetype, an *Ortsarchetypus*, that is surging into appearance.

For our modern, rationalistic world such thoughts are, of course, nothing less than heresy. But it is not only Klages who entertains these ideas: Jung, too, is attracted to them, most often with reference to the comparative religious concept of 'mana', which is 'not confined to Melanesia, but can also be found in Indonesia and on the east coast of Africa', and 'still echoes in the Latin *numen* and, more faintly, in *genius* (e.g. *genius loci*)'.[74] If, as Jung says, the psyche is 'a reflection of the world and man'; if, precisely for this reason, it is 'the only phenomenon that is given to us immediately and, therefore, is the *sine qua non* of all experience';[75] and if, on the level of this world-related psychic experience there is much to indicate that the 'mana'-concept is not quite so unreal as today we moderns would like to believe, then 'we hesitate to accept [these statements] and begin to look around for a comfortable theory of psychic projection'.[76] For 'the question', Jung continues, is this: 'Does the psychic in general – the soul or spirit or the unconscious – originate in *us*, or is the psyche, in the early stages of conscious evolution, actually outside us in the form of arbitrary powers with intentions of their own, and does it gradually take its place within us in the course of psychic development?'[77]

However one chooses to answer this question, whether one sees in the site of a 'genius loci' the presence of an archaic and originary (mana) spirit or whether one sees it simply as a surface onto which something is projected,[78] one thing remains certain: it is not only in relation to the path of life in the biography of an individual that a certain originary magic becomes noticeable, but in relation to important locations in the social space of life as it is lived. Indeed, one could take those lines from Hesse – 'For *every beginning* has a special magic / That nurtures life and bestows protection' – and, in the light of the concept of the 'genius loci', modify them as follows: 'For *certain places* have a special magic, / That protects it and helps us to live.' For such archaic and originary places really do help us: they renew our connection with our natural, social, and personal, life-historical roots, and so make possible a firm foundation for a future, in which continuity and transformation are not seen as opposites, but as reciprocally interrelated and as establishing the place we call our home.

Notes

1 H. Hesse, *Gesammelte Werke in zwölf Bänden*, Frankfurt am Main: Suhrkamp, 1987, vol. 9, p. 483. The poem appears in Hesse's novel *The Glass Bead Game* (*Das Glasperlenspiel*).

2 The Romans believed that, while each man had his 'genius', every woman had a similar spirit or 'iuno', but the Romans were by no means systematic about this distinction, and sometimes attributed a 'genius' to a woman. For the sake of clarity, then, I shall speak here about the masculine form of the 'genius'.

3 'The Genius alone knows – that companion who rules our star of birth [*natale comes qui temperat astrum*], the god of human nature, though mortal for each single life, and changing in countenance, white or black' (*Epistles*, book 2, epistle 2, ll. 186–89, in Horace, *Satires, Epistles and 'Ars Poetica'*, tr. H. R. Fairclough, London: Heinemann; New York: Putnam, 1926, p. 439).

4 Thus, for example, Plautus speaks with reference to the pleasure enjoyed at a meal of the soul, and not – as is usually the case in such circumstances (see below) – of the genius: *facite vostro animo volup* (*Casina*, 784; see *Plautus*, vol. 2, tr. P. Nixon, London: Heinemann; New York: Putnam, 1932, p. 85).

5 See T. Birt, 'Genius', in W. H. Roscher *et al.* (eds), *Ausführliches Lexikon der griechischen und römischen Mythologie*, 6 vols, Leipzig: Teubner, 1886–1937, vol. 1.2, cols 1613–25 (here: col. 1614).

6 W. F. Otto, 'Genius', in G. Wissowa *et al.* (eds), *Paulys Real-Encyclopädie der classischen Altertumswissenschaft*, 1st edn, 20 vols, Stuttgart: Metzler and Alfred Druckenmüller, 1894–1963, vol. 7.1, cols 1155–70 (here: col. 1156). According to the OED, the Latin *genius* derives from the base of *gignere*, 'to beget', and in turn the Greek *gignesthai*, 'to be born' or 'to come into being'.

7 See Plautus, *The Persian*, l. 263: 'Here's where I both prosper a friend and contribute a lot to my personal comfort' (*nun et amico prosperabo et genio meo multa bona faciam*), and *Stichus*, l. 622: 'For you really won't increase your personal comfort in here' (*nam hac quidem genium meliorem tuom non facies*), in *Plautus*, tr. P. Nixon, vol. 3, London: Heinemann; New York: Putnam, 1924, p. 453; and vol. 4, London: Heinemann; New York: Putnam, 1938, p. 73.

8 For this reason Virgil can later speak of 'the winter that delights the genius' (*genialis hiems*): 'In cold weather farmers chiefly enjoy their gains, and feast together in merry companies' (*frigoribus parto agricolae plerumque fruuntur / mutuaque inter se laeti convivia curant*) (*Georgics*, 1, 299–302, in Virgil, *Eclogues; Georgics; Aeneid I–VI*, tr. H. R. Fairclough, rev. G. P. Goold, Cambridge, MA: Harvard University Press, 1999, p. 131.

9 Plautus, *The Pot of Gold*, l. 725: 'I've lost all that gold [. . .] I've denied myself, denied my own self comforts and pleasures' (*tantum auri / perdidi . . . egomet me defraudavi / animumque meum geniumque meum*), in *Plautus*, tr. P. Nixon, vol. 1, London: Heinemann; New York: Putnam, 1937, p. 309.

10 See Birt, 'Genius', cols 1623–24.

11 Seneca, *On Anger (De Ira)*, book 2, chapter 31, l. 6: 'how serpents crawl in harmless course among our cups and over our laps', in Seneca, *Moral Essays*, tr. J.W. Basore, vol. 1, London: Heinemann; New York: Putnam, 1928, p. 235.

12 Otto, 'Genius', col. 1162; see *The Attic Nights of Aulus Gellius*, vol. 2, tr. J. C. Rolfe, London; Cambridge, MA: Heinemann; Harvard University Press, 1968, book 6, part 1, 2, p. 3; see Livy, book 26, chapter 19, l. 7: 'It revived the tale previously told of Alexander the Great [cf. Plutarch, *Alexander*, book 2, chapter 4] and rivalling it as unfounded gossip, that his conception was due to an immense serpent, and that the form of the strange creature has very often been seen in his mother's chamber, and that, when persons came in, it had suddenly glided away and disappeared from sight' (in Livy, *Books XXVI–XXVII*, tr. F. G. Moore, London: Heinemann; Cambridge, MA: Harvard University Press, 1970, p. 75).

13 Horace, *Epistles*, book 2, epistle 1, l. 144: 'and with flowers and wine [propitiate] the Genius who is ever mindful of the shortness of life' (*floribus et vino Genium memorem brevis aevi*), in Horace, *Satires, Epistles and 'Ars Poetica'*, p. 409.

14 Horace, *Odes*, book 4, ode 11, l. 8 and book 3, ode 17, ll. 14–15, in Horace, *Odes and Epodes*, ed. and tr. N. Rudd, Cambridge, MA: Harvard University Press, 2004, pp. 249 and 187.

15 See O. Jessen and W. Schmidt, 'Genethlios' and W. Schmidt, 'Genéthlios heméra' (in which, cols 1142–49, the Roman equivalent, the 'Natalis', is described), in *Paulys Real-Encyclopädie der classischen Altertumswissenschaft*, vol. 7. 1, cols 1133–35 and cols 1135–49; see also Otto, 'Genius', cols 1158–60; and Birt, 'Genius', col. 1614.

16 See the image reproduced in R. J. Kozljanič, *Der Geist eines Ortes, Kulturgeschichte und Phänomenologie des Genius Loci*, 2 vols, München: Albunea, 2004, vol. 1, p. 45, picture 1.

17 Horace, *Epistles*, book 2, epistle 1, l. 16, in *Satires, Epistles and 'Ars Poetica'*, p. 397. See Birt, 'Genius', col. 1617; and Otto, 'Genius', col. 1164.

18 Horace, *Epistles*, book 1, epistle 7, ll. 91–95: 'by your genius, by your right hand and household gods, I implore and entreat you, put me back in my former life' (*quod te per Genium dextramque deosque Penatis / obsecro et obtestor, vitae me redde priori!*), in Horace, *Satires, Epistles and 'Ars Poetica'*, pp. 301–03; see Seneca, *Letters to Lucilius*, epistle 12, §2, in Seneca, *Ad Lucilium Epistulae Morales*, tr. R. M. Gummere, vol. 1, Cambridge, MA: Harvard University Press; London: Heinemann, 1934, p. 67.

19 Horace, *Epistles*, book 1, epistle 1, l. 87 (in *Satires, Epistles and 'Ars Poetica'*, p. 259); Book 6, §603, in Servius Grammaticus, *In Vergilii Carmina Commentarii*, ed. G. Thilo, vol. 2, *In Vergilii Aeneidos Librum Sextum Commentarius*, Leipzig: Teubner, 1884, pp. 83–84 (*non geniales proprie sunt qui sternuntur puellis nubentibus, dicti a generandis liberis*); and Arnobius, *Adversus Nationes*, book 2, chapter 67: 'When you enter wedlock, do you spread the couches with a toga and do you invoke the genii of married couples?' (*cum in matrimonia convenitis, toga sternitis lectulos et maritorum genios advocatis*), in Arnobius of Sicca, *The Case Against the Pagans*, tr. G. E. McCracken, 2 vols, Westminster: Newman Press; London: Longman, Green, 1949, vol. 1, p. 180.

20 Catullus, *Poems*, no. 64, l. 47: 'See, the original marriage bed is being set for the goddess in the midst of the palace' (*pulvinar vero divae geniale locatur / sedibus in mediis*), in *Catullus, Tibullus, Pervigilium Veneris*, 'The Poems of Gaius Veneris Catullus', tr. F. W. Cornish, London: Heinemann, 1912, pp. 1–183 (here: p.101).

21 Juvenal, *Satires*, no. 6, l. 22: 'It's an ancient and established practice, Postumus, to pound someone else's bed, belittling the spirit of the sacred couch' (in Juvenal and Persius, *Satires*, ed. and tr. S. M. Braund, Cambridge, MA: Harvard University Press, 2004, p. 237). See Birt, 'Genius', col. 1615; and Otto, 'Genius', col. 1160.

22 Otto, 'Genius', col. 1160.

23 Petronius, *Satyricon*, §57, 1.2, in *Petronius; Seneca, 'Apocolocyntosis'*, tr. M. Heseltine, rev. E. H. Warmington and W. H. D. Rouse, Cambridge, MA: Harvard University Press; London: Heinemann, 1975, p. 119. See also the following inscriptions in the *Corpus Inscriptionum Latinarum*, full bibliographical details for which are available at

http://cil.bbaw.de/dateien/cil_baende.html, and which is henceforth abbreviated as CIL; here, CIL, vol. 6, nos 216 and 777, pp. 40 and 138; vol. 13.1, no. 440, p. 58.

24 Birt, 'Genius', col. 1625. In Birt, col. 1624, there is a reproduction of the picture from Herculaneum.

25 Calpurnius Siculus, *The Eclogues*, ed. C. H. Keene [1887], London: Bristol Classical Press, 1996, pp. 119–20; see Birt, 'Genius', col. 1622.

26 See Y.-F. Tuan, 'Geopiety: A Theme in Man's Attachment to Nature and to Place', in D. Lowenthal and M. J. Bowden (eds), *Geographies of the Mind: Essays in Historical Geosophy in Honor of John Kirtland Wright* [American Geographical Society Special Publication 40], New York: Oxford University Press, 1976, pp. 11–39 (p. 19).

27 Cato, *De agricultura*, 139, in *Marcus Porcius Cato, 'On Agriculture'; Marcus Terentius Varro, 'On Agriculture'*, tr. W. D. Hooper, rev. H. B. Ash, London: Heinemann; Cambridge, MA: Harvard University Press, 1934, pp. 1–157 (here: §139, p. 121).

28 See R. J. Kozljanič, *Der Geist eines Ortes*, vol. 1, pp. 52–58.

29 For example, *genio domi suae* [. . .] *aram* (CIL, vol. 8, no. 2597, p. 307); ⟨Io⟩vi o. m. et ⟨He⟩rculi et ⟨Sil⟩vano et ⟨Ge⟩nio ⟨do⟩mus (CIL, vol. 13.2, no. 8016, p. 539); cited in Otto, 'Genius', col. 1167.

30 See, for example, the references to *genius vici* (CIL, vol. 8.1, nos 2604 and 6352, pp. 308 and 591) or *pagi* (CIL, vol. 5.1, no. 4909, p. 515); *genius curiae* (CIL, vol. 8.1, no. 1548, p. 189); *genius conservator horreorum* (CIL, vol. 6, no. 236, p. 46) or *tutelae horreorum* (CIL, vol. 2, no. 2991, p. 406); *genius portorii* (CIL, vol. 3.1, nos 751–52, p. 142); *genius area frumentariae augustus* (CIL, vol. 8.1, no. 6339, p. 590); *genius macelli* (CIL, vol. 2, no. 2413, p. 339); *scholae* (CIL, vol. 8.1, no. 2601, p. 308); *thermarum* (CIL, vol. 8.1, no. 8926, p. 761); cited in Birt, 'Genius', cols 1621–22.

31 H. Kunckel, *Der römische Genius*, Heidelberg: Kerle Verlag, 1974, pp. 54–55.

32 Arnobius, *Against the Pagans*, book 3, chapter 29, in vol. 1, p. 215. See S. Pescarin, *Rom: Antike Bauwerke der ewigen Stadt*, pp. 36–37; and W. Eisenhut, 'Iuturna', in K. Ziegler and W. Sontheimer (eds), *Der kleine Pauly: Lexikon der Antike*, 5 vols, Munich: Deutscher Taschenbuchverlag, 1979, vol. 3, cols 25–26.

33 CIL, vol. 6.4.3, no. 36781, p. 3762.

34 See R. Schilling, 'Genius', in T. Klauser (ed.), *Reallexikon für Antike und Christentum*, 23 vols, Stuttgart: Hiersemann-Verlag, 1950–2009, vol. 10, cols 52–83 (here: col. 68).

35 See S. Pescarin, *Rom: Antike Bauwerke der ewigen Stadt*, pp. 26 and 31.

36 Varro, *On the Latin Language*, book 5, §71, in Varro, *On the Latin Language*, tr. R. G. Kent, vol. 1, London: Heinemann; Cambridge, MA: Harvard University Press, 1938, p. 69.

37 Servius, *Commentarii*, vol. 2, *In Vergilii Aeneidos Librum Sextum Commentarius*, §139, p. 591.

38 Ovid, *On the Roman Calendar* or *Festivals*, book 1, l. 706, in *Ovid's Fasti*, tr. J. G. Frazer, London: Heinemann; New York: Putnam, 1931, p. 53; and Dionysius of Halicarnassus, *The Roman Antiquities*, tr. E. Cary (and E. Spelman), vol. 3, London: Heinemann; Cambridge, MA: Harvard University Press, 1960, book 6, chapter 13, pp. 277–79.

39 Ovid, *Metamorphoses*, book 3, 157–62, in Ovid, *Metamorphoses*, tr. F. J. Miller, vol. 1, *Books I-VIII*, Cambridge, MA: Harvard University Press; London: Heinemann, 1936, p. 135.

40 Ovid, *Troubles* or *Sorrows*, book 5, chapter 10, l. 22; *Letters from the Black Sea*, book 2, chapter 1, l. 52 and book 4, chapter 7, l. 22, in Ovid, *Tristia; Ex Ponto*, tr. A. L. Wheeler, Cambridge, MA: Harvard University Press; London: Heinemann, 1939, pp. 247, 323 and 447.

41 For further information, see chapter 'Die antike Lehre des Ingenium', in E. Grassi, *Macht des Bildes: Ohnmacht der rationalen Sprache*, 2nd edn, Munich: Wilhelm Fink Verlag, 1979, pp. 174–78.

42 Cicero, *Catiline Orations*, speech 3, l. 19, in Cicero, *In Catilinam I-IV; Pro Murena; Pro Sulla; Pro Flacco*, tr. C. Macdonald, Cambridge, MA: Harvard University Press; London: Heinemann, 1977, p. 123; Catullus, *Poems*, no. 64, l. 134 and no. 76, l. 4, pp. 107 and 155; and Caesar, *The Gallic War*, tr. H. J. Edwards, London: Heinemann; Cambridge, MA: Harvard University Press, 1970, book 6, section 16, l. 3, p. 341.

43 Nonius Marcellus, *De Conpendiosa Doctrina*, ed. W. M. Lindsay, vol. 1, Leipzig: Teubner, 1903, §173, l. 27 to §174, l. 1, p. 255.

44 Cicero, *Concerning Divination*, book 2, chapter 124, in Cicero, *De Senectute; De Amicitia; De Divinatione*, tr. W. A. Falconer, Cambridge, MA: Harvard University Press; London: Heinemann, 1971, p. 511. See F. Pfister, 'Numen', in G. Wissowa *et al.* (eds), *Paulys Real-Encyclopädie der classischen Altertumswissenschaft*, 1st edn, 20 vols, Stuttgart: Metzler; Alfred Druckenmüller Verlag, 1894–1963, vol. 17.2, cols 1273–6.

45 Virgil, *Aeneid*, book 3, l. 437, and book 9, l. 661, in Virgil, *Eclogues; Georgics; Aeneid I-VI*, p. 401, and Virgil, *Aeneid VII-XII; Appendix Vergiliana*, tr. H. Rushton Fairclough, rev. G. P. Goold, Cambridge, MA: Harvard University Press, 2000, p. 161.

46 Pfister, 'Numen', p. 1279.

47 Virgil, *Eclogues; Georgics; Aeneid I-VI*, p. 98.

48 Virgil, *Eclogues; Georgics; Aeneid I-VI*, p. 536.

49 Horace, *Odes*, book 4, no. 5, l. 35, in *Odes and Epodes*, p. 234.

50 Ovid, *Art Of Love*, book 3, l. 548, in Ovid, *The Art of Love, and Other Poems*, tr. J. H. Mozley, London: Heinemann; New York: Putnam, 1929, p. 157; *Metamorphoses*, book 6, l. 392, in *Metamorphoses, Books I-VIII*, p. 315; *Metamorphoses*, book 1, l. 320, in *Metamorphoses, Book I-VIII*, p. 25; and *Love Poems*, book 3, no. 1, ll. 1–2, in Ovid, *'Heroides' and 'Amores'*, tr. G. Showerman, London: Heinemann; New York: Putnam, 1931, p. 445.

51 Virgil, *Aeneid*, book 12, ll. 181–82, in Virgil, *Aeneid VII-XII; Appendix Vergiliana*, p. 313.

52 Ovid, *Fasti*, book 3, l. 264: 'a lake [. . .] hallowed by religion from of old' (*Ovid's Fasti*, p. 139).

53 Wilhelm Henzen (1816–1887), *Acta Fratrum Arvalium quae supersunt*, Berlin: Riemer, 1874, p. 146.

54 Pfister, 'Numen', col. 1281; for references, see note 23 above.

55 See chapter on the 'locus amoenus' entitled 'The Pleasance' (or in German, 'Der Lustort') in E. R. Curtius, *European Literature and the Latin Middle Ages*, tr. W. R. Trask, London: Routledge and Kegan Paul, 1979, pp. 195–200.

56 Seneca, *Letters to Lucilius*, epistle 41, in *Ad Lucilium Epistulae Morales*, vol. 1, pp. 273–75.

57 R. Otto, *The Idea of the Holy: An Inquiry into the non-rational factor in the idea of the divine and its relation to the rational*, tr. J. W. Harvey, London, Oxford, New York: Oxford University Press, 1958, pp. 12–13.

58 See R. J. Kozljanič, *Antike Heil-Ort-Rituale: Traumorakel, Visionssuche und Naturmantik bei den Griechen und Römern*, Munich: Albunea, 2004; see also the chapter entitled 'Das daimonische Erleben eines numinosen Ortes', in Kozljanič, *Der Geist eines Ortes*, vol. 1, pp. 109–21.

59 U. von Wilamowitz-Moellendorff, *Der Glaube der Hellenen*, 2nd edn, 2 vols, Darmstadt: Wissenschaftliche Buchgesellschaft 1955, vol. 1, pp. 22 and 140.

60 L. Klages, *Der Geist als Widersacher der Seele*, 6th edn, Bonn: Bouvier, 1981, p. 1103; see also pp. 1255–57; and L. Klages, *Vom kosmogonischen Eros*, 9th edn, Bonn: Bouvier, 1988, pp. 112–14.

61 See H. Schmitz, *Der unerschöpfliche Gegenstand*, 2nd edn, Bonn: Bouvier, 1995, pp. 6–7, 196–99 and 292–96.

62 See H. Schmitz, *System der Philosophie*, 5 vols, 2nd edn, Bonn: Bouvier, 1981, vol. 3.2 (*Der Gefühlsraum*), §149, pp. 98–133; and H. Schmitz, *System der Philosophie*, 5 vols, Bonn: Bouvier, 1977, vol. 3.4 (*Das Göttliche und der Raum*), §212ff., pp. 12–15.

63 H. Schmitz, *System der Philosophie*, vol. 3.4, p. 130.

64 *The Description of the World*, book 1, chapter 13, §72 and §74–75, in Pomponius Mela, *Chorographie*, tr. A. Silberman, Paris: Belles Lettres, 1988, pp. 21–23; cf. H. Schmitz, *System der Philosophie*, vol. 3.4, p. 132.

65 H. Schmitz, *System der Philosophie*, vol. 3.2, p. 128.

66 H. Schmitz, *System der Philosophie*, vol. 3.4, pp. 133–34.

67 H. Schmitz, *Der unerschöpfliche Gegenstand*, p. 439.

68 H. Schmitz, *System der Philosophie*, vol. 3.4, p. 149.

69 H. Schmitz, *System der Philosophie*, vol. 3.4, p. 150.

70 H. Schmitz, *System der Philosophie*, vol. 3.4, p. 146.

71 For further discussion, see the chapter 'Das ausdrucksphänomenologische Konzept der "Bilder" und "Elementarseelen" eines Ortes von Ludwig Klages', in Kozljanič, *Der Geist eines Ortes*, vol. 2, pp. 262–300.

72 Klages, *Der Geist als Widersacher der Seele*, p. 1263.

73 Klages, *Der Geist als Widersacher der Seele*, p. 1264.

74 C. G. Jung, 'On the Nature of the Psyche' (*Theoretische Überlegungen zum Wesen des Psychischen*), in C. G. Jung, *Collected Works*, ed. Sir H. Read, M. Fordham, G. Adler, and W. McGuire, 20 vols, London: Routledge and Kegan Paul, 1953–1983, vol. 8, paras. 343–442, pp. 159–234 (here: para. 441, p. 233).

75 C. G. Jung, 'The Structure of the Psyche' (*Die Struktur der Seele*), in *Collected Works*, vol. 8, paras. 283–342, pp. 139–58 (here: para. 283, p. 139).

76 C. G. Jung, 'Archaic Man' (*Der archaische Mensch*), in C. G. Jung, *Collected Works*, vol. 10, paras. 104–47, pp. 50–73 (here: para. 140, p. 60).

77 Jung, 'Archaic Man', para. 140, p. 60.

78 On the depth-psychological theory of projection, see M.-L. von Franz, *Spiegelungen der Seele: Projektion und innere Sammlung in der Psychologie C. G. Jungs*, Stuttgart: Kreuz Verlag, 1978, p. 11; see also C. G. Jung, 'Archetypes and the Collective Unconscious', in *Collected Works*, vol. 9/i, paras. 1–86 (pp. 3–41), esp. paras 53–54, pp. 25–26. This theory has been applied to the problems of landscape and urban development; see T. Abt, *Fortschritt ohne Seelenverlust: Versuch einer ganzheitlichen Schau gesellschaftlicher Probleme am Beispiel des Wandels im ländlichen Raum*, Bern: Hallwag, 1983, p. 149. A similar theme is discussed (although on the basis of Freudian psychoanalysis) in A. Lorenzer, 'Städtebau: Funktionalismus und Sozialmontage? Zur sozialpsychologischen Funktion der Architektur', in H. Berndt, A. Lorenzer and K. Horn (eds), *Architektur als Ideologie*, Frankfurt am Main: Suhrkamp, 1968, pp. 51–104 (p. 83); see also C. Majunke, *Der Genius loci: Geist des Ortes oder verorteter Geist: Landschaftsplanung zwischen dem Wunsch nach Ganzheit und moderner Subjektivität*, Berlin: Eisel, 1999, p. 89; and Kozljanič, *Der Geist eines Ortes*, vol. 2, pp. 370–83.

Chapter 4

The archaic and the sublimity of origins

Alan Cardew

My abyss *speaks*, I have turned my ultimate depth into the light.
(Nietzsche, *Thus Spoke Zarathustra*)

The primordial nothing

Robert Fludd commences the first volume of his *History of the Macrocosm and Microcosm* (1617) with a representation of Nothing, a black square. On each side of the square is written *Et sic in infinitum*. It is not strictly an illustration because there is no lustration, or light in it. It depicts the *Mysterium Magnum*, or the 'Great Darkness', which preceded creation and which stretches for infinity in every direction – however, there is no direction, it is ungraspable, there are no properties, inclinations, no quantity, no thing.

Fludd took the term *Mysterium Magnum* from Paracelsus, and Paracelsus gave the original Nothing the name 'the Iliaster'. In a sense this is not very informative, as 'Iliaster' is intended to be the ultimate oxymoron, compounded from *hyle* (i.e. matter) and *aster* (i.e., star, or, better, spirit; as the stars were thought to be spiritual substances). For Paracelsus, to understand the meaning of 'Iliaster', the intangible original condition which was both soul and matter, was to understand the nature of the Philosopher's Stone. To comprehend the secret of the *Mysterium Magnum*, 'the true principles which obtained in the universal genesis', was 'enough to possess anyone with a full and practical illumination concerning the arcanum of philosophy'.[1] Not only is the primordial state a challenge to expression, a state to be approached through paradox and contradiction, it has the added difficulty that it is a secret, a great mystery which requires initiation. The idea of mystery is a constant and persistent association with the primordial.

Logically, working our way back to the Beginning, we are stranded in a temporal version of the 'regressus argument' which is customarily illustrated by the apocryphal Indian myth that the world is supported by an elephant which is supported by a turtle – what supports the turtle? The same argument appears in John Locke, Henry David Thoreau, Lewis Carroll, Bertrand Russell, William James, and

Figure I Robert Fludd, 'The Great Darkness or Mysterium Magnum', *Utriusque cosmi maioris scilicet et minores metaphysica, physica atque technica historia,* Oppenheim: Hieronymus Galler for Johann Theodor de Bry, 1617, pt. I, I.

Stephen Hawking, beginning the latter's *A Brief History of Time* (1988). These problems are anticipated by Immanuel Kant (1724-1804) in his first critique, the *Critique of Pure Reason* (*Kritik der reinen Vernunft*) (1781; 1787), in his discussion of the Antinomies of Reason, which occur when discussing any system of cosmological ideas. It is difficult to arrive at a notion of an original cause, a beginning to space and time, when our understanding is made up of notions of space and time, and our concept of causality always demands a preceding cause in which each effect depends on a preceding cause, and that cause depends on a preceding cause, and that cause on a preceding cause and so on ad infinitem. There is always an antecedent in our understanding, and we cannot arrive at the original cause, however assiduously we pursue a regressive synthesis; we are ultimately at an impasse in which there must be an origin but at the same time there cannot be one. Kant posits the possibility of an absolute spontaneity which begins of itself (see A

446; B 474), and this must be an act of absolute freedom as it has not been pre-determined by a pre-existing cause. Kant anticipates the 'singularity' of the Big Bang of current cosmology.

To contemplate Fludd's Mysterium is to contemplate everything and nothing. The blank black square is like Borges' 'Mirror of Ink', which reflects, not the world, but the soul of the man gazing into it; not just the personality, but the very depths of that soul. The way to the primordial may lie, not beyond, but beneath the rational mind. In his essay 'The Mirror of Enigmas', Borges examines the verse from St Paul: *Videmus nunc per speculum in aenigmate: tunc autum facie ad faciem. Nunc cognosco ex parte: tunc autem cognoscam sicut et cognitus sum* ('Now we see in a mirror, in darkness; but later we shall see face to face. Now I know in part: but later I shall know as I am known') (1 Corinthians 13:12).

This version reflects Borges' preferred translation by Cipriano de Valera, rather than the familiar King James version: 'For now we see through a glass, darkly; but then face to face: now I know in part; but then shall I know even as also I am known.' Borges includes in his essay a number of reflections on this passage by Léon Bloy, including the following:

> The statement by St Paul [. . .] would be a skylight though which one might submerge himself in the true Abyss, which is the soul of man. The terrifying immensity of the firmament's abysses is an illusion, an external reflection of our own abysses, perceived 'in a mirror'. We should invert our eyes and practice a sublime astronomy in the infinitude of our hearts, for which a God was willing to die. [. . .] If we see the Milky Way, it is because it actually exists in our souls.[2]

The very obscurity of the primordial, like the darkest night and the profoundest mystery, engenders a strong reaction, even though – and perhaps because – it is beyond thought and the horizon of the manifest world. Edmund Burke (1729–1797), in *A Philosophical Enquiry into the Origin of our Ideas of the Sublime and Beautiful* (1757), analyses such responses and makes the following, rather alarming statements:

> It is one thing to make an idea clear, and another to make it affecting to the imagination. [. . .]
>
> To make any thing very terrible, obscurity seems in general to be necessary. [. . .]
>
> A clear idea [. . .] is another name for a little idea.[3]

The primordial and the archaic are certainly obscure and affecting, and they are most certainly threats to clear reason.[4] They are sublime in Burke's realization of the term, in that the mind is overawed and the individual overwhelmed. The feeling of the sublime, whether it is prompted by storms, darkness, mountains, gulfs, or the vast unknown produces a tremendous affect; it 'hurries us on with an irresistible force'. It is best understood through our emotional response. Burke assigns various passions to the experience of the sublime, the chief of which is 'Astonishment', defined as 'that state of the soul, in which all its motions are suspended, with some degree of horror. In this case the mind is so entirely filled with its object, that it cannot entertain any other' (Burke, 57).

Horror is excited by the terrible; which includes poisonous serpents, limitless space, and great dimensions. Obscurity, 'that condition when we are unable to make out the possible dangers that confront us', is linked by Burke to the dread experienced in despotism, and, in particular to 'the unfathomable obscurity of religion':

> Almost all the pagan temples were dark. Even in the barbarous temples of the Americans at this day, they keep their idol in a dark part of the hut, which is consecrated to his worship. For this purpose the Druids performed all their ceremonies in the bosom of the darkest woods, and in the shade of the oldest and most spreading oaks.
>
> (Burke, 59)

Burke anticipates Nietzsche's observations on the architecture of antiquity:

> [An] atmosphere of inexhaustible meaningfulness hung about [an ancient] building like a magic veil. Beauty entered the system only secondarily, without impairing the basic feeling of uncanny sublimity, of sanctification by magic of the gods' nearness. At most the beauty tempered the dread – but this dread was the prerequisite everywhere.[5]

Other associations with the sublime which may equally attach to the primordial are Privation with which are associated Vacuity, Darkness, Solitude and Silence, and Vastness, linked to tremendous gulf and heights; an Infinity which exceeds the senses. At the start of his analysis of the Sublime and the Beautiful, Burke states his belief that 'an attempt to range and methodize some of our most leading passions, would be a good preparative' to his ensuing discourse (Burke, 52). He thereupon develops a palette of qualities, of feelings, which enable us not only to recognize the Sublime but also to articulate and recognise the response to the Primordial; although it defies definition we can describe the emotions which are associated with it. They emerge from its very obscurity.

Throughout the descriptions of the primordial and the archaic there is conceptual darkness, the original condition is rendered in the darkest chiaroscuro. Hegel famously criticised the 'abyss of vacuity' of Schelling's identity philosophy in the Preface to his *Phenomenology* (1807) which dealt with an Absolute about which nothing could be said; or, in other words, 'the night in which, as we say, all cows are black'.[6] Yet it is not the case that this is an empty nothing; a blank. In contradiction of King Lear's 'nothing comes of nothing', *everything* comes of nothing. There may be a lack of determinate and complete knowledge, but the forces associated with it are overwhelming.

To contemplate the primordial and the archaic is to gaze far into the black waters of the Lake at Nemi, to uncover what lies in the deepest mines, to fathom the profoundest gulfs, and to search 'the dark backward and abysm of time'. It is a *katabasis*, a descent, not into the Underworld, but deeper, before the creation of the Underworld. Nothing can be said of such depths, there is no time and space, no where, no thing. However, the feelings which the primordial engenders, its dark sublimity, are one with those very qualities which engender the Creation. Terror and dread, obscurity, vastness, the uncanny awaken forces, 'the pregnant causes' which are the beginnings of experience and are the dim, unconscious shadowings of order. It is also an exploration of the innermost abyss of the soul. Those who have grasped the ebon bough – Friedrich Creuzer (1771–1858), Jakob Böhme (1575–1624), Friedrich Wilhelm Joseph Schelling (1775–1854), Petar Petrović-Njegoš (1813–1851) – write of affect, of longings and passions, violent, beautiful, and terrible which precede the World and which drive the emergence of God. It is these responses which we will examine in what follows.

Cancelled cycles: Before the beginning

In *The Myth of the Eternal Return* (1954), Mircea Eliade famously argued that in pre-modern societies Man's actions are given force and meaning through repetition; re-enacting what was laid down by gods and the ancestors in the Beginning, 'in those days' (*in illo tempore, ab origine*).[7] When such beginnings impart such authenticity and value, it seems rather perverse, if not irreverent, to enquire what preceded the beginning. Yet so strong and inveterate is the idea of a causal chain that when a beginning is identified it is hard to resist the idea that there must have been something before it. How did it get there? And, if creation has some sort of narrative sequence, was there some sort of ontological prequel, was there a beginning before the beginning, or was there a whole string of previous beginnings? Did there exist some state or entity that was more primordial than the primordial, and when the archaic was new was there another, more ancient archaic? Is there something earlier and more deeply sunk than Atlantis?

There exist in various traditions fragmentary accounts of cancelled cycles, previous creations and pre-primordial entities; remains of which may even persist in our own time and universe. For example, in what appears to be an unexceptional passage in the Book of Genesis (35: 31ff.), there is a list of Esau and his successors, known as the Kings of Edom. In the Kabbalistic tradition, in the Zohar, this was interpreted in the profoundest terms as an 'unbridled flow of life'.[8] It signified 'the emergence of God from the depths of Himself into creation' through an initial series of unsuccessful creations. Thus, inscribed in Genesis is a record of previous geneses. According to Gershom Scholem's *Kabbalah* (1974), the Zohar interpreted Edomite Kings (*malkhei Edom* or *malkin Kadma'in*) as being primordial but unstable realms that preceded the primordial Adam, Adam Kadmon. They lacked his self-sustaining balance of parts.[9] Each creation involved a 'breaking of vessels', or *Kelim*, the violent irruption of the Godhead into a new sphere, each of which crystallised an aspect of God's being. The fragments of these vessels, *Kelippot* or *Qlippoth*, cosmic leftovers, either persisted in the present creation, or were the material from which the present creation was formed.

Similar ideas found mythopoetic speculations in the Romantic period, notably in works by Byron (1788–1824) and Shelley (1792–1822). In *Cain: A Mystery* (1821) by Lord Byron, Cain is tempted by Satan into becoming the first murderer. As part of this process, Cain is taken on a malign tour of Creation which includes the Underworld. In the 'Realm of Death', 'shadowy and dim', Cain sees shades or phantoms which exhibit neither the intelligences of heaven nor the forms of man:

> And yet they have an aspect, which though not
> Of men nor angels, looks like something, which
> If not the last, rose higher than the first,
> Haughty and high, and beautiful and full
> Of seeming strength, but of inexplicable
> Shape; for never I saw such. They bear not,
> The wing of seraph, nor the face of man,
> Nor form of mightiest brute, nor aught that is
> Now breathing; mighty yet and beautiful
> As the most beautiful and mighty which
> Live, and yet unlike them, that I scarce
> Can call them living.[10]

These intelligent, great, glorious things were once inhabitants of the same, though different earth; the present earth being 'too little and too lowly' to sustain such creatures. Thus, the place of Man is diminished in the order of creation(s), and Cain is given a vision of the contingent, evanescent nature of existence; to murder a man is insignificant in the scale of things, for even these beings, living high above mankind, were simply wiped away:

> By a most crushing and inexorable
> Destruction and disorder of the elements,
> Which struck a world to chaos, as a chaos
> Subsiding has struck out a world: such things,
> Though rare in time, are frequent in eternity.[11]

Byron and Shelley had been reading and discussing the *Discours sur les revolutions de la surface du globe et sur les changemens qu'elles ont produits dans la régne animal* (1812) by Georges Cuvier (1769–1832), and its argument for 'catastrophism'. Cuvier argued that different species such as the 'megathereum', whose bones had been recently identified, had existed on the globe at earlier times and had disappeared as the result of huge convulsions that racked creation from time to time.[12]

Shelley's *Prometheus Unbound: A Lyrical Drama* (1818) also exhibits these ideas. The final vision of the work uncovers the secrets of the Earth's deep heart and the '[m]elancholy ruins of cancelled cycles':

> The wrecks beside of many a city vast,
> Whose population which the earth grew over
> Was mortal, but not human; see, they lie,
> Their monstrous works, and uncouth skeletons,
> Their statues, homes and fanes; prodigious shapes
> Huddled in grey annihilation, split,
> Jammed in the hard, black deep.[13]

Such notions of different races of 'men' replacing one another clearly inform Mary Shelley's *Frankenstein* (1818), in which Frankenstein seeks to fashion a new race of more perfect men, or in her novel *The Last Man* (1826), in which mankind is reduced to one remaining survivor. There may be future cycles and future beings to replace us as the inhabitants of Earth, something which is developed in Bulwer Lytton's novel *The Coming Race* (1871). In this book, dedicated to the mythologist Max Müller, Lytton's protagonist, a mining engineer, descends to a subterranean utopia inhabited by the 'Vril-Ya'. The latter are a race more perfect than man, taller, stronger, winged and matriarchal. The women are larger, more imposing than the men, and they command a mysterious life energy called Vril. One day, the coming race will emerge from their glowing golden underworld to replace mankind. A cross between Valkyrie and the Nibelungs of Wagner's contemporary *Ring*-cycle who toil in the depths, they share the potential of destiny. Just as the depths are the repository of cancelled cycles and the relics of other pasts, they also hold the future which emerges from the depths, like Wagner's Erda, prophesying the art work of the future or the ring of power.

A further, but altogether more sinister, poetic account of the time before origin may be found in the work of the Montenegran epic poet Petar Petrović Njegoš. Njegoš, 'the poet of Serbian cosmic misfortune', was the last Vladika, or Prince Bishop, of Cernagora. He succeeded his uncle Bishop Petar I at the age of 17, and became Petar II, ruler of a savage and remote country sandwiched between the independent Turkish rulers of Mostar and the Pireus. A mountain Sparta, Montenegro (Cjrna Gora; literally, 'black mountains') was a beleaguered wilderness, which had for centuries been a place of constant war and blood feuds. Vengeance, or *Osveta*, plundering and raiding were the chief stays of the economy.[14] Its landscape was not less exacting; in appearance it has been described by Milovan Djilas as 'the crucified land':

> The land is one of utter destitution and forlorn silence, its billowing crags engulf all that is alive and all that the humane hand has built and cultivated. [. . .] All is stone. Even all that is human is of stone. Man himself is made of

it – without an ounce of fat, honed down by it all and with his sharp edge turned outward to the whole world. Every evil assails him, and he uses evil to ward off evil, on a soil where even the wild beast has no lair.[15]

An old story said that, after the creation, God had some rubble left over and he dumped it on Montenegro, and if there was anywhere composed of 'broken vessels' it was there. Certainly, it was a high water mark of European culture, where all manner of cultural fragments, pagan, Christian, Orthodox and heretical, had washed up and caught fast in the rocks.

Behind Njegoš' Episcopal Palace at Cetinje was a ruined tower topped by Turkish heads which corresponded with a tower behind the palace of the Turkish vizier of Mostar which was topped with Montenegrin heads.[16] Montenegro owed its existence to a harsh balance of forces which constantly had to be maintained, and these forces were measured in heads.[17]

Njegoš wrote his mystical poem, *The Ray of the Microcosm* (*Luca Mikrokozma*) in 1845, and the dualist elements of the poems almost certainly reflect the unceasing effort of the poet at the time to battle against both the Turks and the feuding tribes of the mountains. Djilas observed that the existence of Montenegro was a constant act of Will and that Njegoš not only created a country, but lived in that creation' (Djilas, 50). This endless conflict is fiercely depicted, and, indeed, distressingly so, in Njegoš' great epic 'play', *The Mountain Wreath* (*Gorski Vijenac*) (1847), which is as unsparing as the *Iliad* in its portrayal of war. A far more dreadful conflict, however, appears in *The Ray of the Microcosm*.

For the poem not only describes a Hobbesian state of nature on a metaphysical scale, it is infused with a Gnostic pessimism. Its sources are a matter of a great deal of speculation: the dualism of the Eastern Cathars, the Bogomils, Neoplatonism, Origen, the widespread circulation of heretical texts in the Bulgarian Empire and the European part of the Byzantine world, the survival of Dionysian mystery religions in Thrace, the uncertain canon of Biblical texts in early Serbian Orthodoxy.[18] What is certain is that there are influences of Byron and Shelley, whose works were extensively read by Njegoš, and the tradition of 'Merkabah' mysticism. In this tradition, which has its roots in the the Old Testament books of Ezekiel, Chronicles and Ecclesiastes, an inspired individual goes through a passage of ecstasy and ascension, escaping from the miseries of this world, and flying upwards through the Heavenly Spheres (*Hekhalot*; literally, 'palaces'), past a series of *Archons* and eventually leading to a vision of the Throne of God, His *Merkavah* or 'Throne-Chariot'. This became an important element of the Jewish Midrash; an active, theurgical process of the interpretation of biblical texts. Examples of Merkabah mysticism are found in a number of Apocryphal texts such as the 'Testament of Levi', the 'Book of the Secrets of Enoch', the 'Greek Apocalypse of Baruch', and the 'Ascension of Isaiah' and the 'Book of Enoch', many of which were extant in the Byzantine Empire.[19] In *The Ray of the Microcosm* the protagonist is lost in the darkness of the world. He is addressed by the Ray who seeks to reawaken the memory of the heavenly temple within him; the whole is reminiscent of Gnosticism, particularly the Mandean Gnosis. The personified 'Ray' calls to him with an angelic voice:

> Of thine own darkened soul I am the Ray
> Of flame immortal; and it is through me
> That thou rememberest still what thou hast lost.
> What are the shapes of goddesses, created,
> By fiery-minded poets of the past?
> 'Tis I alone that penetrate the dark,
> 'Tis I alone that reach the heaven's gates.[20]

Men are themselves fallen angels sunk in the dark vale of sorrow, subjects of the Creator's wrath; 'the victims of unceasing torments sad':

> O'erpowered by oblivion's sullen sleep,
> Thou didst forget thy early bright abode,
> The fountains of immortal happiness,
> The fields of Paradise, eternal bliss
> And thy Creator's life-maturing glance.
> Escape, O spark divine from the embrace
> Of the dark power, soar up on radiant wings,
> Ablaze with flames of immortality.[21]

Responding to the call the poet is led by the spark/ray; part guide and genius, part guardian angel, and part Platonic daemon, across six circling heavens, and through as many gleaming galaxies, and in Canto 2 he reaches the Plain of Heaven, which is the result of the vanquishing of ancient foes; the 'Cradle of Eternity' is not without memories of 'ancient conflict':

> The powers of darkness, here they once were broken
> And split asunder the giant blocks,
> From here they fled out of the realm of being
> And sank into the state of woeful death.[22]

Eventually, the protagonist is vouchsafed a vision of God's throne and His crown:

> Above the throne of thrones there turns on high
> In circling, dazzling motion through the air
> A ring, like other bright aerial rings
> That turn above the thrones of the Archangels;
> But this is far greater far a million times.
> Myriads of suns are hung on shining locks,
> And shed their light throughout the circling ring.
> This is the crown of God.[23]

> [. . .] Around the throne-sustaining mountain
> Four mounts of sparkling adamant arise,
> From where there burst four fountains to the skies
> Like giant rays of fire; their columns rise
> As high as from our earth to touch the moon,
> And reach the utmost heaven-sustaining sphere,
> Then, vaulted into rainbows, they spread forth,
> O'erarching the four borders of the skies,
> And fall into the ocean of the air.[24]

From this throne stream millions of orbits, the constant creation of ideas, music, poetry and light.

As a striking contrast to the beauty and perfection of this realm, in Canto 3 we hear about Creation and more about the powers of darkness that preceded it. God speaks to the Archangel Michael, telling him:

> To me alone is subject all that is.
> And if my sacred duty I deserted
> The reign of darkness should remain for all
> Eternity. And orbits numberless,
> Unknown to anyone but me, they should
> Remain in heavy sleep under the clouds
> Of chaos, buried in the depth of gloom.
> How much I had to toil ere I freed time
> Out of the bondage of the gloomy powers,
> And let fly in liberty through space,
> The clear bright region of eternity![25]

Before Creation there was only this:

> A frightful huge empire had once extended,
> On far, the sad dominion of its gloom.
> Its monstrous legions entered heaven's fields;
> The horrid ugliness of such abortions
> No one can ev'n imagine but myself.
> And only to the sacred mount sublime
> And to my throne they did not dare approach –
> My fiery looks with horror crushed them all.
> Those masses huge, in senseless motion heaving,
> Have ever yet been subject to my will.
>
> They often used to sink in horrified
> Precipitated flight into th' abyss,
> Their gloomy cradle, on their broken spines.
> At times, though rarely, they escaped the rule,

Until I raised the crown above the throne.
This blow, the first and brightest, crushed their might,
And split asunder their dark empire.
Thus, vanquished by the brightness of my crown,
They hid themselves in crowded multitudes,
And sank into the void and cold abyss,
Creating there such dead and woeful shapes
With taste repulsive and the help of death.[26]

These entities are defeated, but yet not completely vanquished, and they are only held in place by an almighty act of sustained will. Gradually, God extends the boundaries of the Universe and pushes back the borders of 'their dumb empire'.

It is intriguing to speculate whether this disturbing account of the forces which existed outside of God, and which persist even now somewhere in the abyss, is simply a straightforward allegory of the constant struggle of Montenegro against the forces of the Ottoman Turks; or whether, more disturbingly, the poem and the historical conflict are themselves perhaps manifestations of a deeper ontological struggle buried deep in the Balkan psyche, whose wellsprings can be found in the traditions of the Thracian mysteries and Eastern heresy.

The poem then introduces Satan and his rather secondary account of creation, portrayed as a universal decline and confinement by a tyrannous Divinity rather than the steady victory of light. Having passed through the history of Satan and Adam the poem concludes with prophecy of the complete and utter redemption of the universe; though it is yet to be achieved.

Archaic man: The Pelasgians

In Njegoš' *Ray of the Microcosm*, the primordial precedes Adam who is a comparative latecomer in the cosmic drama, yet his role as the original Man is unchallenged. Things are more difficult in the classical world, whose legends of origin, particularly of the Greeks and their cities, is always a matter of contention. Were the first Greeks autochthonous or did they migrate from elsewhere? Were the founders of Greece Greek at all? Attempting to untangle the provenance of the legendary founders of Athens, Kekrops, Erechtheus and Neth (who all appear to be Egyptian), Bulwer Lytton sadly observed that, 'by the evidence of all history, savage tribes [including the Greeks] appear to owe their first enlightenment to foreigners.'[27]

Any enquiry into the origins of Greek culture comes across an exemplary group of aboriginals who, in many ways, set the standard for a mysterious, archaic race. These were the Pelasgians. Their influence on nineteenth-century mythography was profound, and they are important for the way in which myth, mystery religion, and metaphysics developed the idea of the primordial unconscious. Their elusiveness made it possible to develop all manner of connections, and to point to a time when religion, and the nature of gods or God in particular, had not evolved into dogma or any kind of clear theology.

They appear a number of times in Herodotus, from whose *Histories* it remains quite unclear whether they were barbarous and alien forerunners of the Hellenes, ancestors of the Hellenes, or a race who brought mystery religion and the alphabet to the Aegean from Phoenicia.[28] The Pelasgians exist somewhere between the migratory and the autochthonous, the deepest tradition or its first disruption. They play an essential role in the development of ideas of the primordial and the archaic, as they appear in Creuzer and Schelling, who will be discussed below.

History begins, according to Hegel, with a transgression, and Herodotus begins his history with the abduction of a king's daughter, Io, and the other women of Argos by Phoenician traders, which led to the war between the Hellenes and the barbarians (here the Persians). Some Hellenes appear to have avenged the abduction by proceeding to Tyre and abducting that king's daughter. This, in turn, was followed by the abduction of Medea from Asian Colchis, and Paris's abduction of the Hellene's Helen.[29] This violent interchange certainly represents the crossing of cultural and social barriers, and the breaching of walls and mores. Most interestingly, this is associated with merchants and trade.[30] Such disruptions are frequently identified with the beginnings of narrative, as Hayden White has argued:

> To raise the question of the nature of narrative is to invite reflection on the very nature of culture and, possibly, even on the nature of humanity itself. So natural is the impulse to narrate, so inevitable is the form of narrative for any report of the way things really happened, that narrativity could appear problematical only in a culture in which it was absent – absent or, as in some domains of contemporary Western intellectual and artistic culture, programmatically refused.[31]

Hegel, in the introduction to his *Lectures on the Philosophy of History* (1837), defined history, not as a record of what happened, but as the narration of what happened and said that 'the uniform course of events is no subject of serious remembrance'. Only in a state cognizant of laws can distinct transactions take place, 'accompanied by such a clear consciousness of them as supplies the ability and suggests the necessity of an enduring nature':

> The reality which lends itself to narrative representation is conflict between desire, on the one side, and the law on the other. Where there is no rule of law there can be neither a subject nor the kind of event which lends itself to narrative representation.[32]

The boundary between Hellene and barbarian breaks down, as does that between myth and history, and truth and conjecture. Herodotus distances himself from what he records, and in Book Seven when dealing with the question as to whether the dealings of the Argives with Xerxes were shameful or not, states: 'I may be obliged to tell what is said, but I am not at all obliged to believe it' (Her. 7.152).[33] The evidence for the Pelasgians is even more equivocal. In just a few lines we are told that the Athenians were a Pelasgian people and the Pelasgians once lived with the Athenians; they were distinct in that they spoke a barbarian language but they learned a new language when they became Hellenes (Her. 1.57). In Book 6, Herodotus says that the Athenians gave the Pelasgians land as a payment for having built the Cyclopean wall around the Acropolis. Following this, a dispute broke out and the Pelasgians violated and then abducted Athenian women, and were driven out to Lemnos. There is both difference, the building of social and physical separation, and the rupture of linguistic and racial boundaries.

Added to the uncertain racial status of the Pelasgians, there are their links with the Oracle of Dodona and the Mysteries of Samothrace, which further increase doubts and uncertainties. In the case of Dodona, it was the result of yet one more abduction. The Phoenicians this time abducted two Egyptian priestesses from Thebes. One was taken to Libya where she began the Oracle of Ammon at Siwa, and the other to Western Greece (Her. 2.54–2.56). According to Herodotus, the names of the gods came from Egypt, and it was from the priests at Dodona that Herodotus himself learned the following:

> Long ago the Pelasgians, when they made sacrifices, used to pray to the gods, but they could not address prayers to any one of them by name or epithet because they had not yet learned or heard of these. They called them gods because they believed that it was the gods who had ordered and allotted all things. But after the passage of quite a long time, they did learn the names of the gods from Egypt [. . .]. Then they consulted the oracle at Dodona about these names. This oracle, which is thought to be the most ancient of all Greek oracles, was in fact the only one that existed at the time. When the Pelasgians asked this oracle whether they should agree to those names that came from the barbarians, the oracle answered agreeably that they should indeed use them.
>
> (Her. 2.52, 2–3)

Later, the Pelasgians passed on these names to the Hellenes. Thus the gods gained an identity, but nothing more was known of them, their origins or whether they always existed, and what they looked like. The latter qualities were quite recent achievements, according to Herodotus – the contribution of the poets, Hesiod and Homer, who composed a Theogony for the Hellenes.

Thus the primary differentiation of the gods came about on the very fringes of history, before the poets, in the period of the Pelasgians. To understand archaic Hellenic culture and the source of its heroes, gods, and customs, demanded of later classicists that the origin and nature of the Pelasgians should be fully under-stood as they were somehow at the root of things, and this became a source of dispute in later mythography. Did Herodotus' account of the Pelasgians mean that

Hellenic culture came from elsewhere, such as Phoenicia, or were the Phoenicians the conduits which led back to Egypt?

Deep etymology: Creuzer and myth

One way of establishing the origin of names, gods and peoples was by examining the origin of that name. In antiquity, this was a hit-and-miss method which strained the principle of the similarity of words, paronomasia, to breaking point; for example, the Medes took their name from Medea or the Persians from Persius. This similarity of sounds was thought to be a clinching argument, particularly in the case of peoples who must derive from a founding hero; so the Hellenes were named after Hellen, son of Deucalion. Hellen had two sons, Aeolus and Dorus, from whom came the Aeolians and the Dorians, and two grandsons, Ion and Achaeus, from whom came the Ionians and the Achaeans.

Some ancient commentators, such as Plutarch or Athenaeus, were happy to entertain a number of possible explanations, whereas others, such as Euhemerus or Macrobius, were eager to establish a single principle – that the gods were derived from famous men or were solar in nature.[34] One explanation of the Pelasgians was that they were named after Pelasgus, who was possibly the first Man. This is mentioned by Pausanias, in the eighth book of his *Travels*. Pelasgus was the first king, the inventor of shelter, clothes and food in their most primitive and unattractive manifestation: sheds, sheepskins, and acorns. Pausanias cites the poet Asuis as follows:

> The godlike Pelasgus on the wooded mountains
> Black earth gave up, that the race of mortals might exist.[35]

The etymological derivation of the origin of Pelasgians was a particularly important but equally tricky business, as they (in their Phoenician aspect) were associated with the origins of written Greek. Herodotus notes that the Pelasgians had their own language (Her. 1.57), but he can shed little light on it, besides giving it a wide distribution. Because of its claim to primacy, spoken by the Mediterranean version of Adam, it had, and still has, some considerable consequence, and it continues to play a political role, as it is currently claimed to be the ancestor of Illyrian and modern Albanian.

From a fragment of Hyginus, *Fabulae* 277, it seems that Cadmus, a notable Pelasgian, brought the alphabet to Greece from Egypt – Hermes, an Egyptian deity at Saïs, having invented the letter shapes, basing them on the shapes that cranes take in flight. Amongst the numerous etymologies for Pelasgian are included everything from 'springing from the earth' and 'hairy', to 'the flat sea' and 'neighbour' or 'near land' (πελάς γή). The most repeated is the derivation from another bird, the stork; deriving Πελασγός from 'pelargos' (πελαργός).[36] The 'Third Discourse' of Sir William 'Oriental' Jones (1746–1794), presented to the Asiatic Society in Calcutta in February 1786, argued that Sanskrit, Latin,

Greek, and the Gothic and the Celtic languages had a close lexical and grammatical relationship and came from a common source. For Jones, Sanskrit was the more perfect, more finely inflected, and also a sacred language, the language of the Vedas. The origin of the Indo-European language and its gods lay in the East.

This perception was reinforced by Friedrich Schlegel (1772–1829) in his *On the Language and Wisdom of the Indians (Über die Sprache und Weisheit der Indier)* (1808), following extensive research on Sanskrit manuscripts in Paris. As the Pelasgians were, admittedly, close to the origin of Greek language and culture, philologically one might expect to find traces of an Indian origin. There was, of course, a strong Semitic case, as there was evidence in Herodotus for Phoenician and Egyptian origins. All such influences were rejected by Karl Otfried Müller (1797–1840), who championed the unique nature of Greek culture. Müller had the temerity to condemn Herodotus for his orientalism, and in his Apollonian *History of Greek Tribes and Cities (Geschichten hellenischer Stämme und Städte)* (1820–1824), Müller's Caucasian Dorians were quite distinct from the Pelasgians and their mystery cults, and were quite free from 'derivation'. In France, Guillaume Baron de Sainte-Croix (1746–1809) favoured an Egyptian derivation of all rites and mysteries, as did Jörgen Zoega (1755–1809), the Danish-born, German-educated archaeologist.[37]

In his *Symbolism and Mythology of the Ancient Peoples, especially the Greeks (Symbolik und Mythologie der alten Völker, besonders der Griechen)* (1810–1812), Georg Friedrich Creuzer brought together and synthesized the contending ideas that the Pelasgians had brought, via Phoenicia, both language, myth and mystery.[38] As has already been suggested, the significance of the Pelasgians lay in their position on the edge of language and history, and in a profound sense that their undifferentiated obscurity became their value rather than a source of contention; a ground for the emergence of cultural practices, and also a common point of origin. For, once one had accepted that beginnings were unclear, then the archaic was a source rather than the horizon of investigation. Indeed, such shadowy beginnings became a form of legitimization. Pelasgus, perhaps the first man, and the Pelasgians, became identified with the time where man and culture, language, religion, and myth emerged from Nature. The autochthony was not from a particular geographical location, but from the Earth itself. The subterranean, the primitive, the obscure was for Creuzer the place where matter and spirit were merged, their *Urquell* or primordial spring. From it emerged the symbol, which is primarily an expression of sudden illumination, a mid-point between the World and the Spirit (Creuzer, vol. 1, 63). It acts on the beholder like a lightening flash (*wie ein Blitzstrahl*) that lights up the gloomy night. It is like the play of colours in a rainbow which manifests both sunlight and the Idea against the darkest cloud. It combines unity and differentiation and acts instantaneously; boundless and compact, it is literally a pregnant moment, for it recapitulated the instant of creation with all its potency, it conveyed the meaning of creation, and enjoyed the exuberance of the spring of all things. It was the essence of the revelation of the mysteries which was reputedly so sudden and so complete. Other forms of religious manifestation – narrative, legends, myth and allegory – were, for Creuzer, inferior. They were slower and less effective.[39] Just as the world is, in Hegel's words, the Spirit emptied out in time, so the allegory, the myth, was the symbol emptied out in time.

The symbol was, for Creuzer, the essence of a secret priestly teaching which had been brought from the East by a chain formed by the priests of the Pelasgians, the Phoenicians and the Egyptians from the East. It had the most profound associations – a secret doctrine in its highest concentration, imparted at night in hidden subterranean locations, the inverse of the Platonic cave, light in darkness and light from darkness; a perfect instance of the primordial sublime which is one with the depth of our being. It was something for initiates who understood and, indeed, experienced the true meaning of myth.

That the Pelasgian gods were originally nameless, almost undifferentiated, was, Creuzer argued, of the greatest significance; it was both a mystery to be penetrated, and also a sort of negative evidence that they were close to the original ground of religion. Near the commencement of his *Mythologie* (vol. 1, 4) Creuzer draws a parallel between a passage in Herodotus (discussed above) and a section of Plato's *Cratylus* (397 c-d), in which Socrates and Hermogenes are discussing the aptness of the name '*theoi*' for the gods. Socrates says that this is truly an example of the right name for things 'that by nature always are':

> It seems to me that the first inhabitants of Greece believed only in those gods in which many foreigners still believe today – the sun, moon, earth, stars, and sky. And seeing that these were always moving or running, they gave them the name '*theoi*' because it was in their nature to run (*thein*). Later, when they learned about the other gods, they called them all by that name.[40]

From the outset, Creuzer emphasises the source of religious symbols in the forces of Nature, the sun, stars, winds, waves. Yet there is a unified power behind these manifestations and in this respect, as we can see in the interpretation of Plato, there is a movement towards abstraction.

Creuzer then moves on to Proclus' great interpretive work on Plato, his *Platonic Theology*:

> For those who treat of divine concerns in an indicative manner, either speak symbolically and fabulously, or through images. But of those who openly announce their conceptions, some frame their discourses according to science, but others according to inspiration from the Gods. And he who desires to signify divine concerns through symbols is Orphic, and in short, accords with those who write fables concerning the Gods. But he who does this through images is Pythagoric.[41]

What is significant here is the Platonic – indeed, Neoplatonic – cast of Creuzer's thought. Myth is explained as an allegory of the most philosophic and transcendent forces which may be more precisely imparted through reason (*nous*) and an exalted philosophic experience. Creuzer is very happy to consider myth through the works of the last commentators and scholiasts of antiquity such as Damascius, Iamblichus, Hermeias, and Syrianus, though he does admit an unease about the anti-Christian sentiments of Plotinus, Proclus, and Porphyry (vol. 1, 52). Creuzer also echoes an essential and increasingly important feature of Neoplatonism, namely Plotinus' 'identification of metaphysical realities with states of consciousness'.[42] Creuzer manages to reconcile philosophic ideas emanating from the highest spheres with illumination rising from the depths, largely by drawing on Neoplatonic commentaries on the mysteries, but in his contemporary and friend, Schelling, this was to become a significant problem. In both Creuzer and Schelling, all of these elements – primordial obscurity, the archaic, the symbolic, the transcendent – were at their most concentrated in the interpretation of the Mysteries of Samothrace.

The Mysteries of Samothrace I: Creuzer

Herodotus, when discussing the Pelasgians and the naming of the gods, refers to the link between the Pelasgians and the mysteries of Samothrace. We encounter a double veil: the lack of certainty about the past, and a reluctance to divulge the secrets of the Mysteries. Herodotus is famously reticent concerning the 'nocturnal rites' held in the sanctuary of Athena at Sais in Egypt, which he connects with the ritual of the Thesmophoria held annually in honour of Demeter in Athens:

> Now, although I know all the details of these rites, may my reverence ensure that they remain unspoken. I feel the same way about the rite of Demeter which the Hellenes call the Thesmophoria, so may my reverence ensure that they also remain unspoken, except for that which one can say without offence to religion.
>
> (Her. 2.171)[43]

On other matters, however, Herodotus is a little more forthcoming. In Greece, the Athenians were the first to make statues of Hermes with an erect phallus, taking this from Egypt via the Pelasgians, who also took this custom to Samothrace. Herodotus says that whoever has been initiated into the secret rites of the *Kabeiroi* (or Cabiri) will know what he means: 'There exists a secret story about Hermes that was told by the Pelasgians; its details are revealed in the mysteries at Samothrace' (Her. 2.51).

The silence demanded from those who had witnessed the Mysteries was an absolute in antiquity; there are no known accounts of what took place, what was said, what was done, or what was shown at Eleusis, Sais, or Samothrace – only a collection of fragments, rumours, and canards in Patristic works. Alcibiades was one of the few condemned by law for infringing the sacred silence, but such was the strength of the oaths taken and the profound effect of the occasion, or so it seems, that secrecy was guaranteed. The Homeric Hymn to Demeter states the absolute nature of the Mysteries which cannot in any way be transgressed, questioned or expressed: 'For the awe of the god stops the voice', and the importance of the rites of Eleusis is clear: 'Happy is he among men upon earth who has seen these mysteries; but he who is uninitiated and has no part of them never has the lot of like good things once he is dead, down in the darkness and gloom.'[44]

The secrecy may have been partly because what was enacted and the experiences gained were inexpressible, ineffable.[45] The Mysteries had a double effect on later commentators. The first was the desire to penetrate into their profound depths by piecing together clues from fragmentary texts, and then, later, clues from archaeological traces and the evidence of excavation. The second was to invoke secrecy and mystery as a characteristic style of writing, affecting what Edgar Wind once described as a 'cryptic pomp'. The commentator could suggest that he or she was an initiate and write in a veiled and occult manner; a style that was particularly important in the Renaissance. Often the two were brought together in the composition of elusive fragmentary writing. Whereas, in antiquity, initiation into the Mysteries (particularly on Samothrace) and the Orphic rites was for the many – for slaves as well as for emperors – later there was a turn towards exclusivity; the meaning of the Mysteries was now for the few to understand. Secret meanings and symbols, it was suggested, were passed on from one initiate to another, rather in the manner of Creuzer's Pelasgian priests.[46]

For his part, Creuzer approached the Mysteries of Samothrace primarily through the Cabiri; by means of etymology and analogy, it was possible to demonstrate their Eastern origin, and its transmission to the Greeks. The primary derivation for Κάβιροι, *Kabiroi*, could be the Semitic *kabirim*, 'the Great', close to the Arabic *Ghebr* (Creuzer, vol. 2, 313 et seq.).[47] In many inscriptions on Samothrace later than Creuzer, but also in sources known to Creuzer, the gods of Samothrace are referred to simply as Μεγάλοι Θεόί, 'Great Gods', pointing to a link with *kabirim* and to the archaic, undifferentiated nature of the deities of the Pelasgians and Hellenes. The Semitic word would point to a possible Phoenician origin, and Creuzer accepts this, tracing it further back and further East to the *Chobar* of Mesopotamia.

Their undifferentiated nature allows Creuzer to explore a primal link between the Cabiri and the forces of nature; they are primordial *Urzustände*, elementary spirits. They are sons of the water – there are particular links between the Cabiri on Samothrace and the Dioscuri, the saviours of seamen. The force of the tide, the wind, fire and the volcanic earth are worshipped in the Cabiri, they are 'potencies' (Creuzer, vol. 2, 309) which energise and create, and they may find an analogy in the Deii Potes of archaic Rome (Creuzer, vol. 1, 13).[48]

The Cabiri were also worshipped at Lemnos, where they were associated with Aetna and Hephaistos. Creuzer develops a special identification with Herculean, telluric potencies (Creuzer, vol. 2, 309) and the powerful gods of the forge (*die starken Schmiedegötter*), who usually appeared in the form of dwarves. For Creuzer, they are the causal principle and they trigger events (*Anlässe*), they emerge from the life-engendering warmth of the earth and its mines, to smelt ores and create the weapons for heroes which engender the affairs of men. They are part fire, part mineral, part smith, part energy.

Their effective, urgent nature puts them close to Plato's derivation of the gods *theoi* from *thein* (to 'run') – they are always running, they make things run – rather than to Herodotus' derivation of *theoi* from *theutes*, which anticipates the definition of Aristotle, setting in order and allotting. Creuzer connects the Cabiri with a whole range of Dactyls, Cobolds, Telchines, Kuretes, *Untergötter* across the ancient world – fire brings him to the Parses, demiurgic forces to Egypt and the god who fashioned the world, 'Phtah' (or *Ptah*). Although Creuzer does not develop the link, certainly Ptah as 'primeval artificer god' was accompanied in his task of fashioning the sun egg and the moon egg by eight fellow earth-spirits, the Khnûmû, at the site of the 'primordial mound'.[49]

Creuzer also makes the link with the seven sons of Syndaxl – or perhaps eight, including Esmur – of Babylonian myth. The numbers are important, or rather their inconclusiveness is, as they point to the vexed problem of the numbering of the Cabiri which appears in Goethe's *Faust: Part Two*. Were they so undifferentiated that they could not even be counted?

Creuzer's ideas on the primordial energies emerging from the underworld reflect the fascination with mines and a sort of mystical mineralogy in the Romantic period in Germany. It may have received its initial impetus from Emmanuel Swedenborg, but it can be found in, for example, E. T. A. Hoffmann's 'The Mines at Falun' (*Die Bergwerke zu Falun*) (1819), Novalis's *Heinrich von Ofterdingen* (1802), and Ludwig Tieck's 'The Old Man of the Mountain' (*Der Alte vom Berge*) (1828). Both Goethe and Novalis had mining responsibilities in their career, and the mining region of Ilmenau features in *Wilhelm Meisters Wanderjahre* (1821; 1829). As Theodore Ziolkowski has extensively shown in his *German Romanticism and its Institutions*,[50] the mine became an image of the soul, and as such it appears later in Rilke:

> That was the strange unfathomed mine of souls.
> And they, like silent veins of silver ore,
> were winding through its darkness. Between roots
> welled up the blood that flows on to mankind,
> like blocks of heavy porphyry in the darkness.[51]

The underworld is filled with rude antediluvian bones, a treasure house of ore and precious gems, and forests of silver and gold trees rooted in crystal. There was a proto-scientific belief that mineral nature had powers of growth, veins of ore were like veins of blood in animal life, leaf-shaped patterns could be found in rocks, and crystals were akin to stars. The secrets of the primal plant (or *Urpflanze*), the primal animal (or *Urtier*), and the primal stone (or *Urstein*) could be discovered in the depths. Descending into the black labyrinths of the earth and exploring caverns was a journey back to the archaic, which was still at work with a daemonic magical force; just as the eyes of the explorer become accustomed to the dark, gradually one could 'acquire the power of reading in the stones, the gems, and the minerals, the mirroring of secrets which are hidden above the clouds'.[52] There was a dark hermetic equivalence, the principle of 'as below, so above':

> You miners are almost astrologers in reverse. [. . .] Whereas they gaze inces-santly at the heavens and stray through those immeasurable spaces, you turn your gaze into the earth and explore its structure. They study the powers and influence of the constellations, and you investigate the powers of rocks and mountains and the manifold effects of the strata of the earth and rock. To them the sky is the book of the future, while to you the earth reveals monuments of the primeval world.[53]

Creuzer's Cabiri more than merely inhabit this underworld: they are its active forces, emerging from its rocks and stones to smelt and hammer out the world. Recent excavations at Samothrace have confirmed Creuzer's intuitions. Close by the central hall of the Mysteries, the Hieron, has been found the site of a forge and a place for smelting iron. Those who had been initiated in the Mysteries were given an iron magnetic ring to wear, whose rhapsodic, magnetic powers are described by Proclus and in Plato's *Ion* (533d–534e). These rings were sacred to the cult of the Mother of the Rocks, whose places of worship are particularly associated with outcrops of porphyry such as are found on Samothrace.[54] The ore for the iron rings was mined near the Sanctuary of the Great Gods.[55]

There existed only one real clue to the identity of the Great Gods and this is found in a note by an Alexandrian Scholiast on Apollonius Rhodius. Mnaseas of Patrae, an historian of the third century BCE, is said to have recorded that the gods' names are Axieros, Axiokersa, Axiokersos, with a fourth, Kasmilos.[56] Mnaseas identifies the first three with Demeter, Persephone and Hades.[57] Creuzer was aware of this source (Creuzer, vol. 2, 320), though his interpretation is different. For him, Axieros (Αξίερος) was the 'magnipotens' Hephaistos; Axiokersos (Αξιόκερσος), 'the great fecundator/inseminator', is associated with Ares and the Egyptian planet 'Hertosi'; Axiokersa (Αξιόκερσα), the 'magna fecundatrix', is Aphrodite. Finally, Kasmilos (Κασμίλος) is the 'perfecte sapiens' (or, *der Allweise*), related to the heroes Cadmus and Hermes. Why there is this divergence from Mnaseas is unclear, but the fourth figure is almost conjectural, though in line with the views of Herodotus on the Mysteries and the phallic Hermes. Creuzer identifies the four figures with potency, both sexual and physically creating, and says that they are part of the secret teaching of religious symbols (*die Priestlehre esoterischen Charakters*) (Creuzer, vol. 2, 312–13). The lack of differentiation and mystery enables a wide (and sometimes ungraspable) series of associations, and the Great Gods are the force that is manifest in that diversity.

The Cabiri appear in the scene entitled 'Classical Walpurgisnacht' (*Klassische Walpurgisnacht*) in Goethe's *Faust: Part Two*, featuring amongst a host of marginal, vestigial mythological entities: lamiae, sphinxes, sirens, nereids, emmets, and dactyls. Their uncertain nature conveys what Anthony Phelan has described as the 'ontologically unstable' characterisation in this scene.[58] In it, the Cabiri are associated with Cranes and Pygmies, and Goethe anticipates more recent archaeological discoveries at Thebes on the mysteries of the Cabiri where pottery fragments show the Cabiri and the Pelasgian cranes and pygmies.[59] Goethe's setting for the passage on the Cabiri is a marine one – 'The Rocky Inlets of the Aegean Sea'. Besides alluding to the role of the Cabiri as the gods of sailors, rescuing them from a watery grave and assuaging Neptune's rage, it also supports their Protean nature. Not only is there the problem of their uncertain quantity (three or four? – seven or eight?), there is also the problem of their indeterminate being. The Sirens reflect that the Cabiri of Samothrace

> Are Gods in the strangest setting,
> Ever evolving themselves anew,
> And never aware of their own nature.

> *Sind Götter! wundersam eigen,*
> *Die sich immerfort selbst erzeugen*
> *Und niemals wissen, was sie sind.*[60]

Their qualities are very powerful; they are entities who are tiny, mighty, and eternally unappeasable:

> These incomparable ones
> Ever wanting
> Yearning, hungry
> For the unattainable.

> *Diese Unvergleichlichen*
> *Wollen immer weiter,*
> *Sehnsuchtsvolle Hungerleider*
> *Nach dem Unerreichlichen.*[61]

In a letter to Creuzer dated 1 October 1817, Goethe reproached the mythographer for having forced him to confront 'Orphic darknesses' and

> to glimpse a region which I am ordinarily at pains to avoid. [. . .] When the passion for delving into these mysteries goes so far as to link the Hellenic sphere of god and man with the remotest regions of the earth [. . .] our suffering becomes acute and we hasten to take refuge again in Ionia.[62]

Creuzer was certainly the source of much of the material on the Cabiri in *Faust*, but the notion that they were 'shadows of the shining Olympians', their yearning, ever emerging and willing force, and their uncanny setting on Samothrace owe much to another thinker – to Friedrich Schelling.

The Mysteries of Samothrace II: Schelling

Schelling begins his *On the Deities of Samothrace* (*Über die Gottheiten von Samothrake*) (1815) with the island itself; it has a primordial life of its own like Delos, rising out of the northern part of the Aegean. It is the embodiment of volcanic processes, the product of 'a great convulsion of nature' (*große Naturerschütterungen*), the site of the Flood where the waters of the Mediterranean poured through to inundate the area which is now the Black Sea. The oldest legends of the island express 'the ever present awe of a vast and mighty nature' and 'the terror of its memories' (DS, 15). From the single peak of the island Poseidon, himself a god of earthquake and subterranean thunder, surveyed the convulsion of the Trojan War (see *Iliad*, book 13, 10–14); the human, mythic, geological energies merge as a single entity.

The description of the island manifests the central ideas of *Naturphilosophie* which Schelling had developed in his earlier *Ideas for a Philosophy of Nature* (*Ideen zu einer Philosophie der Natur*) (1797). In this work, Schelling put forward an organic, interrelated vision of Nature in which all phenomena and forces, observer and observed, spirit and matter, were parts of the same process of development. The argument sought to overcome (or, in a profound sense, undermine) the transcendent separations, identified by Kant in his *Critique of Pure Reason*, which appeared to have severed forever the link between man and things (*Ding an sich*) and his own innermost Self. Schelling attempted to look beneath and within phenomena, and to penetrate the noumenal world that lay beyond the categories of conscious perception. Both the world and the mind that observed it were generated from a single source whose manifestations they were; enjoying a sort of ontological kinship, they had *au fond* a single, underlying identity. This generation of the cosmos was active and ongoing, and here Schelling anticipated the notion of evolution, the 'infusion of time into Nature'. Drawing on Spinoza's notion of nature as a process, *natura naturans*, Schelling developed a dynamic idea of the cosmos in which Nature and Nature's God – the two are one – emerged from a dark origin driven by a longing for spiritual awareness and consciousness; the inchoate always seeking form and substance. Necessarily, this view meant a turn towards a consideration of archaic pre-conscious forces which emerged in dreams and in myth; and it involved a consideration of the irrational which underlay all things, both mental and physical, and was on its way to becoming rational and self-aware.

The development of the argument itself (and, indeed, the philosopher) also followed this pattern of universal *Bildungsroman*, and was part of the cosmic process. In fact, Ernst Benz in his essay 'Theogony and Transformation in Schelling' (1964) argues that the entire Schelling opus should be treated as a single organic process of Bildung, and that his work represents an attempt to 'encompass the whole development of mankind in metaphysical terms'.[63] Robert J. Richards, too, in his *The Romantic Conception of Life* (2002), suggests that the early titles of Schelling's works – from *Ideas for a Philosophy of Nature*, via *On the World Soul: A Hypothesis of the Higher Physics for the Clarification of Universal Organicity* (*Von der Weltseele: Eine Hypothese der höheren Physik der Erklärung des allgemeinen Organismus*) (1798), to *First Sketch of a System of Nature Philosophy* (*Erster Entwurf eines Systems der Naturphilosophie*) (1799) – map such a process: 'These tracts reflected the rapid evolution of Schelling's changing conceptions of nature and the philosophical approach to it,' he writes, for 'each title suggests an effort still in the process of becoming (*Ideas for. . . a hypothesis of. . . First sketch of. . .*), a conception not having achieved completion, indeed, a Romantic adventure.'[64]

Schelling's virtual exploration of the island of Samothrace was one such adventure, a textual equivalent of Goethe's Italian journey and Alexander von Humbolt's exploration of South America: a process of self-discovery. In the Jena circle of Schleiermacher, the Schlegels, Tieck, Novalis, and Schelling there was a great interest in 'spiritual geography'. Novalis explored the islands of the mind with the same rapture as Keats did his 'realms of gold':

> I have been on a journey of discovery, or on my pursuit [. . .] and have chanced upon promising coastlines which perhaps circumscribe a new scientific continent. This ocean is teeming with fledgling *islands*.[65]

Samothrace itself is a little more forbidding. Nevertheless, however dark its energies, it does, in Schelling's account, manifest the same compound of elements found in the Romantic vision of Nature, in which all the sciences and arts, chemistry, magnetism, botany, philology, geology, mythology and theology, are all parts of the same process of emerging. In his celebrated equation from the Introduction to *Ideas for a Philosophy of Nature*, Schelling stated: 'Nature is visible Spirit, Spirit is invisible Nature.'[66] Here, in this compressed fragment of *Identitätsphilosophie*, we find an expression of the unity of the material and the spiritual; physical substance is a concatenation of the spirit, and spirit is a volatilised substance.[67]

The island is the site in which one is transformed into another, and Schelling is drawn to the 'mysterious' and 'most ancient cult' of Samothrace. 'Together with the secret history of the gods,' he wrote, 'Greece first received from the forests of Samothrace the belief in a future life' (DS, 15). The 'most sacred island' was a place of confession, purification, and the evolution of the spirit, associated with Pythagoras and the initiation of both Emperors (such as Hadrian) and slaves. Demeter, Dionysos, Hermes and Zeus were all worshipped as Cabiri.

Like Creuzer, Schelling relies on the fragmentary reference to the island gods by the Scholiast Mnaseas for the actual naming of the Cabiri, but he favours a different allocation of gods, or goddesses, than Creuzer: Axieros/Demeter; Axiokersa/ Persephone; Axiokersos/Hades; and the conjectural fourth, Kasmilos/ Hermes. Schelling develops a Phoenician/Egyptian derivation of the names, though he shows great caution when contemplating 'the hazardous path' of philology. His criticism of past philogical exercises reads like an anti-*Naturphilosophie*; a malign, rather than benign, identification: 'A new frenzy of linguistic derivations struggled to produce everything from everything and mixed all things together is a crazy manner, even in the old stories of the gods' (DS, 18).

For the word 'Cabiri' itself, Schelling goes along with Creuzer's Phoenician/Hebrew origin, *Kabirim*, signifying 'the great', fitting in with the notion of 'the Great Gods'. Axieros has as its meaning 'yearning', 'hunger', 'need' (*Armuth, Schmachten, Sucht*). Axieros is the first being, desolate and yet promising the world in abundant fullness; it has 'the greatest need insofar as it has *nothing* to which it can communicate itself', for 'in the concept of *every* beginning lies the concept of a lack' (*daß im Begriff jedes Anfangs der Begriff eines Mangels liegt*) (DS, 18).

Schelling links this observation with Diotima's account of the birth of Eros in the *Symposium* (203 a–d). Eros is the child of Poros (Resourcefulness) and Penia (Poverty); Love will find a way to fulfil its need. Schelling finds the raw material for Plato's dialogue in one version of the Orphic myth of creation – Eros being the first god to emerge from the World Egg. Before the Egg there is and was only Night, and the most archaic foundation is Night (DS, 18):

> For it was the teaching of all peoples who counted time by nights that the *night* is the most primordial of things in all of nature [. . .]. But what is the essence of night, if not lack, need, and longing [*Mangel, Bedürftigkeit und Sehnsucht*]? For this night is not darkness, not the enemy of the light, but it is the nature looking forward to the light, the night longing for it, eager to receive it.
>
> (DS, 18)

Schelling then turns to another image of first Nature (*erste Natur*), whose essence is desire and passion. This is fire – a consuming fire, whose essence is nothing but a hunger which draws everything into itself. Schelling now reverses the usual allegorical philosophical interpretation of mythology, tracing back the gods to physical forces, a practice that was common in antiquity (for example, the *Saturnalia* of Macrobius, in which every myth was traced back to the Sun). Instead, Schelling starts with the nameless forces of Night and fire, and marks their manifestations as gods and goddesses; there is no space for derivation.

Fire is embodied in the figure of Hestia, the goddess of the hearth, which represents the subduing of fire. The philosopher also makes a link between these forces and Demeter, as she appears in the first part of the Homeric Hymn to Demeter, frantically searching for her lost daughter, Proserpine (or Persephone), who has been abducted by Hades 'in deep and insatiable hunger'. Demeter appears at Eleusis obscured by a dark veil and disguises herself as an old nurse. She is offered hospitality by the daughter of Keleos, to whom she identified herself as 'Dorso'.[68] Schelling, here, rather relies on the reader's knowledge of the poem, and the specific occasion when Demeter begins to nurse the infant son of the household, Demophoön, whom she passes each night through the fire of the hearth in order to make him immortal.

He proceeds to an identification of fire, yearning, night and the goddess Demeter. Hunger is the strongest aspect; it characterises the shades, 'the seeking-ones' in the underworld, who are desperate for life: 'For the burning desire must precede the satisfied desire, and the greatest receptivity, thus consuming hunger, must precede the abundant fullness of fecundity' (DS, 19). The ravenous world is condemned to starve, before the return of Proserpine (Persephone), and the re-emergence of Demeter. Schelling proceeds to connect this overmastering emotion with a Phoenician creation myth of 'great antiquity':

First there was the breathing of a dark atmosphere and a turbid chaos, in itself entirely boundless. But when the spirit of love was kindled in the presence of the special beginning, and a contraction [of the two] resulted, this bond was called longing and it was the beginning of the creation of all things.

(DS, 19–20)

How might this be linked to Samothrace? Pliny mentions a statue of Pothos, or 'longing', on the island (see *Naturalis Historia*, book 36, chapter 5, §25), and so Schelling finds some evidence as to the nature of Axieros/Demeter. At the same time, it should be noted that the distinction between forms of desire, Eros and Pothos, is as important as the pagan and Christian distinction between the two forms of love: Agape and Eros.[69] In a long footnote, Schelling carefully observes the distinction and criticises Creuzer for confusing Eros and Pothos. Pothos is a longing for an absent good, a loss, whereas Eros is the initial kindling of the desire which precedes possession.[70]

Schelling seems to be developing a theogony of desire – a temporalised sequence of the emergence of yearning that matches the generations of the gods in Hesiod. It is a deixis of desire; Eros is the desire for something in the future, Pothos for something in the past, and Himeros (Ἵμερος) a desire for that which lies in the present – thus anticipating Bergsonian vitalism and a present made up of retention and protention. In this connection, Schelling cites *De primis principiis*, one of the last works of antiquity, written by Damascius, the last Head of the Platonic Academy in Athens before it was closed by the Emperor Justinian in 529 CE. In his account of creation myths, Damascius could look back over the entire world of antiquity and all of its mythic and religious systems. According to Damascius, the Sidonians posited Time before all else, followed by Desire and Gloom; a truly sublime obscurity. In a remarkably long footnote – and in his preface, he apologises for the fact that the footnotes are longer than the main text which rises above them like a flower above a rhizome – Schelling conjectures the whole development of the nature of Time, moving from Damascius back to Zoroastrianism. According to the Zend-Avesta, 'the true creator is *time*, which knows no limits, has nothing over it, has no root, and both has been and will be eternal' (DS 33, n. 44). In his gloss on this passage, Schelling entertains the idea of a sort of temporal pantheism, arguing that 'Time' which permeates all things is not a 'mere eternity'; it is a limitless time in which unity and difference are established as one (DS 33, n. 44). One of Kant's major forms of intuition is harnessed to identify the categories with the noumenal itself.

In his interpretation of the second, third and fourth Cabiri, Axiokersa, Axiokersos, and Kasmilos (or Kadmilos), Schelling appears to be moving towards a Neoplatonic and Orphic understanding. For he identifies the first three as follows: Axieros is Demeter, Axiokersa her daughter, Persphone, and Axiokersos, Dionysos; thus, the major gods of the Eleusinian and Orphic mysteries are brought together. Cadmilus is, for Schelling, an adjunct to the primary three, an interstitial entity which links the primary three with the next three. This makes seven Cabiri in all, an upper and lower triad with Kasmilos/Hermes as messenger of the gods who mediates between different realms of creation, the heavens, the earth and the underworld. The vestigial eighth Cabir may be a further linking principle, though this is not made clear.

The very real difficulty of Schelling's text at this point appears to stem from the unacknowledged problem which stems from attempting to reconcile three different accounts of creation from antiquity, which may be gleaned from the discussion of time and the typology of desire.

First, there is the idea of an underlying primordial hypostasis which is unchanging, ever present in phenomena as a founding principle. This idea is found in such pre-Socratics as Thales, who argued that the material principle of water is the arche of everything; or Anaximenes, who claimed it is air; or even the later Stoic notion of fire, the physical world being but 'curdled smoke'. Though superficially changing, the universe is at least ever present or, in some accounts, eternal.[71]

Second, according to the Neoplatonic concept of emanation, which proceeds downwards from the ineffable transcendent Principle, there is a movement of increasing differentiation, through a descending series of triads. It should be emphasised that the transcendent principle has no need, being utterly contained and self-sufficient; it does not desire creation in any way (see Plotinus, 'On the Three Hypostases', in his fifth *Ennead*). On an abstract level there is the Triad of the primary hypostases: the One, the Intellect (*nous*), and Soul (*psyche*). Below this, Proclus identifies another five descending levels.[72]

And third, there is an evolution upwards from a primal matrix towards greater and greater differentiation and clarity. Thus Hesiod's *Theogony*, like other theogonies, was seen as a pattern of increasing clarification and refinement from the horrors of Chaos, through Erebos and Night, the monstrous Titans, Cronos, and up to Zeus and the shining Olympians – the first in beauty should be first in might. Following this notion of an upward path, the Homeric world is seen as a model of order and clarity compared with the obscure mystery which preceded it.

Now it would seem that the Cabiri are an early stage in the third of these ontologies, energies embodying a primitive yearning for creation and expressing the primordial lack. Yet, in the comparatively short section allocating the second, third, and fourth Cabiri, Schelling describes the Cabiri as primitive reflections of the eternal gods. He seems to be agreeing with Goethe, with whom he certainly discussed the matter. Kasmilos moves between ontologies as much as he traffics between levels:

> As servant of the great gods, Camillus [or Kasmilos] is not necessarily servant of those first three. But it is established that he simultaneously served the lower and the higher gods. Hence he served the former only insofar as he was the *mediator* between them and the higher, and thus himself was higher than they. [. . .] This is the *most essential* concept of Hermes, to be the governing bond between the higher and lower gods.
>
> (DS, 22)

It seems that, for Schelling, it is the principle of identity itself that mediates cosmically, epistemologically, and theologically across those rifts that *Naturphilosophie* sought to bridge – thought and matter, observer and observed, the gods and Man. Here, Schelling confronts the following problems: Why are the gods plural, and why is their differentiation important? Why not just believe in a god of gods, a transcendent unity rather than mere fractions? While admitting that it is necessary when researching archaic gods to harmonise the multiplicity of divine natures with the indelible idea of the unity of God, he draws back from emanation cosmology:

> But it is neither suitable and clear in itself to represent the diverse gods as merely emanations of One into them, as a primal power propagating itself into diverse rays; nor can its indeterminateness and boundlessness be compatible as well with the determinateness and sharpness of the outlines of every individual form, as also with the limited number of these forms.
>
> (DS, 23)

Schelling asserts that, if such a view of an irradiating One were conveyed at the Mysteries, then this would be inimical to belief; why worship the 'radiations' when it is possible to direct one's reverence to the immediate One? Schelling reverses the process, so that, instead of top-down, the movement is bottom-up:

> It is an entirely different matter if the various gods be not downward-proceeding, ever more self-attenuating emanations of a highest and superior deity; if instead they be gradations of a lowest power lying at the basis, which are all finally transfigured in One highest personality.
>
> (DS, 23)

The gods are thus rungs on a ladder, which rises up ever higher.

To conclude Schelling's study, there follows a double movement, one towards abstraction and the other towards energy. Following a Platonic-Pythagorean model, Schelling considers the mathematical aspect of the Cabiri. The four Samothracian deities form a series ascending from below, as numbers do. Kasmilos is the way to a doubling of the series; and the following group of three or four are known collectively as Haephestos, combining as a sort of indefinite group demiurge. Rather like Creuzer, he links the Cabiri to the Phoenician and Egyptian pygmy gods. The latter are essentially phallic in nature, as Schelling modestly hints, referring to 'crude yet candid idol sculptures' (DS, 23). The dwarf, *Zwerg* in German (Old German: *Tuwerg*), is from the same root as the Greek *theurgos*, which denotes a being of magical power. The Cabiri thus can stand as energies or potencies that, in part, create and, in part, open the way to higher levels as they 'practise the magic by which the transcendent is drawn into reality' (DS, 28). The Cabiri, or so it seems in Schelling's interpretation, make the three levels of interpretation possible; as changeless mathematical forms, as the access to higher levels, and as energised forces burning upwards from below. But the preferred way is from the ground upwards.

Schelling even hazards a guess as to the nature of the initiate meaning of the Mysteries of Samothrace:

> The holy, revered teaching of the Cabiri, in its profoundest significance, was the representation of insoluble life itself as it progresses in a sequence of levels from the lowest to the highest, a representation of the universal magic and of the theurgy ever abiding in the whole universe, through which the invisible, indeed the super-actual, incessantly is brought to revelation and actuality.
>
> (DS, 29)

Here, the rising and falling motions (and, perhaps, no motion at all) become in Schelling's thought interfused, as in the higher doctrine of Neoplatonism; Proclus had observed of the original triad that they all have the qualities of the uniform and intelligible: 'the abiding, and the proceeding, and the returning' (*On the Theology of Plato*, book 3, chapter 14).[73] Schelling always favours the proceeding, as he strongly favours a positive theology, which says what the godhead is, rather than a negative theology, which says what it is not. He also favours process over frozen and abiding, permanent *a priori* truths (as Robert Brown emphasises in his 'Philosophical Interpretation' of the treatise) (see DS, 58). Always there is an enthusiasm for effectiveness which, at its highest, is disinterested ordering, and, at its lowest, urges and desires.

All of these levels again come together in Schelling's idea of the 'Potencies' in his dynamically redrafted, necessarily incomplete work *The Ages of the World (Die Weltalter)*, which occupied the philosopher from 1811 to 1815. His treatise *On the Deities of Samothrace* was in some sense a parallel text, an illustration, a concretisation of the abstract ontology of the world ages. But, before turning to *Die Weltalter*, it is helpful to consider C. G. Jung's encounters with the Cabiri.

Jung and the Cabiri

The appearance of Jung's *Symbols and Transformations of the Libido* (*Symbole und Wandlungen der Libido*) (1911–1912) established the grounds for his break with Freud. One important element of this break was a re-evaluation of the libido; was it simply sexual energy, or something wider and greater? Jung's considera- tion of the meaning of the Cabiri is an essential part of this re-evaluation, and prefaces the analysis of the transformation of the libido which forms the second part of the book. These passages were taken over, virtually unchanged, into the revised second edition, published in 1952 as *Symbols of Transformation* (*Symbole der Wandlung*) (whereas other sections underwent drastic revision). In his treat- ment of the Cabiri, Jung clearly draws on Creuzer, whom he cites later in the text, but there are also clear signs of the influence of Schelling (and, through Schelling, Jakob Böhme). Werner Leibbrand has asserted that Jung's teachings are 'not intelligible if they are not connected with Schelling', and perhaps, with hindsight, it might be possible to argue that Jung is really a late member of the Schelling School of philosophy.[74]

Jung examines the nature of the phallus and its relation with creative divinity. The phallus is the creative dwarf which toils away in secret, working in darkness. Jung associates this with Faust's 'descent to the Mothers' and also Hephaistos, Wieland the Smith, dactyls, Tom Thumb, and the Cabiri. These mysterious chthonic gods are represented as ugly and deformed – the limp of Hephaistos – and Jung illustrates his text with vase paintings from the Cabirion at Thebes which portray the Cabiri as grotesque pygmies. Jung here draws attention to Herodotus' disclosure that the Pelasgians brought the cult of the ithyphallic Hermes to Samothrace, and he moves from the creative (and patently sexual) force of the Cabiri and their uncertain number to argue that the libido is not exclusively a sexual force, as maintained by Freud.

To do so, Jung quotes Cicero's *Tusculan Disputations* which distinguishes between delight concerned with present good and desire associated with future good; if the pursuit of the good is pursued with moderation and prudence this is called will, 'but when it is divorced from reason and is too violently aroused, then it is "libido" or unbridled desire'.[75] Lasciviousness, pleasure, want, wish – Jung finds a whole series of associations and variations of the libido, and digs up a host of philological parallels (Sanskrit, Gothic, Old Bulgarian) to show that the mean- ings of libido can arise from violent longing. These are rapid and violent semantic shifts. Yet Jung's sudden conclusion that it all has to do with psychic energy[76] becomes less hurried and, indeed, more intelligible when read in tandem with Schelling's differentiation of desire (Eros, Pothos and Himeris), but, as I shall argue below, the true source of the varieties of longing may be found in the visionary writings of Böhme.

The Cabiri appear in the 'Liber Secundus' of Jung's *Red Book* or *Liber Novus*. This secret book – built up over many years from material revealed in dreams, meditations, active imagination, visions, and material circulated through three preceding Black Books, with its mandalas, calligraphy, and paintings – is not simply a record; it is an activity, like the Great Work of the alchemists, lasting over the second decade of the twentieth century. The Red Book is truly theurgic in Schelling's sense, building up from subterranean darkness to greater and greater articulation, not through conscious decision, but through the use of energies, symbols, and potencies.[77]

In the 'Liber Secundus' of the *Red Book* there is an encounter with the Cabiri, who are described as having 'delightful misshapen forms' and as 'young and yet old'.[78] They are interrogated as to their emergence from the 'egg of the gods', and the narrator figure specifically asks the Cabiri if they are the earthly feet of the godhead, in response to which they declare:

> We carry what is not to be carried from below to above. We are the juices that rise secretly, not by force, but sucked out of inertia and affixed to what is growing. We know the unknown ways and the inexplicable laws of living matter. We carry up what slumbers in the earthly, what is dead and yet enters into the living. We do this slowly and easily, what you do in vain in your human way. We complete what is impossible for you.[79]

The Cabiri create a foundation, they form the tower which grows up from inside from the labyrinths, from the chthonic. In Creuzer and Schelling, the Cabiri are potencies, energies of a mysteriously productive Nature, the shadowy adumbrations of separate names and specific identities. Similarly, in Jung, they are the vital forces of the unconscious which underlie the separation of spirit and matter, nature and reason. In Jung, they are slightly more conscious than Goethe's Cabiri, 'never aware of their own nature' and 'ever evolving themselves anew'.[80] Schelling had already attempted to articulate the full signification of such mythic potencies and their primordial ground in his *Die Weltalter*; a task that was to prove almost impossible.

Schelling: The ages of the world

In the Romantic group at Jena there was a particular interest in employing mathematics as a philosophic and symbolic system in the manner of Pythagoras or Proclus' Commentary on Euclid. The mathematical notion of potentisation, raising to a higher power, could be – and, in Jena, was – a metaphor of metaphysical ascent. In a famous passage by Novalis, we find the equation between potentisation and romanticising:

> The world must be romanticized. This yields again its original meaning. Romanticizing is nothing else than a qualitative potentization. In this operation the lower self becomes identified with a better self. Just as we ourselves are a potential series of this kind. This operation is still entirely unknown. By giving the common a higher meaning, the everyday a mysterious semblance, the known the dignity of the unknown, the finite the appearance of the infinite, I *romanticize* it – For what is higher, unknown, mystical, infinite, one uses the inverse operation – in this manner it becomes logarithmicized. It receives a common expression.[81]

This 'reciprocal raising and lowering' follows the Neoplatonic trajectories of descent and ascent known as 'the process'. Such potententisation can act on a humble word, raising it to a proposition, which can, in turn, be raised to a science. Similarly (and significantly), thinking and dreaming can be subject, according to Novalis, to qualitive potentisation:

> Thinking in the ordinary sense of the word, is thinking of thinking – Comparing etc. different specific thoughts. Direct dreaming – reflected dreaming – potenticized dreaming.[82]

The central causative idea of potency, the idea of an energetic force that brings about qualitative as well as quantative change, and possesses the power to create, has its foundation in Book Delta of Aristotle's *Metaphysics*. Aristotle's 'potential' is dynamic, life-giving and associated with process. Without such a capacity, things are inert, compressed, and they cannot be raised to a final end (*telos*). Such potencies are essentially transitive, working and having the capacity to work in a created world of acts, events, things, and individual entities. There subsists a causal chain of interlinked activity, but it is difficult to trace this back to an original cause that contained the capacity for the emergence of the whole to time and the cosmos. This tracing is essentially negative, for one is returning to that which essentially requires no capacity of its own, no movement, no change, no place. . . It is potentially all things and nothing.

Schelling takes on this challenge in his *Die Weltalter*, a work which was also a product of a process of reworking over a number of years, from 1811 to 1815, and it remained unfinished as it attempted to express the inexpressible, and describe that which precedes language and the world.[83] The treatise *On the Deities of Samothrace* was the mythic side of a complete project, the realisation of the ideas of *Die Weltalter* in a more plastic form.

The study of the world is a matter of time, and science, Schelling asserts, is a species of history, seeking to present the actual living essence, the primordially alive, the essence preceded by no other, the first, the oldest of essences. Philosophy, for its part, tries to overstep the boundaries of present time in seeking that essence, and retracing 'the long path of developments from the present back into the deepest night of the past' (W, 114). Man alone seeks to do this, because he is drawn from the source of things; he is akin to it, and to the eternal essence.

In a way similar to Rousseau in his *Discourse on the Origin of Inequality* (1755), Schelling returns to the original ground by avoiding the impossible task of tracing back through time; instead, the archaic and, indeed, the primordial are accessible through introspection. The primal man is within and 'the unfathomable, prehistoric age rests in this essence' (W, 114). We must turn to 'the inner oracle'; the only witness to 'a time before the world' (W, 114). The 'archetypal images of things' (*Urbilder*) slumber within and can be awakened through a process of inner dialogue which seems a cross between anamnesis and Jung's interrogation of the Cabiri; indeed, according to Schelling, one comes to one's inner identity through the conscious interrogation of the unconscious, and 'everything remains incomprehensible to man until it has become inward for him' (W, 116).

Thus, to move backward and inward and to the essential are one and the same; and by doing this, one is renewed by the primordial, which seems to be part hypostasis, and part primordial unconscious. New life comes from the most archaic, for 'Man is rejuvenated time and again and becomes blessed anew through the feeling of unity of his nature' (W, 117). Schelling's own treatise is partly scientific, partly philosophical, and partly theosophic; and these studies are also revivified by the return to the ontological roots:

> Science has no longer begun with thoughts drawn from far away in order to descend from these to the natural; rather, it is now the reverse: beginning with the unconscious presence [bewußlosen Daseyn] of the Eternal, science leads up to the supreme transfiguration in a divine consciousness. The supersensible thoughts now acquire physical force and life, while nature for its part becomes ever more the imprint of the highest concepts.
>
> (W, 119)

The entire enquiry thus becomes itself a potentiation, and the physical world becomes worthy of the highest consideration; it cannot be passed over and scorned; indeed, 'the stone the builders rejected becomes the cornerstone' (W, 119).[84]

Nevertheless, there is still a place for rejection; Schelling asserts that it is only the man who has the strength 'to rise above himself' who is able to create a true past and becomes able 'to savour' the present and look forward to the future. This is difficult both to perform and to understand, but resistance is the secret of change, for it is the negativity which makes the variety of the world possible. Once accomplished, it becomes clear that

> [e]verything that surrounds us points back to a past, to a past of incredible grandeur. The oldest formations of the earth bear such a foreign aspect that we are hardly in a position to form a concept of their time of origin or of the forces that were then at work. We find the greater part of them collapsed in ruins, witness to a savage devastation. More tranquil eras followed, but they were interrupted by storms as well, and lie buried with their creations beneath those of a new era. In a series from time immemorial, each era has always obscured its predecessor, so that it hardly betrays any sign of an origin; an abundance of strata – the work of thousands of years – must be stripped away to come at last to the foundation, to the ground.
>
> (W, 121)

We are reminded here not only of Shelley and Byron and their cancelled cycles, but also of the more melancholy sensibility of an age that habitually visited, painted, mourned over, and wrote on and in ruins; an age infused as it was with the Comte de Volney's *Les Ruines, ou Méditation sur les révolutions des empires* (1791), and which found itself, in Peter Frizsche's phrase, 'stranded in the present'.[85]

For Schelling, things human and physical must be uncovered and traced back to their source. All phenomena demand an 'immense investigation' to unravel all of the multifarious intricacies and folds that went into their growth and development; even a grain of sand possesses a determination within itself that 'we cannot exhaust until we have laid out the entire course of creative nature leading up to it' (W, 121–22).

The vastness of the task and the resistances and difficulties involved become part of the process of uncovering. Undaunted, Schelling seeks not only the sources of the beginning of phenomena, but what was before that beginning. 'The spirit' – a sort of metaphysical explorer – finds a presupposition of a time before the last visible thing when there was nothing but 'the one inscrutable, self-sustaining essence, from whose depths all has come forth' (W, 122). However, it is possible – and here there is a sense of the Gothic sublime – for the spirit to probe even deeper, and venture even further into 'new abysses':

> It would not be without a kind of horror that spirit would finally recognize that even in the primordial essence itself something had to be posited as a past before the present time became possible, and that it is precisely this past that is borne by the present creation, and that still remains fundamentally concealed.
>
> (W, 122)

Whilst noting that he, Schelling, has personally had insufficient time to pursue the theme of time, past, present, future; and that time is a constant of human experience, it is nevertheless possible, it would seem, to collapse the development of time and history and perceive 'everything as one in life and deed':

> We have a presentiment that one organism lies hidden deep in time and encompasses even the smallest of things. We are convinced (but who is not?) that each great event, each deed rich in consequence, is determined to the day, the hour – indeed to the very moment – and that it does not come to light one instant earlier than is willed by the force that stops and regulates time.
>
> (W, 123)

Schelling discovers two forces present in time: one of development, the flow of time itself, and one that holds back, retards and indeed inhibits and stops the flow of time. Without the latter, negative force, the universe (so Schelling claims) would be over 'in a flash'. Negativity is essential striving against the positive force which drives forward, and without contradiction and struggle there would be 'a deadly slumber' of forces. We are close to Blake's Proverb of Hell ('Damn braces, bless relaxes'), and this energetic, even infernal, opposition can be traced back in both Schelling's and Blake's case to the visionary works of Jakob Böhme.

'Contradiction', Schelling writes, 'is in fact the venom of all life and all vital motion is nothing but the attempt to overcome this poisoning' (W, 124). In the grand economy of dynamic tension, even the negative and seemingly destructive and sinful has a positive role to play. Few people have realised what force lies in limitation, not in expansion.[86]

Everything in time yearns for the peace of non-contradiction, but opposition can be found even in the Highest, not just in creation, even though we would expect the Highest to be the Unconditioned. Schelling says that, although this state of contradiction is well-nigh universal, and philosophers just have to accept this, he himself struggles to find some sort of difficult 'peace' by identifying 'what is' (*Seyendes*) with 'being' (*das Seyn*), linguistically aligning the stative and the dynamic (and thus reflecting the joining of Spinoza's *natura naturata* and *natura naturans*). Yet the problem remains: how can there be duality in unity?; how is the timeless One realised as the temporal many? Here Schelling follows a Neoplatonic model of triads of Hypostases, distinguishing three stages of generation. Each of these stages or Potencies encapsulates the next. As stages of generation, they are far from being merely the emergence of the possible, for they are expressions of the lawful freedom of creation, not the spawning of every possible contingency; they represent an increase. As such, they form part of Schelling's concern with a radical and underlying freedom, as considered in his earlier treatise *On the Nature of Human Freedom* (*Über das Wesen der menschlichen Freiheit*) (1809), and his concern to reconcile order and the unchecked free energy represented by liberty.

The first potency A^1, the affirming principle, must contain B, the negating principle, in order to freely create that which is independent from it. The self-affirming, expanding agency is A^2. The unity of A^1 and A^2 is A^3. This triad was later revised a number of times by Schelling, being rendered as a more deferred process of differentiation: A^1 is the Unlimited, A^2 the Limited, and A^3 the Self-limiting, or as a more experimental triad: Possibility, Impossibility, and Transpotentiality (the latter being a higher form in exactly the same manner as, in the treatise on Samothrace, Hermes is above the preceding Cabiri, in that it leads, not only to reconciliation, but to a superior order of things from the lower forces of creation).[87]

It is difficult to finely equate the Potencies with the Cabiri as, *ex hypothesi*, all enjoy a defining but undefined nature. However, Schelling finds the same quality of yearning in the first of the Potencies A^1 as he identified in the first of the Cabiri, Demeter: there is a longing in Eternity, a yearning which comes about without Eternity either helping or knowing – a longing to come to itself, to become conscious, '*of which Eternity itself does not become conscious*' (W, 136). Schelling likens this condition to the polar opposites of a magnet, which experience a state of endless longing, and it would be interesting to know if some distant image of the magnetism of Samothrace lay behind Schelling's analogy. This primordial longing, unlike that of the Goddess, is utterly unaware: 'This is a seeking that remains silent and completely unconscious, in which the essence remains alone with itself, and is all the more profound, deep, and unconscious, the greater the fullness it contains in itself' (W, 137).

This 'resting will' generates or (it would seem) gives birth to the second Potency, more peaceful perhaps than the first, but equally unconscious, actively self-seeking will. It is unconditioned, omnipotent, and self-producing. This will is independent of Eternity; it longs for Eternity, but cannot become so itself. From it comes 'the great process of the whole'; probably, Schelling means this in the Neoplatonic sense of process (πρόοδος), in which the One advances through the cosmos back to itself, although (in Schelling) the trajectory is almost at once reversed. There is right from the start a greed in nature and matter for the eternal. This is reflected in the will which springs from a lack (*Mangel*), a need for essence which it has to posit outside itself. 'Unconscious longing is its mother', Schelling writes of the will, 'but she only conceived it and it has *produced* itself', so the will 'produces itself not *out* of eternity, but rather *in* eternity' (W, 137).

Here, there are two points of tremendous importance. The first is that primacy is given to the feminine, and this is (it seems) not merely metaphoric; Schelling later says that a germinal (potential) life precedes the active, and that the 'receptive sex' is present first and on its own, in the 'supposedly asexual' lower species of animals (W, 148). The second is that there is an immediate parallel made between the free will of the Origin and an individual person's will, which unconsciously produces itself in a man's mind – he simply finds it, and only when it is found does it become 'a means for him to externalize what lies innermost within him' (W, 137). Schelling's notion of a free will is not at all, it would appear, a matter of rational choice, but rather the force of realising one's innermost nature, one's potentiality. It is volition, but not in the usual, utterly untrammelled sense of voluntary. Its very obscurity, its unconscious nature, gives it both power and authenticity without external limitation.

Man, nature, matter, all burn for release and the fulfilment of a need for articulation, and to overcome the force of negation which keeps them in their preset condition. Dreams are seen by Schelling as a window into these inner forces, or rather the inner Potency, and the search for the Philosopher's Stone is analogous to this quest for potency in material form. Schelling speculates whether an inner force in all things desires to be released to change into a higher spiritual essence, to feed on their own spiritual pabulum. In a wonderful passage he argues that 'things do not seem fully completed by what constitutes their existence in the strictest sense', for 'something else in and around them first grants them the full sparkle and shine of life':

> There is always an overflow, as it were, playing and streaming around them, an essence that, though indeed intangible, is not for that matter unremarkable. But this essence that shines through everything – is this not just that inner spiritual matter which still lies concealed in all things of this world, only awaiting its liberation? Among the most corporeal of things, metals in particular have always been regarded as individual sparks of light from this essence, glimmering in the darkness of matter. A universal instinct divined the presence of this essence in gold, which seemed most closely related to the

> spiritual-corporeal essence by virtue of its more passive qualities, its almost infinite ductility, its softness and tenderness, which render it so similar to flesh and result in the greatest indestructibility.
>
> (W, 151)

It is because of this spiritual essence, its association with Golden Ages, and the fact that it was a sign of 'that blessed primordial time' that the alchemists 'wanted – never gold, but rather the gold of gold' (W, 152). Schelling further speculates whether it would be possible for humans to get their hands on this inner spiritual essence; if so, then they would be able to transform baser metals into a more noble one through a series of steps. By the inner essence he seems to suggest the Philosophers' Stone; and, once attained, its inner force would be stronger than the outer restricting force.

   Just as matter might be 'released', so in dreams, which have the greatest form of intensity, we have the release of the boundary of the soul. In a similar way, alchemy sets free the interior, thus 'initiating true transformations' (W, 160–61). In that inner world of matter and dreams – the gold of mines, the green of plants – we are close to the archetypes, energies which are different from the transcendent Ideas of Plato. For Schelling, Plato was only a reporter and interpreter of a doctrine, a sacred legacy, which was by Plato's time already lost in deepest antiquity. The archetypes, energies, *Urbilder*, are essential to matter, to the destiny of the individual, and to life itself:

> The production of such archetypes or visions of future things is a necessary moment in the overall development of life, and even if these archetypal images cannot be understood as physical natures in precisely the normal sense, they certainly cannot be thought apart from all physicality. They are neither merely universal concepts of the understanding, nor fixed models; for they are Ideas precisely because they are eternally full of life, in ceaseless motion and production.
>
> (W, 161)

These archetypes stream out from the innermost part of creative nature and the individual has access to them through inner vision. Sleep, the mystery of the depths, the primal springs of the deepest night of antiquity, and the obscure, unconscious generation of the cosmos present, for Schelling, the same energies and sources. The primordial at its most obscure, the archaic at its most primeval, consciousness at its most occluded, matter at its densest, and the deepest, darkest recesses of Nature: these are the places where the potencies and archetypes may be encountered (although, like Fafnir, at their most sluggish, albeit most rich). Schelling conveys all this in an exemplary passage of the primordial sublime, which contains echoes of Kant's early essay *Observations on the Feeling of the Beautiful and Sublime* (*Beobachtungen über das Gefühl des Schönen und Erhabenen*) (1764):

Darkness and concealment are the dominant characteristics of the primordial time. All life first becomes and develops in the night; for this reason, the ancients called night the fertile mother of things and indeed, together with chaos, the oldest of beings. The deeper we return into the past, the more we find unmoving rest, indistinction, and indifferent coexistence of the very forces that, though gentle at the beginning, flare up later into ever more turbulent struggle. The mountains of the primordial world seem to look down on the animated life at their feet with eternally mute indifference; and likewise with the oldest formations of the human spirit. We encounter the same character of concealment in the mute solemnity of the Egyptians, as well as the immense monuments of India, monuments that seem built for no time but rather for eternity. Indeed, this character even emerges in the silent grandeur and sublime tranquillity of the oldest works of Hellenistic art, works that still bear within themselves the (albeit softened) force of that pure, noble age of the world.

(W, 179–80)

Force and energy are essential, in that they vivify the density of these serene but largely static edifices, and these energies must have, for mortals, a perilous fierceness, for they are the very forces of nature, the unextinguishable driving-upward into a higher resolution. Only an exceptional individual can embark on such an exploration, or on 'the Odyssey of the spirit', as Schelling liked to call it.[88] Such an individual must overcome the limitations of his being in order to become (in philosophical terms) '*posited-outside-himself*' (W, 163). He must be in contact with the 'eternal in his soul' or enjoy the intercession of a daemon, 'a genius that nature has given us as a companion, and that alone is capable of serving as an instrument to being, to the extent that the conscious spirit lifts itself above the genius' (W, 163).

Just as the world in its substance is a coagulation of subterranean energies, *Die Weltalter* itself is a plutonic geology of cultural forces. Here gleams a vein of Neoplatonism, there the triads of Plotinus and Proclus; there is a glint of Iamblichan theurgy, now we find buried in passages on magnetic vision the Ideas of Plato and the Daemons of the Chaldean Oracles; and, in the depths, we encounter the Mysteries of Samothrace and hear the rumble of Spinoza's *deus sive natura*. At the same time, we find energies for future developments, especially dynamic psychology: the unconscious, the archetypes, dreams, inhibitions, the importance of desire and its restriction, the individuation process. It is as if the cultural matrix were becoming ready for the emergence of Freud and Jung.[89]

Ultimately, the exploration of God and the spirit is dangerous and terrible, because of the sheer force of negation which keeps the boundaries of the universe. Only the inner spirit of God, love, can counter such forces:

This force is the white heat of purity, intensified to a fiery glare by the pull of nature. It is unapproachable, unbearable to all created things, an eternal wrath that tolerates nothing, fatally contracting but for the resistance of love.

(W, 171)

The fierceness of this wrath burns up from a deeper, largely unacknowledged layer of Schelling's cultural matrix: the work of the seventeenth-century German mystic, Jakob Böhme.

Böhme: The unconscious of the unconscious

Schelling was quite unforthcoming about his reading and his sources, but the link with the books of the visionary of Görlitz is well attested. He discussed Böhme with Coleridge, he was acquainted with Franz von Baader (who attempted to reconcile Böhme and Catholicism), and there is evidence that Ludwig Tieck introduced the thought of Böhme to the Jena Circle as early as 1799.[90] Like Schelling, Böhme sought to understand not simply the origins of the world, but also the nature of God, and in his works the archaic sublime is truly at its most terrible. These tremendous subjects appear primarily in *Aurora* (1612), *Signatura Rerum* (1622), and *Mysterium Magnum* (1623).

In *Aurora* Böhme gives us an account of his principal vision. He tells us how he was seized with melancholy 'as with the presence of evil and good in the world'.[91] He overcame his apprehensions and made an assault upon God and upon the Gate of Hell, so that his spirit burst through 'even into the innermost birth and geniture of the Deity'.[92] This resurrection of his spirit allowed him to see 'through all, and *in* all and *by* all the creatures, even in herbs and grass it knew God' (the echo of this moment in Schelling's *Weltalter* is already clear). Böhme continues that, in that light, his will was set on by 'a mighty *impulse*, to describe the *being of God*'.[93]

Böhme's experience recapitulates God's own emergence, and the Dawn of the title is both the visionary's and the Deity's, so there is a deep parallel between the development of the two. The violence of Böhme's breakthrough, the rupture of the barrier, anticipates the dangerous energies and the strife of the origin of the world, and again anticipates Schelling's discussion of boundaries and overcoming the negative force of containment.

In *Aurora*, the qualities of the sublime – obscurity, terror, awe, power, associated with the primordial and the archaic – become active qualities, active and potent agents which produce the world. They are the Qualities of alchemy, partly chemical, partly spiritual, in the crucible from which God, the Cosmos, and the forces which drive it are formed. These forces are not just attributes of God, they reveal the nature of God, and in a deep sense they *are* God, and are working throughout the world. Böhme describes the Qualities as the seven sources, species, kinds, manners, circumstance, powers, operation, faculties of a thing. The Quality is 'the mobility, boiling, springing, and driving of a thing', and the seven Qualities are the seven Spirits of God, each combining two species of good and evil. *Qualität* is identified by Böhme with both *Quelle* (or 'source') and *Qual* (or 'suffering').[94] These Qualities are in the innermost nature of all things and constitute their 'deep Mystery': 'In the deep the Power of all Stars, together with the Heat and Lustre of the Sun are all but one Thing, a moving, hovering like a Spirit or Matter [*eine bewegende Wallung gleich eines Geistes oder einer Materia*].'[95]

Adjectives, verbal nouns (gerunds), and qualifiers (adjectives) are of supreme significance here; nouns tend to be petrified, dead. It is a vision of change, indeed transfiguration, which is reflected later in the style of Schelling's writing, though it is not just style, it is the force itself. Buffon's observation, *le style c'est l'homme même*, here takes on a profound significance.

These Qualities are working even in the deepest, most material of locations, 'as in the corrupted, murderous den or dark valley and dungeon of the earth, there spring up all manner of earthly trees, plants, flowers, and fruits'.[96] We can already see the ground for the later Romantic fascination with the mine for, as Böhme tells us, within the Earth grow curious precious stones, silver and gold, and these are a type of 'Heavenly Generation and Production'.[97]

In the *Mysterium Magnum: An Exposition of the First Book of Moses* (1623) Böhme returns to the subject of spiritual forces working in and through nature; nature taking on the likeness of the invisible, spiritual world as the soul in the body. 'God is nigh unto All and through All.'[98] From material creation we can find out a way to understanding, and indeed, witnessing, not only the forming of the world, the emergence of the Deity out of Nothing, but what was before the world and God. Just as the unappeasable longing of Demeter/Axieros precedes creation so too longing in Böhme precedes everything: 'The Nothing hungereth after something, and the hunger is a Desire.'[99] This 'Nothing' (*das Nichts*) is potentially all, the 'Desire' is the Fiat of the creating power which comes into being in order to know itself: 'For the Desire has nothing that it is able to make or Conceive – it concentratreth it selfe – it impresseth it selfe – it coagulateth it selfe – it draweth it selfe into it selfe and comprehends it selfe'.[100] So the Abyss becomes 'the Byss' (foundation, bottom), the *Ungrund* becomes the *Grund*. The Abyss becomes 'the Eye of the Abyss', and 'the Eternall Chaos brings it selfe into Nature'; from this forms a series of successive interacting and contending Properties.

The First Property which arises from obscurity is that of Desire which is 'Astringent, Harsh and Eager', overshadowing itself it first creates 'the Darknesse of the Abyss'.[101] This property is the 'Will of the Abyss' and Böhme identifies it with God the Father.[102] The Second Property is 'the Delight or Impassion of the Will', it has 'a magneticall attraction' and it is a 'Compunction, a Stirring, a Motion', and it is the Son.[103] Böhme gives it the name 'Lubet' which signifies a pleasing or pleasurable delight.[104] 'Lubet', according to Jung's philological analysis in *Symbols of Transformation*, is part of the etymological history of the libido: 'Libido or *lubido* (with *libet*, formerly *lubet*), "it pleases"; *libens* or *lubens*, "gladly, willingly"; Skr. *lúbhyati*, "to experience violent longing" [etc.].'[105] The Third Property is the Spirit, and whereas the Second Principle, the Son, is 'a driving together', the Third is 'the Severable', the rising spring of distinctness which moves towards the 'Outbirth of Creation'.[106] It is 'Anguish' (*Angst*).[107]

There is a dark, infernal side to the Properties which is Wrath, burning, fire, corrosiveness, and, although the Holy Ghost reigns in calm and meekness, there is fierceness, also Fury (*Grimmigkeit*).[108] The three Properties are identified with the primary alchemic ingredients: Sulphur, Mercury, and Sal, which boil, contend, and solidify in the self-created alembic of creation. Substance and matter, where they occur, are arrested process, things are and stand in place only as a balance of negative and positive forces, again as in Schelling. Böhme's sentences or phrases errupt with a mix of hypotaxis and parataxis, like vast jets of fire in the sun's photosphere. There is no peaceful emanation of the godhead, matter gradually becoming dense and less good with the privation of the spirit, as there is in Neoplatonism:

> For the Hardnesse causeth Substance and weight: & the Compunction giveth Spirit and the 'Active Life': these both mutually Circulate in themselves and out of themselves, yet cannot go any whither [parted] what the Desire: viz. the Magnet maketh Hard, that the attraction doth again breake in pieces: & it is the Greatest unquietnesse in it selfe; like a Raging madnesse: and in it selfe an horrible Anguish [*Angst*] [. . .].[109]

This caldera of seething affect is an expression of an active Will, and the all-creating Word is a tremendous Speech Act, a breath before what Böhme calls 'the Coagulation of the Syllable':

> The Creation of the outward world, is a manifestation of the inward Spirituall Mystery, viz or the centre of the Eternall Nature with the holy Element and was brought forth by the Eternall Speaking Word through the motion of the inward world as a spiration; which Eternall Speaking Word hath expressed the Essence in the Speaking.[110]

Böhme enlarges on these themes in his *Signatura Rerum* (1622). Once again, there is a violent anticipation of Romantic expressivism, vitalism and magnetism in his description of the cosmic agon: 'Sicknesse and paine arise when an Essence destroyeth another & there is contrariety and combate in this Being of Beings, how one doth oppose, poyson, and kill another.'[111]

All is driven by a fierce convection that springs from the *Ungrund* of the original Nothings, and seeks rest and self-knowledge through creation. It is a much rougher ride than the flight of the alone to the alone:

> We understand that without Nature there is an Eternall Stillness & Rest, and then we understand that an eternal Will ariseth in the Nothing, to introduce the Nothing into a Something that the Will might find, feel & behold it selfe.[112]

In chapter 3 of *Signatura Rerum*, Böhme imparts the hidden 'Arcanum', 'the Great Mystery of All Beings':

> We give you to understand This of the Divine Essence; without Nature God is a Mystery understood [a sort of active hypostasis] in the Nothing, which is an Eye of Eternity an Abyssal Eye, that standeth or seeth in the Nothing for it

is the Abyss, & this same Eye is a Will, understand a longing after manifestation, to find the Nothing; but now there is Nothing before the Will, where it might find Something, where it might find a place to Rest; therefore it entreth into it selfe, and findeth it self through Nature.[113]

Nature becomes the foundation (*Grund*), albeit a rather seismic one. But the Nothing, as an undetermined blind force of Will, is perfect freedom. And yet that freedom, that 'Eye', demands the mirror of self-revelation. This yearning desire is a universal force and Böhme gives it the name *Abgrund* – the source of life in Anguish (*Angst*).[114]

Thus Böhme develops a vision of tremendous energies, qualities, and potencies, which emerge from the primordial, and of the blind search which is the emergence of God. The link with Schelling's development of a notion of a striving unconscious, and the ideas of 'lack' and 'yearning', are already there. These same drives are to be found within the individual and all phenomena, although in Böhme the sheer affect is at a greater level, and potencies such as the Cabiri take the form of Qualities. If we seek for the ground or origin of the notion of the unconscious, then perhaps it may be found in Böhme's *Ungrund*, even if this immense proto-concept had not yet been fully developed in cultural terms.

The will, the sublime, the uncanny and the tremendum

In Böhme, too, we find the direct ancestor of Arthur Schopenhauer (1788–1860), and the idea that behind all creation and destruction of phenomena lies the turbulent Will. For Schopenhauer, Böhme offered a model of the development of creation, and creatures; a development through a universal *principium individuationis*, which was the same for Man, the Cosmos, and God.

Here, as in Schelling, a possibility is opened up to penetrate beyond the horizon of the Kantian categories to the noumenal which underlies both subject and object – the depths of the self and the thing-in-itself (*Ding-an-sich*). Phenomena composed the great book of nature which revealed an under-meaning, the *hyponoia*, of things. Through them, in Böhme, one can trace the lineaments of God and discover what prevailed before God and creation. Schopenhauer sees in Nature the manifold grades and modes in which the Will manifests itself.[115] Whereas Böhme's Will, the yearning of origin, works towards self-knowledge, and *Angst*, Wrath, Bitterness, and Evil are the engines which energise the quest for rest, redemption, wisdom and love, Schopenhauer's Will sweeps all things on to their ultimate destruction – creating and devouring individuals in a swirling vortex of creation and obliteration: *Apparent rari, nantes in gurgite vasto* ('Singly they appear, swimming by in the vast waste of waves').[116]

Schopenhauer's pessimistic sublime overwhelms the final distinctions of Nature, and ultimately of rational philosophy. But even in Kant's fine apparatus of the mind we can discern the tremors of a stirring affect. In an early work,

Observations on the Feeling of the Beautiful and Sublime (1764), Kant considers Burke's essay *A Philosophical Enquiry into the Origin of our Ideas of the Sublime and the Beautiful* (1757), along with its raging storms, horror, soaring mountains, and Milton's portrayal of Hell. Kant carefully categorises the sublime into the terrifying sublime; great depths and chasms, and the noble, great heights and the 'splendid', as seen in vast monuments such as the pyramids. In his later third Critique, the *Critique of Judgement (Kritik der Urteilskraft)* (1790), the sublime exceeds the boundaries of cognition and so 'to contravene the ends of our power of judgement'(§23). The sublime is 'an outrage of the imagination', and, although he is discussing aesthetics and not theology, Kant like Böhme makes a link between sublime greatness and suffering; for the pleasure of the sublime is 'only possible through a medium of pain' (§27).[117]

The might of Nature is, for Kant, an example of the 'dynamical sublime', being superior to all hindrances. This excites fear, and we can only appreciate sublimity from a safe vantage when we observe

> [b]old, overhanging, and as it were threatening rocks; clouds piled up in the sky, moving with lightening flashes and thunder peals; volcanoes in all their violence of destruction; hurricanes with their track of devastation; the boundless ocean in a state of tumult; the lofty waterfall of a mighty river, and such like.
>
> (§28)

In the same section, Kant engages in a discussion of the sublimity of war and conflict, as a contrast to peace, which – though it brings peace and a predominant commercial spirit – can also debase mankind, bringing about 'low selfishness, cowardice and effeminacy'. From this, he immediately turns to the fear of God and the outbursts of his wrath appearing in the tempest, the storm, and the earthquake. These operate on man as a moral lesson, and he is awakened to a stern judgement on his own faults. Kant's own judgement quickly re-asserts its authority, and he turns to the abstraction of the mathematical sublime and a discussion of the beautiful, which is of such importance in Neoplatonism, but less so in the primordial sublime of Böhme and Schelling. A similar shift occurs, it would seem, in the later Schelling, where there seems to have been a retrenchment, as he sought to contain the energies of the *Frühromantiker* of Jena.

In the early and middle writings on *Naturphilosophie* and *Identitätsphilosophie* there is an underlying conception of the unconscious, an eternal unconscious (*ewig Unbewußtes*) which exceeds the object/subject distinction. In a work of 1809, *On the Nature of Human Freedom* (*Über das Wesen der menschlichen Freiheit*), Schelling identified the irrational principle (*das irrationale Prinzip*):

> After the eternal act of divine self-revelation, everything in the world as we now see it is according to rule, order, and form. But still there always lies in the ground (*im Grunde*) that which is unruly (*das Regellose*), as if it could someday break through again.[118]

Madness is divine and a source of freedom, available for the higher individual to use, although it must be carefully kept in check, as it represents the 'inner-laceration of Nature', and so the negative force of constraint must not be entirely abandoned.[119]

In his much later work, *Philosophy of Mythology* (*Philosophie der Mythologie*) (1842), Schelling returns once again to the primordial consciousness (*Urbewußtein*), the *Urreligion* of the most primitive mythologies. But, like Goethe, Schelling now seems to favour an Olympian classical 'Homeric clarity' over the archaic mystery religions. Indeed, in his treatment of the two faces of Dionysos, at once savage and benign, he seems to favour a discreet veiling. In a celebrated passage from this work, Schelling anticipates Nietzsche's *The Birth of Tragedy* and Erwin Rohde's *Psyche*, and he develops the quality of the 'uncanny' (*das Unheimliche*). Schelling suggests there are some things which are too unsettling, so freedom comes from these forces being held in check:

> Greece had a Homer precisely because it had mysteries, that is because it succeeded in completely subduing that principle of the past, which was still dominant and outwardly manifest in the Oriental systems, and in pushing it back into the interior, that is, into secrecy, into the Mystery (out of which it had, after all, originally emerged). That clear sky which hovers above the Homeric poems, that ether which arches over Homer's world, could not have spread itself over Greece until the dark and obscure power of that uncanny principle which dominated earlier religions had been reduced to the Mysteries (all things are called uncanny which should have remained a secret, hidden, latent, but which have come to light); the Homeric age could not contemplate fashioning its purely poetic mythology until the genuine religious principle had been secured in the interior, thereby granting the mind complete outward freedom.[120]

Freud cited this passage at the start of his essay 'The Uncanny' (*Das Unheimliche*) (1919), an analysis of E. T. A. Hoffmann's short story 'The Sandman' (*Der Sandmann*) (1816). In the course of his discussion, Freud associates the perception of anguish and dread with hidden fears of castration and conflict with the father that is kept down or repressed.[121]

A more positive although arguably more terrible account of affect, appeared two years before Freud's essay; namely, Rudolf Otto's study of *The Idea of the Holy* (*Das Heilige*) (1917), an exploration of 'the depths of the divine nature'.[122] Otto attempts to analyse how the numinous grips and stirs the human mind. At its most profound, it may be characterised as a *mysterium tremendum*, whose effects may be the deepest worship or intoxicated frenzy, spasms and ecstasy. Otto examines the quality of these feelings: awfulness, fear, majesty, the overpowering power, the shudder, energy and urgency. Much is reminiscent of the terrors of the sublime in Burke and the awakening *Ungrund* in Böhme. In fact, in a discussion of the numinous in Martin Luther, Otto talks about the 'non-rational energy' and 'awefulness of God' in Böhme – the ferocity, fierce wrath, and the Will which are all ideograms of the *tremendum*.[123] Thus, once more, we are brought to the immense force of the primordial as a religious experience; the irruption of forces which are at once irrational and transcendent.

Dark with excessive bright

In Plato's dialogue *Parmenides*, the young Socrates is interrogated by Parmenides about his notion of the Ideas. How is it possible, he asks, that an Idea may be present at once in a number of earthly manifestations? Is the Idea most like a sail that could cover the heads of all the crew of a ship at once? This may certainly be asked about the primordial, which is blacker than the sail of Theseus and whose coverage envelops the whole of Creation. It would seem that, in Pope's words, 'a universal darkness buries all',[124] and that the obscurity verges on the meaningless, were it not for the sublimity of the passions it arouses; the potencies which rend its veil, and ultimately illuminate it. Burke illustrated his discourse on the sublime with quotations from Milton and from the Book of Job; Milton, for his visions of chaos and the infernal regions, and Job, for his terrors and sufferings.[125] As, for example, a passage in Job which he describes as 'amazingly sublime', principally due to the terrible uncertainty of what is described:

> In thoughts from the visions of the night, when deep sleep falleth upon men, fear came upon me and trembling, which made all my bones to shake. Then a spirit passed before my face. The hair of my flesh stood up. It stood still, *but I could not discern the form thereof*; an image was before mine eyes; there was silence; and I heard a voice, – Shall mortal man be more just than God?
> (*Job*, 4: 13–17 [Burke, 63; Burke's italics])

In his *Answer to Job* (1952), Jung grapples with the dangerous, wrathful side of God, the non-rational energy which, at its worst in the Book of Revelation, becomes 'a veritable orgy of hatred, wrath, vindictiveness, and blind destructive fury'.[126] The wrath of God is tempered and moralised through his relationship to Job, to Man. Indeed, Yahweh, 'an archaic god',[127] following the pattern of the development we have seen in Böhme and Schelling, emerges into consciousness and differentiation through Nature and Man. He *needs* conscious Man, though he resents that consciousness; he is no friend to critical thought, but he needs the tribute of recognition, for without Man there would be 'a withdrawal into hellish loneliness and the torture of non-existence, followed by a gradual reawakening of an unutterable longing for something which would make him conscious of himself'.[128]

For Man, this is a perilous business, exposed to omnipotent 'hatred, wrath, vindictiveness', and so Man must, in his encounters with archaic unconsciousness, always favour the light: 'The encounter between conscious and unconscious has to ensure that the light which shines in the darkness is not only comprehended by the darkness, but comprehends it.'[129] The light, too, can be overpowering:

Burke turns to Milton to describe the terrible beauty, the overwhelming light of God; a 'light which by its very excess is converted into a species of darkness', so that: 'Dark with excessive light thy skirts appear' (Burke, 80).[130]

In the primordial, darkness and light are one and yet to be manifest as separate. Damascius, in *De principiis*, says that '[o]f the first principle the Egyptians said nothing, but celebrated it as a darkness beyond all intellectual conception, a thrice-unknown darkness'.[131] It was observed of this darkness that it was the 'brilliancy of the primal veil' which was too strong even for spiritual sight.[132]

Another Hellenistic philosopher, Porphyry, in his commentary on book 12 of the *Odyssey*, entitled 'Concerning the Cave of the Nymphs', says that the ancients consecrated the cave as 'the representative of every invisible power: because as a cave is obscure and dark, so the essence of these powers is unknown'.[133] However, these powers are formative, and so the cave is the way to greater light. At the climax of the Mysteries, after a period of subterranean wanderings, the light emerged from darkness and obscurity. This was, according to Plutarch, an anticipation of the entrance of the the newly dead into the underworld:

> Then [at the point of death] it [i.e., the soul] suffers something like what those who participate in the great initiations (τελεταί) suffer. Hence even the word 'dying' (τελευτᾶν) is like the word 'to be initiated' (τελεῖσθαι) [. . .]. First of all there are wanderings and wearisome rushings about and certain journeys fearful and unending (ἀτέλεστοι) through the darkness, and then before the end (τέλος) all the terrors – frights and trembling and sweating and amazement. But then one encounters an extraordinary light, and pure regions and meadows offer welcome, with voices and dances and majesties of sacred sounds and holy sights; in which now the completely initiated one (παντελής . . . μεμυηένος) becoming free and set loose enjoys the rite, crowned, and consorts with holy and pure men.[134]

Plutarch's play on words deriving from *telos* ('end') brings together the ideas of death, mystery, and initiation. If this move is applied to notions of the primordial, then the originating darkness has something deathly about it, but linked with a desire for a self-created initiation; it reminds us of the blind teleology of the Potencies. The experience of the initiate is the same as that of the emerging Godhead, an aspiration towards enlightenment, symbolised by the dazzling light of the torches, and of the redeemed hero of the Golden Ass, initiated in the nocturnal Mysteries of Isis, who sees 'the sun shining as if were noon'.[135] Indeed, something of this experience is, as we have seen, affirmed in Schelling in his treatise on the deities of Samothrace, when he describes the yearning and lack which characterises the origin of things as primarily a desire for the light:

> For it is the teaching of all peoples who counted time by nights that the night is the most primordial of all things in all Nature. [. . .] But what is the essence of night, if not lack, need, and longing? For this night is not darkness, nor the

Figure 2 Robert Fludd, 'The Creation of Light', *Utriusque cosmi maioris scilicet et minores metaphysica, physica atque technica historia,* Oppenheim: Hieronymus Galler for Johann Theodor de Bry, 1617, pt.1, 2.

enemy of light, but it is Nature looking forward to the light, the night longing for it, eager to receive it.

(DS, 18)

As in the Mysteries, there is a reversal in the depths of darkness which converts all to brilliant illumination; the unknown turns into knowledge, and obscurity becomes an epiphany.

Thus the primordial sublime is the coming of light out of darkness, not darkness as a diminution of the light. In *De visione Dei*, Nicholas of Cusa (1401–1464) considers Paul's remark in 1 Corinthians about how we now see 'through a glass, darkly', and places obscurity as a necessary stage on the path to light. If obscurity, cloud, and darkness are not experienced, then there can be no transcendence; and, in a passage which must surely give solace to all those who contemplate the original darkness, he offers the reassuring advice that 'the more densely the darkness is felt, the truer and closer is the approach, by virtue of this darkness, to the invisible light'.[136]

Notes

1 See A. E. Waite's annotation to 'Text One' in 'The Philosophy of Theophrastus Concerning the Generations of the Elements', book 1, in *The Hermetic and Alchemical Writings of Aureolus Philippus Theophrastus Bombast, of Hohenheim, called Paracelsus the Great*, ed. A. E. Waite, 2 vols, London: J. Elliot, 1894, vol. 1, *Hermetic Chemistry*, pp. 201–09 (p. 201).

2 J. L. Borges, 'The Mirror of Enigmas', in *Labyrinths: Selected Stories and Other Writings*, London: Penguin, 1970, p. 245. 'The Mirror of Ink' is included in J. L. Borges, *A Universal History of Infamy*, London: Allen Lane, 1973.

3 Edmund Burke, *A Philosophical Enquiry into the Origin of our Ideas of the Sublime and Beautiful* [1757], ed. J. Boulton, Oxford: Basil Blackwell, 1958, pp. 60, 58, 63; cf. p. xix. Henceforth cited as Burke, followed by a page reference.

4 The archaic and the primordial in terms of definition enjoy equal obscurity, and it is hard to say anything except that they are coeval. Jung, happily, brings the notions together in the first lines of 'Archaic Man' (1931): 'The word archaic means primal, original' (C. G. Jung, *Collected Works*, ed. Sir H. Read, M. Fordham, G. Adler, and W. McGuire, 20 vols, London: Routledge and Kegan Paul, 1953–1983, vol. 10, paras. 104–47 [here: para. 104]).

5 F. Nietzsche, *Human, All-Too-Human*, tr. M. Faber with S. Lehmann, Lincoln, NE: University of Nebraska Press, 1984, p. 31 (vol. 1, §218). For a discussion of this dread, see G. Hersey, *The Lost Meaning of Classical Architecture: Speculations on Ornament from Vitruvius to Venturi*, Cambridge, MA: MIT Press, 1988, to which the Nietzsche quotation serves as an epigraph; and A. Vidler, *The Architectural Uncanny: Essays in the Modern Unhomely*, Cambridge, MA: MIT Press, 1992.

6 G. W. F. Hegel, *The Phenomenology of the Mind* [*Phänomenologie des Geistes*] [1807], tr. J. B. Baillie, New York: Harper, 1967, p. 79. Hegel, it would seem, adapted Friedrich Schlegel's slightly less dramatic criticism of Fichte: 'In the dark all cats are grey'; see R. J. Richards, *The Romantic Conception of Life: Science and Philosophy in the Age of Goethe*, Chicago: University of Chicago Press, 2002, p. 151.

7 M. Eliade, *The Myth of the Eternal Return* [1954], new revised edn, ed. J. Smith, Princeton, NJ: Princeton University Press, 2005, p. 4.

8 G. Scholem, *Kabbalah*, New York: Dorset Press, 1974, p. 293.

9 Scholem, *Kabbalah*, p. 117.

10 *Cain: A Mystery*, Act II, Scene 2, ll. 51–62; Byron, *Complete Poetical Works*, ed. F. Page, rev. J. Jump, Oxford and New York: Oxford University Press, 1970, pp. 520–45 (p. 532).

11 *Cain*, Act II, scene 2, ll. 80–84 (p. 532).

12 See W. D. Brewer, *The Shelley-Byron Conversation*, Gainesville, FL: University of Florida Press, 1994.

13 *Prometheus Unbound*, Act IV, ll. 296–302; *Shelley's Poetry and Prose*, ed. D. H. Reiman and S. B. Powers, New York and London: Norton, 1977, pp. 132–210 (p. 202).

14 See Z. Zlatar, *Njegoš's Montenegro: Epic Poetry, Blood Feud and Warfare in a Tribal Zone 1830–1851*, Boulder, CO: East European Monographs, 2005.

15 M. Djilas, *Njegoš: Poet, Prince, Bishop*, tr. M. B. Petrovich, New York: Harcourt Brace, 1966, pp. 13–14. Henceforth cited as Djilas, followed by a page reference.

16 Njegos 'wearily' explained to an English visitor, Sir Gardiner Wilkinson, in 1845 that the heads could not be removed as Turks would think they were weakening and invade. Wilkinson later discovered the similar tower when he visited Herzogovina to see the Vizier of Mostar, decorated this time with Montenegrin heads. Wilkinson's account was published in *Blackwood's Magazine*, vol. 65, no. 401, 18 February 1849, 212–18.

17 Njegoš' biographer, Milovan Djilas, observes that, in Montenegro, 'the taking of heads was a proof of heroism and victory': 'The severed head was the greatest pride and joy

of the Montenegrin. He regarded the taking of heads as the most exalted act and spiritual solace – having been nurtured in mythical history and the naked struggle for life. He did not feel any hatred for the cut off head, the hatred he was bound to feel for it in its live state, but only esteem and solicitude. He washed it, salted it, combed it. After all, it was a human head and the badge of his own highest merit' (Djilas, 245). Indeed, it was said in Montenegro that, 'had Adam cut off the heads of a couple of Angels, God would never have chased him out of Paradise, for he would have seen that he had created a hero not a good-for-nothing' (Djilas, 247).

18   Compare with the rather inconclusive discussion in Anica Savić-Rebac's 'Introduction', in P. P. Njegoš, *The Ray of the Microcosm*, tr. A. Savić-Rebac, Cambridge, MA: Harvard University Press, 1957, pp. 105–200. For an extensive discussion of the Bogomils and Eastern heresiology, see Y. Stoyanov, *The Other God: Dualist Religion from Antiquity to the Cathar Heresy*, New Haven: Yale University Press, 2000, esp. pp. 166–83.

19   See Stoyanov, *The Other God*, pp. 261ff; and D. J. Halperin, *The Faces of the Chariot: Early Jewish Responses to Ezekiel's Vision*, Tübingen: Mohr, 1988.

20   Canto 1, ll. 214–20; Njegoš, *The Ray of the Microcosm*, p. 157.

21   Canto 1, ll. 232–39; Njegoš, *The Ray of the Microcosm*, p. 157.

22   Canto 2, ll. 586–89; Njegoš, *The Ray of the Microcosm*, p. 167.

23   Canto 2, ll. 596–603; Njegoš, *The Ray of the Microcosm*, p. 167.

24   Canto 2, ll. 610–18; Njegoš, *The Ray of the Microcosm*, p. 167.

25   Canto 3, ll. 774–84; Njegoš, *The Ray of the Microcosm*, p. 172.

26   Canto 3, ll. 785–806; Njegoš, *The Ray of the Microcosm*, p. 172.

27   B. Lytton, *Athens, its Rise and Fall*, 2 vols, New York: Harper, 1852, vol. 1, p. 21.

28   See Herodotus, *The Histories (The Persian Wars)*; cited here from *The Landmark Herodotus*, tr. Andrea Parvis, ed. R. B. Strassler, London: Quercus, 2008, as Her. followed by reference to book and section.

29   Roberto Calasso organises his exploration of the Greek myths around a series of abductions; see R. Calasso, *The Marriage of Cadmus and Harmony*, tr. T. Parks, London: Cape, 1993.

30   For discussion of the relation of money, exchange, and women, see L. Kurke, *Coins, Bodies, Games, and Gold: The Politics of Meaning in Archaic Greece*, Princeton, NJ: Princeton University Press, 1999, esp. chapter 6, 'Herodotus's Traffic in Women'.

31   Hayden White, 'The Value of Narrativity in the Representation of Reality', *Critical Inquiry*, vol. 7, no. 1 (Autumn 1980), 5–27 (p. 5).

32   White, 'The Value of Narrativity', p. 16; citing Hegel, *The Philosophy of History*, tr. J. Sibree, New York: Dover, 1956, pp. 60–61.

33   For a discussion of this passage and the borders between myth and history, see P. Veyne, *Did the Greeks Believe in their Myths? An Essay on the Constitutive Imagination [Les Grecs ont-ils cru à leurs mythes?]* [1983], tr. P. Wissing, Chicago: Chicago University Press, 1988, p. 12.

34   See Plutarch's *Moralia*, Athenaeus's *Deipnosophists*, or Macrobius's *Saturnalia*.

35   Pausanias, book 8, 'Arcadia', chapter 1, §4 (Pausanias, *Description of Greece*, tr. W. H. S. Jones, vol. 3, London; New York: Heinemann; Putnam, 1933, p. 349).

36   See C. Kerényini, 'The Mysteries of the Kabeiroi' [1944], in *The Mysteries: Papers from the Eranos Yearbooks*, tr. R. Manheim, Princeton, NJ: Princeton University Press, 1955, pp. 32–63. See also M. B. Sakellariou, *Peuples préhelléniques d'origine indo-européenne*, Athens: Ekdotiké Athenon, 1977, pp. 101–04.

37   Guillaume Baron de Sainte-Croix, *Memoires pour servir à histoire de la religion secrète des anciens people; ou recherches historiques sur les mystères du paganisme* (1784); and Jörgen Zoega, *Vorlesungen über die Griechische Mythologie* (1817). For

fuller discussion, see Robert F. Brown's commentary in *Schelling's Treatise on 'The Deities of Samothrace': A Translation and an Interpretation*, Missoula, MT: Scholars Press, 1976, pp. 8–9 (henceforth cited as DS, followed by a page reference). For an extensive account of the disputes of mythic origin in early nineteenth-century German classicism, see George S. Williamson, *The Longing for Myth in Germany: Religion and Aesthetic Culture from Romanticism to Nietzsche*, Princeton, NJ: Princeton University Press, 2004, esp. chapter 3.

38 G. F. Creuzer, *Symbolik und Mythologie der alten Völker, besonders der Griechen*, 4 vols, Leipzig and Darmstadt: Heyer und Leske, 1918–1812; second edn, 1819–1821. Cited henceforth (from the second edition) as Creuzer, followed by a volume and page reference.

39 For an interesting linguistic analysis of Creuzer's views of the symbol, allegory, and metaphor, see T. Todorov, *Theories of the Symbol*, tr. C. Porter, Oxford: Basil Blackwell, 1982.

40 Plato, *Cratylus*, tr. C. D. C. Reeve, in *Plato: Complete Works*, ed. J. M. Cooper, Indianapolis: Hackett, 1997, p. 115.

41 Book 1, chapter 4; Proclus, *On the Theology of Plato*, tr. T. Taylor [1816], Frome: Prometheus Trust, 1995, p. 60.

42 See R. T. Wallis, *Neoplatonism*, London: Duckworth, 1972, p. 5.

43 The secrets of the Mysteries proved irresistible to the German Romantic imagination; see, for example, Novalis's *The Apprentices of Sais (Die Lehrlinge zu Sais)* (1802).

44 'To Demeter', ll. 480–84; Hesiod, *The Homeric Hymns and Homerica*, tr. H. G. Evelyn-White, Cambridge, MA; London: Harvard University Press; Heinemann, 1982, p. 323. See also H. P. Foley, *The Homeric Hymn to Demeter: Translation, Commentary and Interpretive Essay*, Princeton, NJ: Princeton University Press, 1994, pp. 26–27.

45 'Two adjectives, *aporrheta* ('forbidden') and *arrheta* ('unspeakable'), seem to be interchangeable in this usage [. . .]' (W. Burkert, *Ancient Mystery Cults*, Cambridge, MA: Harvard University Press, 1987, p. 9).

46 See E. Wind, *Pagan Mysteries in the Renaissance*, London: Faber and Faber, 1958, p. 11. Wind develops this theme at length in this book. See also D. C. Allen, *Mysteriously Meant: The Rediscovery of Pagan Symbolism and Allegorical Interpretation in the Renaissance*, Baltimore and London: Johns Hopkins Press, 1970.

47 See, too, the later discussion of this derivation in Kerényi, 'The Mysteries of the Kabeiroi', in *The Mysteries*, pp. 48–49.

48 A clear link may be made here with the dwarf Tages who emerged from under the plough of an augur and dictated the *libri Tagetici*, the Roman/Etruscan ritual books that dealt with omens, rites, hallows, and appeasing the gods. See J. Rykwert, *The Idea of a Town: The Anthropology of Urban Form in Rome, Italy and the Ancient World*, Cambridge, MA: MIT Press, 1988, pp. 29–30.

49 See the interesting discussion of Ptah in relation to the dwarf figures carried by Phoenician sailors and the 'black dwarves' of Teutonic mythology, in D. A. MacKenzie, *Egyptian Myth and Legend*, London: Gresham, [1913], pp. 80–81. For a discussion of the 'primeval mound' and the emergence of creation, see R. T. Rundle Clark, *Myth and Symbol in Ancient Egypt*, London: Thames and Hudson, 1959, pp. 36 and 40ff.

50 T. Ziolkowski, *German Romanticism and its Institutions*, Princeton, NJ: Princeton University Press, 1990, pp. 18–63.

51 R. M. Rilke, 'Orpheus. Eurydice. Hermes' (from *Neue Gedichte*, part 1 [1908]), in Rilke, *Selected Poems*, tr. J. B. Leishman, Harmondsworth: Penguin, 2001, pp. 39–42 [p. 39]).

52 E. T. A. Hoffmann, 'The Mines of Falun', tr. A. Ewing, in E. F. Bleiler (ed.), *The Best Tales of Hoffmann*, New York: Dover, pp. 285–307 (p. 290).

53 Novalis, *Heinrich von Ofterdingen*, tr. P. Hilty, New York: Ungar, 1972, p. 86.

54 For further details and pictures, see K. Lehman, *Samothrace: A Guide to the Excavations and the Museum* [6th edition, rev. J. R. McCredie], Thessaloniki: Institute of Fine Arts, New York University, 1998, p. 30.

55 For a modern description of the Sanctuary, see S. G. Cole, *Theoi Megaloi: The Cult of the Great Gods at Samothrace*, Leiden: Brill, 1984.

56 Cole, *Theoi Megaloi*, pp. 2–3.

57 Scholiast on Apollonius Rhodius Mnaseas of Patrae (*Schol* A. R. I, 917).

58 A. Phelan, 'The Classical and the Medieval in *Faust II*', in P. Bishop (ed.), *A Companion to Goethe's 'Faust' Parts I and II*, Rochester, NY: Camden House, 2001, pp. 144–68 (p. 160). In addition to Creuzer and Schelling, Goethe drew on Benjamin Hederich's *Gründliches mythologisches Lexikon* (1770).

59 Kerényi discusses these connections in his essay on 'The Mysteries of the Kabeiroi', in *The Mysteries*, pp. 32–63. He explores the Hermes/phallus connection of the Cabiri in *Hermes Guide of Souls: The Mythologem of the Masculine Source of Life*, tr. M. Stein, Zurich: Spring Publications, 1976, pp. 66–67.

60 Goethe, *Faust II*, ll. 8075–77; partly adapted from Alice Raphael's translation in her *Goethe and the Philosophers' Stone: Symbolical Patterns in 'The Parable' and the Second Part of Goethe's 'Faust'*, London: Routledge, 1965, p. 176.

61 Goethe, *Faust II*, ll. 8202–05; partly adapted from Raphael, *Goethe and the Philosophers' Stone*, p.176.

62 Cited in Raphael, *Goethe and the Philosophers' Stone*, p. 176. As Raphael points out, in the same year Goethe went on to write his great poem 'Primal Words. Orphic' (*Urworte. Orphisch*).

63 E. Benz, 'Theogony and the Transformation of Man in Friedrich Wilhelm Joseph Schelling' [1954], in *Man and Transformation: Papers from the Eranos Yearbooks*, tr. R. Manheim, Princeton, NJ: Princeton University Press, 1964, pp. 203–49 (pp. 204–05).

64 R. J. Richards, *The Romantic Conception of Life: Science and Philosophy in the Age of Goethe*, Chicago: University of Chicago Press, 2002, p. 127.

65 Novalis, note to the Jena Circle, cited in Novalis, *Notes for a Romantic Encylopaedia: 'Das Allgemeine Brouillon'*, ed. and tr. D. Wood, Buffalo, NY: State University of New York Press, p. xi. These are the true 'coasts of light' like those explored by Hyperion in Hölderlin's novellea *Hyperion*, and Goethe's rocky coast of the Aegean Sea.

66 See F. W. J. Schelling, *Ideas for a Philosophy of Nature*, tr. E. E. Harris and P. Heath, Cambridge: Cambridge University Press, 1988, p. 42.

67 The material world is seen as a sort of knot of forces: 'One thinks of a stream, which itself is pure identity; where it meets some resistence, it forms an eddy, which is not an object at rest, but with each moment it disappears and then re-establishes itself' (Schelling, *First Sketch of a System of Nature Philosophy* (1799); cited in Richards, *The Romantic Conception of Life*, p. 145). One is reminded, too, of the aphorisms of the Romantic physician and physicist Johann Wilhelm Ritter (1776–1810), published as *Posthumous Fragments of a Young Physician* (*Fragmente aus dem Nachlass eines jungen Physikers*) (1810). 'Each stone comes into existence anew each moment, continuing to generate itself through all infinity'; 'All bodies are petrified electricity'; and 'The light appearing in the process of combustion is, as it were, a hole into another world' (cited in Eudo C. Mason, 'The Aphorism', in S. Prawer (ed.), *The Romantic Period in Germany*, London: Weidenfeld and Nicholson, 1970, pp. 204–34 [pp. 225–26]).

68 Schelling also uses *Dio*, a title by which Demeter (or *Dea*) is called throughout the rest of the Hymn. This is an important link, in that Demeter may come from Deo-Mater (i.e., god/mother).

69 See A. Nygren, *Agape and Eros*, rev. edn, London: SPCK, 1953.

70 Pothos (Πόθος) has a broader meaning, as in Aristotle's 'Hymn to Virtue (*Arete*)' where *pothos* is the longing of heroes for wisdom and virtue, a longing which brought Achilles and Ajax to the House of Hades. Alexander most certainly knew this poem, written by his tutor, and Arrian in his *Life of Alexander* frequently uses *pothos* to characterise the drive which energised the Conqueror's ambition for great deeds. See V. Ehrenberg, *Alexander and the Greeks*, Oxford: Blackwell, 1938, as well as J. M. O'Brien, *Alexander the Great: The Invisible Enemy*, London: Routledge, 1992, p. 50: Pothos 'displays his longing for things not yet within reach, for the unknown, far distant unattained'.

71 The three positions are part of a debate in antiquity as to whether the universe was created, a matter of accident, or eternal; see D. Sedley, *Creationism and its Critics in Antiquity*, Berkeley: University of California Press, 2007.

72 Namely, the Intelligible-Intellectual [Noëtic-noëric], the Intellectual [Noëric], the Supramundane [Supercosmic], the Liberated, and the Mundane [Cosmic]; cf. Proclus, 'Commentary on the *Parmenides* of Plato', tr. T. Taylor, cited in *The Thomas Taylor Series*, 33 vols, Frome: Prometheus Trust, 1994, vol. 5, *Hymns and Initiations*, p. 315. Matter is the least spiritual and marks the nadir of the emanation process, following this the movement is a return, a path upwards through the many levels towards the One; *katabasis* followed by *anabasis*.

73 Proclus, *On the Theology of Plato*, iii, 14; cited in *Hymns and Initiations* [*The Thomas Taylor Series*, vol. 5], p. 315.

74 W. Leibbrand, 'Schellings Bedeutung für die moderne Medizin', *Atti del XIV° Congresso Internazionale di Storia della Medicina* (Rome: 1954), vol. 2, pp. 891–3, cited in H. F. Ellenberger, *The Discovery of the Unconscious: The History and Evolution of Dynamic Psychiatry*, London: Allen Lane, 1970, p. 204. Ellenberger suggests that a characteristic of the Schelling tradition is a search for polarities everywhere, though this would be a vast oversimplification of Schelling's ideas on the Potencies, and would pass over the importance of triads in his work; Schelling constantly seeks to break down the tyranny of binary opposition. In Jung's text, Schelling and Böhme appear only in a footnote and as an illustrative figure.

75 Cicero, *Tusculan Disputations*, book 4, chapter 6, §12; cited in Jung, *Symbols of Transformation* [*Collected Works*, vol. 5], para. 185.

76 The transformation of sexual to creative psychic energy is a mirror image of Plotinus' elegant explanation of why Herms have phalluses – it is because all generation proceeds from the mind (Third *Ennead*, sixth tractate, §19).

77 Theurgy (Θεοργία) has a number of interpenetrating meanings, all of which seem relevant. The term was apparently coined by the Younger Julian, one of the two editors of the *Chaldean Oracles* during the reign of Marcus Aurelius, signifying 'divine work', the making of individuals into gods, the elevation of individuals from the realm of creation to the divine by active effort using the powers of images, hymns, meditation, contemplations of the gods and employing the assistance of semi-divine Dæmons who mediated between the earthly and the transcendent (see J. H. Lewy, *Chaldaean Oracles and Theurgy: Mysticism, Magic, and Platonism in the later Roman Empire*, Cairo: L'Institut Français d'Archéologie, 1956, pp. 461–66). The great surviving work of antiquity on theurgy is that of Iamblichus, entitled *On the Mysteries of the Egyptians: The Answer of the Priest Abammon* (c. 300 CE).

78 C. G. Jung, *The Red Book: Liber Novus*, ed. S. Shamdasani, tr. M. Kyburz, J. Peck, and S. Shamdasani, New York and London: Norton, 2009, p. 320.

79 Jung, *The Red Book*, pp. 320–21.

80 Goethe, *Faust II*, ll. 8075–77.

81 Novalis, *Notes for a Romantic Encyclopaedia*, tr. Wood, p. xvi.

82 Novalis, 'Freiberg Natural Scientific Studies' (1798–99), No. 50, in Novalis, *Notes for a Romantic Encyclopaedia*, tr. Wood, p. 199.

83 F. W. J. Schelling, *Ages of the World* [*Die Weltalter*], tr. J. Norman, published in conjunction with S. Žižek as *The Abyss of Freedom/Ages of the World*, Ann Arbor: University of Michigan Press, 2005 (henceforth cited as W, followed by a page reference). Schelling begins his own text with a confident statement of deictic ontology: 'The past is known [*gewußt*], the present is recognized [*erkannt*], the future is divined [*geahndet*]' (W, 113).

84 Schelling is clearly referring to Psalm 118: 22, and perhaps to the notion of the Stone of the Philosophers, whose initial ordinariness means that it is *spernitur a stultus*. There is also a strong link with Jung and the cubical stone at Bollingen, which he inscribed with a quotation from Arnaldus de Villanova, and to the *lapis exilis* (see Jung's note to 'Dream 32' in *Psychology and Alchemy* [*Collected Works*, vol. 12], para. 246, n. 125 [p. 180]).

85 See P. Frizsche, *Stranded in the Present: Modern Times and the Melancholy of History*, Cambridge, MA: Harvard University Press, 2004, whose epigraph, 'And only where there are tombs are there resurrections', is taken from Nietzsche's *Thus spoke Zarathustra*.

86 Here, Schelling quotes Luther's translation of Ecclesiastes, perhaps thinking of the toils involved in his present work: 'All works that are done under the sun are full of vexation, the sun ariseth, and the sun goeth down, only to rise up and go down again, and all things are full of labour yet do not tire, and all forces ceaselessly labour and struggle against one another' (W, 124).

87 For a discussion of the later development of Schelling's concept of the Potencies, see E. A. Beach, *The Potencies of God(s) in Schelling's Philosophy of Mythology*, Albany, NY: State University of New York Press, 1994, pp. 116–29.

88 See *System of Transcendental Idealism* (*System des transzendentalen Idealismus*) (1800), part 6, §3; in F. W. J. Schelling, *Texte zur Philosophie der Kunst*, ed. W. Beierwaltes, Stuttgart: Reclam, 1982, p. 121.

89 Coleridge, who was a notorious connoisseur of cultural influence, said in his *Philosophical Lectures* that it was a puzzle to enter into any account of Schelling, and that he would have to refer his audience to a host of sources, one of which, of course, could be himself, and notoriously vice versa – the 'borrowings' of *Biographia Literaria*. See the lengthy discussion of Schelling and Coleridge's originality in T. McFarland, *Coleridge and the Pantheist Tradition*, Oxford: Clarendon Press, 1969, pp. 29–35. Paul Tillich argued that the 'supposed fathers of Schelling's thought, Spinoza, Plato, Böhme, Hegel, Aristotle, & c°, were certainly influences, but also part of the "inner progress" of Schelling's development', which led him into proximity with these philosophers, from whom he adopted 'homogenous elements' (Tillich, *Mystik und Schuldbewußtein in Schellings philosophischen Entwicklung* [1912], cited in McFarland, p. 31). Much the same might be said about Jung and Schelling, but this region of thought generates the deepest defence. Jung agues in *Modern Man in Search of a Soul* (1933) that ideas are 'never the personal property of their so-called author', and he continues: 'They arise from that realm of procreative psychic life out of which the ephemeral mind of the single human being grows like a plant and blossoms, bears fruit and seed, and withers and dies' (cited in McFarland, p. 30). Clearly, the *Weltgeist* waives its copyright.

90 See G. L. Plitt, *Aus Schellings Leben*, vol. 1, pp. 245–47; cited in McFarland, *Coleridge and the Pantheist Tradition*, p. 35.

91 Jacob Böhme, *Aurora: That is, the Day-Spring: Or Dawning of the Day in the Orient or Morning Redness in the Rising of the Sun* [1612], tr. J. Sparrow [1656], ed. C. J. Barker and D. S. Hehner, London: Watkins, 1914, p. 486–87 (chapter 19, §8).

92 Böhme, *Aurora*, p. 488 (chapter 19, §11).

93 Böhme, *Aurora*, p. 488 (chapter 19, §13).

94 Böhme, *Aurora*, p. 40 (chapter 1, §4). See the discussion of this disturbing etymology in R. F. Brown, *The Later Philosophy of Schelling: The Influence of Böhme on the Works of 1809–1811*, Lewisberg, PA: Bucknell University Press, 1977, p. 39.

95 Böhme, *Aurora*, pp. 79–80 (chapter 3, §74).

96 Böhme, *Aurora*, p. 92 (chapter 4, §25).

97 Böhme, *Aurora*, p. 92 (chapter, §26).

98 Jacob Böhme, *Mysterium Magnum: Part One: An Exposition of the First Book of Moses called Genesis* [1623], tr. J. Sparrow, London, H. Blunden at the Castle Cornhill, 1654, p. 2 (chapter 1, §8).

99 Böhme, *Mysterium Magnum*, p. 6 (chapter 3, §5).

100 Böhme, *Mysterium Magnum*, p. 6 (chapter 3, §5). Böhme seems to envision gravity before gravity, and certainly Newton knew his works. Newton saw gravity as the unifying power of Christ's body in creation, and through the power of gravity everything, however small, exerts a degree of force on every other particle.

101 Böhme, *Mysterium Magnum*, p. 7 (chapter 3, §9).

102 Böhme, *Mysterium Magnum*, p. 7 (chapter 3, §11).

103 Böhme, *Mysterium Magnum*, p. 7 (chapter 3, §10)

104 Böhme, *Mysterium Magnum*, p. 7 (chapter 3, §11). The idea of the 'Lubet' is developed at length in the initial chapters of *Mysterium*.

105 Jung, *Symbols of Transformation* [*Collected Works*, vol. 5], para. 188 (p. 131).

106 Böhme, *Mysterium Magnum*, p. 7 (chapter 3, §12).

107 Böhme, *Mysterium Magnum*, p. 7 (chapter 3, §12).

108 Böhme, *Mysterium Magnum*, p. 8 (chapter 3, §16).

109 Böhme, *Mysterium Magnum*, p. 8 (chapter 3, §16).

110 Böhme, *Mysterium Magnum*, p. 36 (chapter 10, §5).

111 Jacob Böhme, *Signatura Rerum or the Signature of all Things, shewing the Sign and Signification of the Severall Forms and Shapes in the Creation* [1622] tr. J. Ellistone, London: Calvert, 1651, p. 5 (chapter 2, §1).

112 Böhme, *Signatura Rerum*, p. 6 (chapter 2, §8).

113 Böhme, *Signatura Rerum*, p. 13 (chapter 3, §2).

114 For a fuller discussion, see Brown, *The Later Philosophy of Schelling*, pp. 54–56.

115 See Schopenhauer's note in *The World as Will and Representation* [1819; 1844], vol. 1, §45: 'Jacob Böhme in his book *De Signatura Rerum*, chap. I, §§ 15, 16, 17, says: "And there is no thing in nature that does not reveal its inner form outwardly as well; for the internal continually works towards revelation [. . .] Each thing has its mouth for revelation. And this is the language of nature in which each thing speaks out of its own property, and always reveals and manifests itself [. . .] For each thing reveals its mother, who therefore give the *essence and the will* to the form"' (Schopenhauer, *The World as Will and Representation*, tr. E. F. J. Payne, 2 vols, New York: Dover, 1969, p. 220).

116 Schopenhauer, *The World as Will and Representation*, vol. 1, §49 (p. 236, fn. 36), alluding to Virgil, *Aeneid*, chapter 1, l. 118.

117 I. Kant, *Critique of Judgement*, tr. J. H. Bernard [1892], London: Hafner Press, 1959 (cited with section number). See, too, P. Crowther, *The Kantian Sublime: From Morality to Art*, Oxford: Oxford University Press, 1989.

118 *On the Nature of Human Freedom*, cited by Beach in *The Potencies of God(s)*, p. 53.

119 See Sonu Shamdasani's discussion of this tradition of divine madness which leads to Schelling and is evident in the '*Liber Primus*' of Jung's *Red Book* (fn. 89, p. 238).

120 F. W. J. Schelling, *Philosophie der Mythologie* [1842], 2 vols, Darmstadt: Wissenschaftliche Buchgesellschaft 1966, vol. 2, p. 649; translated by E. R. Miller, and cited in A. Vidler, *The Architectural Uncanny: Essays in the Modern Unhomely*, Cambridge, MA: MIT Press, pp. 26–27.

121 S. Freud, *The Standard Edition of the Complete Psychological Works*, ed. J. Strachey and A. Freud, 24 vols, London: Hogarth Press, 1953–1974, vol. 17, pp. 217–55. It is interesting to note that Ernst Bertram once suggested that Kant's *Critique of Pure Reason* wants 'to delimit and protect a τέμηηοζ, a sacred grove, a prohibited grove'

(see the section 'Eleusis', in E. Bertram. *Nietzsche: Attempt at a Mythology*, tr. R. E. Norton, Chicago, University of Illinois Press, 2009, p. 292).

122 R. Otto, *The Idea of the Holy: An Inquiry into the Non-Rational Factor in the idea of the Divine and its Relation to the Rational*, tr. J. W. Harvey, Oxford: Oxford University Press, 1950, p. xxi (Foreword to the first English edition of 1923, based on the 9th German edition).

123 Otto, *The Idea of the Holy*, p. 107.

124 A. Pope, *The Dunciad*, book 4, l. 656.

125 It is interesting to note that Milton was deeply affected by Böhme; see M. L. Bailey, *Milton and Jakob Böhme: A Study of German Mysticism in Seventeenth-Century England*, New York: Oxford University Press, 1914.

126 Jung, *Answer to Job*, in *Collected Works*, vol. 11, paras 560–758 (pp. 365–470) (here: para. 708 [p. 438]).

127 Jung, *Collected Works*, vol. 11, para. 571 (p. 371).

128 Jung, *Collected Works*, vol. 11, para. 575 (p. 373).

129 Jung, *Collected Works*, vol. 11, para. 756 (p. 468).

130 Burke misquotes Milton's *Paradise Lost*, book 3, l. 380: 'Dark with excessive *bright* thy skirts appear.'

131 Cited in G. R. S. Mead, *Orpheus: The Theosophy of the Greeks*, London: Theosophical Publishing Society, 1896, p. 93.

132 Mead, *Orpheus*, pp. 93–94.

133 Porphyry, *Concerning the Cave of the Nymphs*, tr. T. Taylor, in K. Raine and G. M. Harper (eds), *Thomas Taylor the Platonist: Selected Writings*, Princeton, NJ: Princeton University Press, 1969, pp. 297–342 (p. 301).

134 Plutarch, *On the Soul*, fr. 178, cited in K. Clinton, 'Stages of Initiation in the Eleusinian and Samothracian Mysteries', in M. B. Cosmopoulos (ed.), *Greek Mysteries: The Archaeology and Ritual of Ancient Greek Secret Cults*, London and NY: Routledge, 2003, pp. 50–78 (p. 66).

135 Apuleius, *The Golden Ass*, tr. R. Graves, Harmondsworth: Penguin, p. 286 (chapter 28).

136 Cusanus, *De visione Dei*, chapter 6, 'De visione faciali', cited in E. Wind, *Pagan Mysteries*, p. 221.

Chapter 5

The idea of the archaic in German thought

Creuzer – Bachofen – Nietzsche – Heidegger

Charles Bambach

> For the *arche* that sits enshrined as a goddess among mortals is the savior of all, provided that she receives the honor due to her from each one who approaches her.
>
> (Plato, *Laws*, 775e)

> Beginning at the beginning is always an illusion.
>
> (Nietzsche)[1]

Arche and the archaic

'Every myth is a myth of origin.'[2] In myth what is most essential lies in the departure from, and the return to, the originary. To speak of myth, then, is to give voice to the hope of recovering the origin – not merely as a kind of archaeological excavation of the distant past. Rather, it serves as a way of leaping out of the present in order to transform its future. Such is the power of the origin that within its source lies hidden the full phenomenological unfolding of a process. This is something that the earliest Greek philosophers understood in the force of a single word: *arche*. For them the *arche* holds within itself the potency of futurity. What will come to be can only come forth from what is – and what is, in turn, owes its existence to that which has come forth from the origin. The origin, as it were, rules over all that proceeds forth from it. We can uncover the roots of such an interpretation in the etymology of the Greek term *arche*, or 'origin', which derives from the verb *archein*, meaning 'to rule over' or 'to command'.[3] Seen from within this perspective, the *arche* holds the key to the destiny of a person, a people, a tradition; our very relation to the present turns upon our recovering the concealed power of the *arche*.

In what follows I want to trace the lineage of the German tradition's embrace of the myth of an *arche* as a way of rethinking the traditional neo–Humanist *Griechenbild* that dominated nineteenth-century German university culture. Starting in the eighteenth century with the work of J. J. Winckelmann and continuing on with the contributions of Herder, Lessing, Goethe, Humboldt, Schiller,

Fichte, Hölderlin, and others there emerges a generational faith that 'only through the Greeks can one pass into the heart of the Germans'.[4] As part of this faith, there develops a deep and abiding vision of ancient Greece as a model for German cultural identity. Following Winckelmann's claim that the path to German greatness lay in 'imitating the ancients [. . .] the Greeks in particular', this generation of German thinkers reconstitutes its image of German identity and its possibilities.[5] As the classicist Walther Rehm put it, 'within German cultural life the belief grew ever stronger that without the relation to Greek antiquity it would be impossible to shape one's own life and to attain a genuinely German form of humanity.'[6] For Winckelmann, this image of Greece was marked above all by a 'noble simplicity and tranquil grandeur' that brought with it a timeless serenity that could serve as a model of aesthetic perfection worthy of imitation.[7] Above all, Winckelmann found such an ideal in the marmoreal sculpture of the Greeks embodied in the Apollo Belvedere.[8]

Winckelmann's vision of Greek antiquity became decisive for the succeeding generation of German thinkers, especially Goethe, whose essay 'Winckelmann' attests to the enduring significance that he had for defining a vision of the Greeks as 'classical'.[9] Yet Winckelmann's classicism was deeply contradictory. On the one hand, he offered up a model of the Greeks as a timeless standard against which contemporary Germans were to measure themselves. On the other, he succeeded in historicizing ancient art and developing a system of classification that organized Greek painting, sculpture, and architecture according to its stylistic periods of 'origin, growth, transformation, and decline'.[10] It was this underlying tension within Winckelmann's Philhellenic vision between what the Greek legacy signifies for the Germans versus a historical understanding of Greek art on its own terms that came to shape the debate concerning the archaic in the nineteenth century.

Winckelmann's classicism comes to dominate so profoundly that its fundamental tenets decisively shape German Philhellenism both within the academic disciplines of archaeology, classical philology, law, history, and literature *and* within the aesthetic-philosophical-cultural world at large. Within the newly founded University of Berlin in 1809, Wilhelm von Humboldt formulated a plan of humanistic *Bildung* that sought to overcome what he saw as the rote, mechanical training of the older German pedagogy. What mattered for Humboldt was the inward, spiritual transformation of the individual through immersion in the study of classical antiquity. As the Prussian minister of cultural affairs, Humboldt was able to implement a plan that transformed both secondary and university education. Out of this *Bildungsideal* within the German Gymnasium there emerged a cadre of classically trained scholars who adapted the Humboldtian ideals of *Bildung* to their own professionalized standards of education. For several generations this new ideal dominating the study of antiquity would be rooted in both a positivist and an historicist form of *Altertumswissenschaft*. The effect of the *Bildungsideologie* spawned by this development was a new vision of Germanness which discovered that German originality lay in its unique affinity to the Greeks.[11] As Vassilis

Lambropoulos claims in *The Rise of Eurocentrism*, 'educated Germans saw themselves as the modern Greeks, the inheritors of ancient culture'.[12] It is both within and against this narrow scientific-nationalist construction of a German *Sonderweg* or 'singular path' that the idea of the archaic develops. Already in his notes for 'We Philologists', Nietzsche rebels against the narrowness of this *Bildungsideal*, arguing that it produces not the creative thinker or artist but '*the classicist castrated by objectivity* who, moreover, is a cultural philistine and *Kulturkämpfer* who dabbles in pure scholarship and is, to be sure, a sorry spectacle' (KSA 8, 5[109], 69).

Rejecting both the antiseptic version of Hellas concocted by the castrated philologists, as well as the phantasm of a serene and cheerful 'classical' Greece fashioned by Winckelmann and his successors, Nietzsche unearthed a radically different Hellas – one marked by dark, chthonic forces associated with the world of Dionysian myth rather than that of Phidias's luminous sculptural surfaces and Socrates' lucid syllogisms. In his 1872 work *The Birth of Tragedy* Nietzsche directs his critique against Philhellenic classicism and 'Winckelmann's dream of Apollonian purity', by excavating a *different* Greece, one that is archaic, rather than classical.[13] Winckelmann had understood the archaic stylistically as an aesthetic approach that preceded the classical and served as its precursor. For Nietzsche, however, the archaic came to symbolize much more; it became a form of the 'anti-classical', a way of 'dismantling the artful edifice of *Apolline culture* stone by stone, as it were, until we catch sight of the foundations on which it rests'.[14] It is this form of a deconstructive archaeology of the 'classical' that would serve as a determining cultural force in the German 'discovery of the archaic', a force whose repercussions would sound in the work of classics (Walter F. Otto), literature (the George Circle), philosophy (Heidegger), politics (Alfred Baeumler), and beyond.[15] Before delineating the effect of Nietzsche's influence on later thinkers, however, I would first like to address the significance of the archaic in two of Nietzsche's predecessors, Friedrich Georg Creuzer and Johann Jacob Bachofen.

The romantic idea of the archaic: Creuzer and Bachofen

The philosophical power of ideas all too often resists the logic of genealogical descent. In composing his own genealogy of Western morality, Nietzsche understood that the task of the philosopher differed radically from the practice of the philologist. That is, the philosopher sought not simply to know the past according to the criteria of *Wissenschaft*; rather, the philosopher sought to critique it in order to transform it as a way of interpreting our present situation and our future. 'We speak of the classical for *us*,' Nietzsche insisted, 'for our modern world, not in consideration of the Hindus, the Babylonians and the Egyptians'.[16] On this reading, the very notion of the archaic will be employed as a philosophical lever to raise the question of the use and abuse of the classical for *life*. What comes to presence in this idea of the archaic is a vision of antiquity as a way of explaining 'our whole culture and its development. It is a means *to understand ourselves*, to set our time aright and thereby overcome it' (KSA 8, 6[2], 97). In his effort to excavate the buried terrain of the archaic, Nietzsche turned to two primary sources that he believed had been covered over and concealed: the chthonic roots of tragedy in the myth of Dionysus Zagreus and the archaic roots of tragic thinking in the pre-Platonic philosophers. In the former case Nietzsche declares in a hyperbolic flourish that his excursions into the problem of tragedy 'led me to understand the amazing phenomenon of the Dionysian, the first person ever to have done so.'[17] Yet, of course, Nietzsche had predecessors – thinkers whose preoccupation with the archaic led them to conceive of Dionysus as a crucial figure who could not be accounted for in the general classicism of Germany's 'Second Humanism'.

Nietzsche, then, was not the first to 'discover' the Dionysian character of Greek culture. Philologists and archaeologists such as Friedrich Creuzer, Anselm Feuerbach, and J. J. Bachofen had already pointed to a tension within Greek culture between an Apollonian deity marked by clarity, measure, form, and limit, *and* a Dionysian deity characterized as the dark god of excess and ecstasy who revels in shattering boundaries, limits, and borders. In the earlier, classical interpretations of the Second Humanism, this Dionysian element was harmonized into a balanced union with Apollo. Yet with the Romantic mythological research of Creuzer there emerges 'another' Greece – an archaic Greece whose fundamental character can be traced in the cultic practices of the Orphics and the mystery religions of Eleusis. Creuzer's *Symbolik und Mythologie der alten Völker, besonders der Griechen* (published in 4 volumes between 1810 and 1812) offers us a survey of ancient religion that uncovers a buried, nocturnal Greece of bloody sacrifices, intoxicating orgies, primitive hunting rites, and other liturgies that burst asunder the tranquil grandeur of Winckelmann's dreams.[18] Seizing on the symbol as the primordial expression of the theological urge for union with the divine, Creuzer saw its function as the desire 'to give limits to the limitless and to fathom the unfathomable'.[19] In this sense, 'the symbol is the Idea itself embodied in the sensual'.[20] As part of his exhaustive study of archaic religion, Creuzer came to understand that the Enlightenment principles of rationality, logic, and discursive language were hardly as relevant to these early Greeks as the vast array of symbols that had come to them from the ancient Orient. Indeed, Creuzer posited a deeprooted opposition in Greek religion between the later 'Olympian' gods of Homer and the earlier 'chthonic' deities concerned with the dark powers of the earth and underworld. In this vision of chthonic, Dionysian religion as the genuine *arche* of Hellas, rather than the 'bright' and 'serene' image of Olympus, Creuzer radically challenged the classical Philhellenism of Winckelmann and his heirs. As Alfred Baeumler put it, Creuzer was 'the new, the anti-Winckelmann'.[21] In uncovering this chthonic realm of the archaic precisely in the myth of Dionysus Zagreus, the dismembered god torn to pieces by the Titans and restored whole by Apollo, Creuzer found a new Greece – an archaic one formed not as harmony, but as generative opposition and duality. This Dionysus was not simply the gregarious god of wine and sexual union; rather, he became 'the *Weltseele*, the demiurge who oversaw the creation of the material world and became intertwined with its suffering'.[22] In the initiatic mysteries of the cult, this Dionysus transformed suffering, dismemberment, and death into redemption, restoration, and rebirth.

Nietzsche's colleague at Basel, J. J. Bachofen, was powerfully influenced by Creuzer's reinterpretation of Greek religion in light of the influence of Dionysus on archaic culture and sought to offer his own narrative of Dionysian myth and its cultural impact. Deeply critical of the optimistic, modernist culture of progress that he saw everywhere around him, Bachofen turned self-consciously to archaic myth as a way of critiquing modernity. Like Creuzer, Bachofen perceives religion as the key to understanding culture – and he sees the depths of culture coming to presence in symbols, particularly the symbols of funerary monuments that express the primal tensions within a cultural order.[23] In works such as *Versuch über die Gräbersymbolik der Alten* (1859), *Das Mutterrecht* (1861), and *Die Sage von Tanquil* (1870), Bachofen presented a sweeping vision of the origins of human society in terms of the struggle between the cosmic forces of spirit and matter, light and darkness, life and death, masculine and feminine. For him, civilization moves from a *hetaeristic* state of nature marked by sexual promiscuity and lawlessness to a matriarchal order rooted in matrilineal descent which then passes through a new agricultural stage, grounded in the worship of the primal earth goddess Demeter and her fecundity, to a patriarchal society of masculine dominance and power. On Bachofen's reading, this order of history is governed by the alternating periods of sundering and reconciliation between the feminine energies of the earth and the masculine power of the sky. The parallels here to Creuzer's notion of the ongoing antagonism between chthonic and Olympian gods can hardly be overlooked. As part of his wide-ranging narrative of historical development, Bachofen presumes to uncover a line of development from a chthonic, matriarchal religion of the earth to a spiritual, patriarchal religion of the sun. Within the primal religion (*Urreligion*) of the ancient world, Bachofen uncovers a 'Dionysian religion' of bacchantic mysteries whose dissemination (in its literal sense) he holds as 'the most important turning point in the history of antiquity'.[24] As Bachofen viewed it, the spread of the cult of Dionysus represented a kind of utopian interlude in the history of culture where the phallic force of the male flowered in a balanced tension with the generative potency of feminine life. If, for Bachofen, Apollo symbolized life, patriarchy, and uranian strength, Dionysus appeared to him as an intermediary figure who served as a phallic conduit to the telluric forces of mother earth. In the orgiastic excesses of the Orphic mysteries and in their myths of destruction and rebirth, Bachofen unearths a more originary Dionysus, Dionysus Zagreus, who functions as the chthonic god of redemption. This Orphic Dionysus of 'the wild *dithyrambos* [...] full of labyrinths and digressions' stands in originary tension with 'the god of light Phoebus Apollo, the disciplined, well-ordered Paean'; this Dionysus Zagreus is 'the riddling god of the world of becoming in whose honor we spin tales and play with gryphons;

he enters partnership not with order and invariable earnestness but with jest, wantonness, frenzy, and variation'.[25]

In Bachofen's rendering, the nineteenth-century *Griechenbild* of a rational, Socratic Hellenism of clarity and Enlightenment offered an impoverished and tendentious understanding of antiquity. His extensive investigations uncover an archaic primal religion of the earth rooted in Dionysian cultic worship that is nearer to the originary character of Greek culture than any of the philologists' projections of a 'classical' Greece of order and harmony. Again, Bachofen was not the first to uncover such a realm. Creuzer, Joseph Görres, K. O. Müller, and the romantic mythologists had all contributed to the growing interest in the chthonic underground of ancient culture.[26] But Bachofen revitalized the myth of the archaic in a radically new way. Creuzer had understood that myth 'derived from a primeval revelation expressed in symbols that were not reducible to concepts and were accessible only to the intuition of those who were innately gifted with a spiritual sensitivity peculiarly adapted to them'.[27] He was excoriated by his colleagues for his lack of scientific rigor. Later, when K. O. Müller in his *Prolegomena zu einer wissenschaftlichen Mythologie* (1825) set out to square the circle by presenting myth scientifically, we encounter the underlying contradictions running through nineteenth-century German *Altertumswissenschaft*. Müller, like so many of his colleagues in philology, archaeology, and art history, wound up dissecting myth by uncovering its historical conditions. Bachofen, however, understood such scholarship as devitalizing in its effects. If *Altertumswissenschaft* set out to historicize myth, Bachofen strove to mythicize history. For him, 'the origin of all development lies in myth. Myth precedes all history, myth determines history'.[28] In his judgment that 'in myth lies the true picture of the most ancient time', Bachofen opened up an original relationship to antiquity rooted in a vision of the origin as *arche* – a myth of the *arche* rooted in the archaic power of myth. Such a vision would have a lasting effect on Nietzsche, even if he harbored deep skepticism concerning Bachofen's matriarchal predilections. Nietzsche shared Bachofen's disdain for the self-congratulatory excesses of historicism and, like him, he was convinced that any 'philosophy without intercourse with works of art [. . .] [remains] a lifeless skeleton'.[29] Yet Bachofen's abiding ties to his Christian faith, ties that made him ambivalent about the hetaeristic excesses of the Dionysian cults, would find no place in Nietzsche's own anti-Christian reading of antiquity. Nietzsche's idea of the archaic would extend far beyond the scientific-historical-religious studies of his colleagues to become a Dionysian cosmodicy of the aesthetic.

Nietzsche's Dionysian *Arche*

If the work of Creuzer and Bachofen helped to provide a prospect for 'overcoming the classicist's conception of antiquity', ultimately it remained foreign to Nietzsche because of its deep academic roots. What ultimately mattered to Nietzsche was less a breakthrough in new scholarly research than a fundamental experience of myth itself as a poetic opening unto the mysteries of being. This he found in his early readings of Hölderlin, especially his epistolary novel, *Hyperion*, a book that attuned the young Nietzsche to the Oriental sources of Greek art in the intensity of ecstatic wonder. More than anything, Hölderlin taught Nietzsche to approach the Greeks in their archaic power through art, not *Wissenschaft*, and to glean from this insight an uncompromising critique of modernity – in short, a critique of the modern from out of the archaic. In this way the Greeks become co-conspirators in undermining the values of the German philistine. 'I can think of no people more divided within itself than the Germans'; so at odds with themselves are they that their poets and artists 'live in the world like strangers in their own house', Hyperion writes to Bellarmin.[30] For both Hölderlin and Nietzsche, 'the' Greeks are no longer the staid, composed figures of Winckelmann's Apollonian fantasy, but the Greeks of Heraclitus, Empedocles, and Pindar before the triumph of Socratic rationalism. It is precisely this affinity that we see in both *The Birth of Tragedy* and *Philosophy in the Tragic Age of the Greeks*: a yearning for a Greece that is 'other', perhaps even for a Greece that never was, or could be. As Henning Ottmann has put it, in this yearning 'Nietzsche is in search of the Hölderlinian homeland of the German soul among the Greeks'.[31]

Where Winckelmann identified the greatness of the Greeks in a homogenous unity of aesthetic style that yielded a spiritual harmony brought to rest in sculptural form, Hölderlin understood the core of Greek culture to lie in the Heraclitean strife of harmonious opposition. It is this Heraclitean principle of an archaic *polemos* that harmonizes all life that marked the composition of *Hyperion*: 'The dissonances of the world are like the quarrels of lovers. In the midst of strife is reconciliation, and all that is separated comes together again.'[32] In this concluding paragraph of *Hyperion* we find a model for the opening sentence of *The Birth of Tragedy* that speaks of the 'perpetual conflict' between the sexes that is 'interrupted only occasionally by periods of reconciliation'.[33] The fundamental tension that Nietzsche identifies here is that between the principle of order, measure, stasis, limit, boundary, *and* that of frenzy, excess, orgy, *ekstasis*, and the boundless. Yet Nietzsche does not grasp these two principles merely as 'concepts' to be applied to a purported 'science of aesthetics'. On the contrary, for him they come to express two oppositional 'drives' that manifest themselves in the human realm, but whose genuine *arche* is *physis* itself. In this way Nietzsche's discussion of Greek tragedy is intended as an aesthetic cosmodicy on the model of Empedocles or Heraclitus, and not as a scholarly contribution to the philology of Greek literary history. Abjuring the scholarly form of footnotes and canonical method, Nietzsche crafts a new myth of German pessimism that understands all life as tragic, even as it presents the authentic task of the philosopher as joyfully affirming life's tragic character. It is this creative-destructive impulse that forms the basis of Nietzsche's Apollonian-Dionysian aesthetic.[34] Or, as Nietzsche frames it, 'the Olympian divine order of joy developed out of the originary, Titanic divine order of terror in a series of slow transitions, in much the same way as roses burst forth from a thicket of thorns'.[35] Out of the mythic tales of Greek religion, cultic practice, and festivals, Nietzsche confects an aesthetic ontology that shatters the tenets of German classicism. Here in the Apollonian-Dionysian tension that yields an aesthetic harmony of opposition, Nietzsche founds what he calls his *Artisten-Metaphysik*, his 'metaphysics of the artist'.[36]

The Birth of Tragedy relates a history of Greece that identifies the struggle between the *principium individuationis* and the principle of division as the hidden ground for the emergence of tragedy. As Nietzsche puts it, 'here for the first time the jubilation of nature achieves expression as art, here for the first time the tearing-apart of the *principium individuationis* becomes an aesthetic phenomenon'.[37] Nietzsche interprets this event of 'timely reconciliation' between these two art deities as 'the most important moment in the history of Greek religion'. Yet it is also, on Nietzsche's reading, the moment that defies historical explication, since it emerges out of the subterranean depths of Greek mythic consciousness. In this sense Nietzsche puts forward a *double* myth. On the one hand, as he views it, Greek art – specifically tragedy – emerges from the depths of Dionysian suffering; yet, on the other, it is worked upon by the dream-forming power of Apollonian representation and channeled into a measured form – as performance. The frenzy and primal lust of the Dionysiac dithyrambs are moderated by the measured pacing of Apollonian images, words, and concepts. The drama – as 'the Apollonian embodiment of Dionysian insights and effects' – displays an Apollonian veneer of beauty that covers over 'the inner, terrible depths of nature' and proffers a sense of healing to 'a gaze seared by gruesome night'.[38]

As he excavates this subterranean, Dionysian layer of Greek ritual and cultic experience, Nietzsche literally dis-covers the chthonic depth-dimension of the archaic world. In this primordial realm of chthonic Greek religion Nietzsche finds infinitely more than Creuzer, Bachofen, Görres, Müller, and their predecessors. This chthonic world of violence, a world where the body of a god is ripped apart from its limbs (*sparagmos*), its torn flesh eaten in its raw state (*omophagia*), and where the frenzied worshippers of this fragmented god lose themselves in the frenzied dance of *ekstatic* dissolution, reveals itself to Nietzsche as the terrifying underbelly of Winckelmann's composed Apollo Belvedere. Hence, Nietzsche can write: 'The ideal of Schiller, Humboldt – like that of Canova's – presents a false antiquity: too varnished and smooth, by no means daring enough to come face-to-face with the hard and hideous truth. One finds noble tones that are priggish, affecting gestures, but no *life*, no genuine blood.'[39] The classical ideal of Germany's Second Humanism offered an illusory vision of the Greeks; at root, it merely reflected the narrow prejudices of the German professoriate who Nietzsche classified as 'culture-philistines'. Classicists had projected their own bourgeois optimism onto the Greeks, thereby cutting themselves off from the genuinely chthonic sources of Greek life. The result for Nietzsche was an 'incongruity between classicists and the ancients'.[40] In an effort to transform modern Germany through a Schopenhaueran-Wagnerian *Kulturkampf* against Bismarck and the German establishment, Nietzsche sought to overcome the present through a retrieval of antiquity. Yet before such a cultural revolution could occur, Nietzsche deemed it essential that we first deconstruct our ideal of antiquity – for only then would such a bold revolutionary gesture succeed. In his notes for 'We Philologists' Nietzsche writes:

> *Our attitude towards classical antiquity is fundamentally the deep cause of modern culture's sterility*: this whole modern concept of culture is something we derive from the Hellenized Romans. We need to *distinguish* within antiquity itself: insofar as we learn to recognize its only creative period, we also *condemn* the whole Alexandrian-Roman culture. *But at the same time we condemn our entire attitude toward antiquity and our classical scholarship along with it*![41]

Not only must classical philology be taken apart and deconstructed brick by brick, but with it the classicists' serene phantasm of antiquity. As it now stands, Nietzsche writes, 'the prototypes of the classical scholar are *the teacher of reading and composition and the proof-reader*'.[42] Yet, as he turns away from the shibboleths of classical scholarship, Nietzsche finds that 'the greatest events in philology are the appearance of Goethe, Schopenhauer, and Wagner. Thanks to them we can manage a glance backwards that penetrates much further now. The fifth and sixth centuries [BCE] are now ready to be discovered'.[43] As he challenges the nostrums and clichés of German Philhellenism, Nietzsche comes to his own mythic version of antiquity rooted in a reading of Greek myth. 'Between our highest art and philosophy and the *truly* understood *archaic* form of antiquity,' Nietzsche claims, 'there is no contradiction: they are mutually supporting and sustaining. My hopes are founded on this.'[44] Nietzsche's own chthonic version of Greek culture expressed a double gesture: even as it rejected the classicists' fetishizing of 'origins', it hastened to confect its own version of an 'other' origin – the *archaic arche* that resists all historical emplacement. The *arche* must exist as *myth*, not as historical fact. No legion of classical archaeologists or ethnographers can succeed in locating the precise historical inception of Nietzsche's Dionysian myth of the archaic. The *arche* functions, rather, as a mythic horizon against which genuine historical inquiry – that is, 'history in the service of *life*' – can take place.[45] In this sense, the meaning of classical studies would be to become radically 'untimely' – that is, 'to act against our time and by so doing to act upon our time and, let us hope, for the benefit of a future time'. This plastic power of fashioning the present on the model of the archaic for the sake of the future would effect the most fundamental change within German Philhellenism – but not until after Nietzsche's death in 1900.

By shifting his gaze from the 'classical' Periclean age of Greek culture to the fifth and sixth centuries BCE, Nietzsche attempts a 'recovery of the ancient soil' of a Dionysian culture.[46] Such a turn would attempt, long before his Zarathustra period, a reevaluation of modern values grounded in a Dionysian reading of both Greek *and* German art and philosophy. The heart of the tragedy book is the recognition that in the sufferings of Dionysus Zagreus, 'the dismembered god', there lies 'the fundamental recognition that everything which exists is a unity'.[47] Here in the ' *doctrine of the mysteries taught by tragedy*' we find the joyful affirmation of existence in 'a premonition of unity restored' (the rebirth of Dionysus from Demeter's womb). With this recognition of the chthonic power of Dionysus, 'Olympian culture is defeated by a yet deeper way of looking at the world'. Now 'Dionysiac truth takes over the entire territory of myth to symbolize its own insights and expresses these partly through the public cult of tragedy and partly in the secretly conducted dramatic mystery-festivals, but always under the old cloak of myth'. In this way the Dionysus Zagreus myth expands its function as merely a religious form of worship to become *the* metaphysical *mythos* of archaic Greek culture. In its form as a homage to Dionysus Zagreus, Greek art – in the form of tragedy – comes to express the tragic nature of the world. Here tragedy becomes 'the highest task and the true metaphysical activity of this life'.[48] In the phenomenon of tragedy Nietzsche discovers and/or confects an aesthetic ontology that finds in art the authentic form of being.[49] This Dionysian optic leads Nietzsche to break down the division between artwork, artist, and the art observer and to see art not as a subject-object relation, but as a phenomenological happening that celebrates the creative-destructive pulse of primordial being. Here Nietzsche comes to understand 'the world as a work of art that gives birth to itself'.[50] This aesthetic ontology will take the form of an aesthetic cosmodicy on the model of pre-Socratic philosophers such as Heraclitus, who interprets the world as a cosmic play of oppositional forces bound in harmonic tension.

In his lectures *Philosophy in the Tragic Age of the Greeks* (1873) and *The Pre-Platonic Philosophers* (1872–1876), Nietzsche builds on his insight that it is 'only as an aesthetic phenomenon that existence and the world are eternally justified'.[51] Drawing on Heraclitus's insight that 'the world is the game Zeus plays',[52] Nietzsche interprets one fragment – 'The time (*aion*) of the cosmos is a playing child, playing draughts here and there – the kingdom of the child' (Diels-Kranz, B 52) – as the very fundament of the aesthetic cosmodicy he proposes:

> Only in the play of the child (or in art) does there exist a coming-to-be and a passing-away without any moralistic attribution. As an un-artistic man, [Heraclitus] reached for a children's game. Here is innocence and yet emergence and destruction. Not one drop of *adikia* [injustice] should remain in the world. The eternally living fire, *aion,* plays, builds up and destroys: the *polemos* [strife] of this confrontation of differing properties, directed by *dike* [justice], can only be grasped as an aesthetic phenomenon. This is a purely aesthetic view of the world.[53]

Being as a whole takes the form of a cosmic game in which the player *becomes* the game even as the game appropriates the player to the 'poesis of cosmic life'.[54] In this philosophical intuition of the tragic age, Nietzsche catches a glimpse of the recursive, Dionysian character of the world as a colossal *Spiel* of inexhaustible energy (the will to power) that emerges, perishes, and re-emerges following a cycle of Anaximandran periodicity and Heraclitean return (eternal recurrence).[55] *Philosophy in the Tragic Age of the Greeks* presents the interplay of these Heraclitean forces as the same harmonically opposing energies as the Dionysian-Apollonian impulses that produce tragic art – Heraclitus's 'aesthetic conception of the play of the cosmos'.[56] As Nietzsche puts it:

> In this world only play, the play of the artist and the child, exhibits coming-to-be and passing-away, building and destroying, without any moral attribution, in the innocence of the eternally same. And as the child and the artist play, so plays the eternally living fire, building and destroying in innocence – such is the game that the *aion* plays with itself.[57]

This ceaseless dynamic of opposition – what Jacob Burckhardt had identified in his interpretation of the *agon* as the core of Greek civic and religious life – becomes for Nietzsche the creative principle for understanding Greek existence, the effect of which is to destructure 'classical' balance to an 'archaic' harmony rooted in *agon*. Drawing from Burckhardt, Nietzsche extended the meaning of *agon* as *Wettkampf* (contest, challenge, duel, feud) and *Gegensatz* (opposition, antagonism, contrast) to a metaphysical principle for understanding the whole cultural world of pre-Platonic Greece: its athletic games in Homer, its cosmic principle of strife in Hesiod and Heraclitus, its rhetorical debates among the sophists, its rhapsodic element in Pindar, and its many different forms and appearances in theological cults, martial competitions, and political struggles. Foremost for him, however, was the significance of the *agon* for philosophy and aesthetics. In both *The Birth of Tragedy* and *Philosophy in the Tragic Age of the Greeks*, Nietzsche comes to think of the Dionysian-Apollonian tension of tragic art as Heraclitus's polar thinking of war/play, destruction/creation, where 'necessity and play, oppositional tension and harmony, must pair to create a work of art'.[58]

At the intersection of *The Birth of Tragedy* and *Philosophy in the Tragic Age of the Greeks* we find the same antipodal play of consonance and dissonance, congruity and contrariety (expressed as the energy of Dionysian emergence that merges all beings) playing against the Apollonian force of individuation that separates and gives plastic form to such boundlessness.[59] Only an epoch that can perceive the deep and necessary connections between these forces, that can think of philosophy *as* art and art *as* the originary and proper form of the philosophical conception of the cosmos, as the game that Zeus plays, can understand the power of myth as the highest form of truth. Nietzsche's greatest hope is to recover this mythic truth of the Greek *arche* for a German cultural revolution. But to undertake such a recovery first requires that we understand the blockages and occlusions that separate us from this Dionysian-Heraclitean world of archaic Greece. This involves above all coming to understand the reasons for the decline of the tragic worldview in Socratic rationalism as the triumph of a moralistic strain of thought that finds expression in Christian ethics and the Enlightenment's metaphysics of *Wissenschaft*, science, scholarship, and optimistic rationality.

Both *The Birth of Tragedy* and *Philosophy in the Tragic Age of the Greeks* offer a cultural diagnosis of this process: of tragedy's death through the conquest of Euripides' de-mythologization and of philosophy's Socratic triumph of 'the imperturbable belief that thought, as it follows the thread of causality, reaches down into the abysses of being, and that it is capable, not simply of understanding existence, but even of *correcting* it'.[60]

By re-thinking the Aristotelian-Hegelian model of Greek philosophy as a progressive development from *mythos* to *logos*, Nietzsche came to see the pre-Socratics as much more than 'precursors' and 'forerunners' of Socrates, Plato, and Aristotle. For him, they were originary thinkers who expressed the profound depths of Greek tragic consciousness. Nietzsche's perception of 'a gap, a break in development' between Socrates and his predecessors contested the academic marginalizing of pre-Socratic philosophy and helped to make it the focus of a new, culturally decisive reading of 'the archaic' – both as a way of understanding ancient Greek culture *and* as a way of re-thinking the aims and directions of modern German culture (KSA 2: 217). Hence, if we have seen that Nietzsche was not the first to 'discover' the irrational, chthonic, ecstatic dimension of the ancient Greeks, yet he was the first to make this reading decisive for the modern era. After the 'Nietzsche Revolution' of the early twentieth century, the old *Griechenbild* of Germany's Second Humanism will be radically altered. In the work of the George Circle, of Walter F. Otto, of Alfred Baeumler and the Weimar-National Socialist generation of artists, poets, thinkers, and scholars, we find the traces of a Nietzschean view of antiquity put to diverse programs of cultural reform and renewal.

In his book *Das Wort der Antike*, Otto will argue that myth stands at the beginning of culture as a primordial experience that 'reigns over all being and is thought in a form that bears a resemblance to the human'.[61] On his reading, the archaic epoch of elemental-chthonic deities will be covered over by the new Homeric gods, ushering in a rational-metaphysical epoch of Olympian enlightenment. In this turn away from the primordial *arche* of Greek culture, Olympian religion will lose touch with the creative sources of divine revelation. This narrative outline of Greek archaic history will, of course, bear a deep affinity with Nietzsche. In the same way we can see Baeumler drawing on Nietzsche's notion of *agon* to present a reading of modern Germany. Baeumler will put forward a forceful account of the chthonic elements that shaped the archaic world of the Greeks yet lay dormant in the modern era. Despite this, Baeumler believed that the struggles of the Great War bore within themselves the traces of the archaic powers of Nietzsche's Dionysian Greeks and pointed to a concealed possibility of cultural renewal. In his essay 'Hellas und Germanien', Baeumler affirms this 'mysterious affinity of the Hellenic and Germanic spirit'.[62] 'The decisive battle has begun,' Baeumler announces; Germany stands in an axial position within the history of the West. 'Our century must provide an answer about which values the West intends to follow to shape its future.' Against the anarchy and nihilism that lay everywhere around him, Baeumler embraces a National Socialist interpretation of antiquity as the source for a new vision of German racial domination: 'The recovery of Hellenic culture for the West is the result of the most violent exertions that the Germanic racial soul has made on the path back to finding itself.' Like Nietzsche, Baeumler insists that 'the discovery of the Hellenic world signifies nothing less than the presentiment of a new epoch'; yet Baeumler's violent appropriation of Nietzsche's idea of the archaic gives voice to the way conceptions of antiquity helped to shape various forms of German self-definition. For Baeumler, Nietzsche's vision of 'the rebirth of the Hellenic world out of the deepest instincts of Germanic being' points towards an 'originary affinity of the Hellenic life-form and the German essence', an affinity that holds forth the hope that Germany can serve as 'the world-historical *Führer*' for saving the West from 'the spiritual catastrophe of Europe'.[63]

Not only in Baeumler, but in so many of the German thinkers and ideologues who heed the call of Nietzsche's archaic Graecomania, we can hear a recurrent refrain: 'We have lost our sense for the originary.'[64]

Heidegger's pre-Socratic Vision of the Archaic

Like Baeumler, Heidegger too confected a vision of Graeco-German affinity rooted in Nietzsche's archaic Greeks and, like Walter F. Otto, he could concur that 'the Greek path is our path'.[65] For Heidegger, Nietzsche's vision of an originary, archaic Greece of pre-Socratic wonder whose insights and discoveries were cut off by the dawn of Platonic-Aristotelian philosophy, would have a powerful effect on his conception of the history of the West. From his very first lecture course in 1919, Heidegger defined philosophy as an *Urwissenschaft*, a discipline focused on inquiry into the origin – 'origin' understood here not in its strictly temporal (chronological) sense, but phenomenologically as a breaking-through to the primordial happening of being.[66] What matters above all to the early Heidegger is that we 'hold onto the beginning by understanding it radically and, while remaining within the beginning, to grasp and retain it, in its own mode, on the basis of our concrete situation'.[67] In his magnum opus *Being and Time* (1927), Heidegger provided a kind of narrative history of Western philosophy going back to Plato, Aristotle, and the early Greeks. But it is in his works of the 1930s, beginning with his famous rectorial address of 1933, 'The Self-Assertion of the German University', that his privileging of the *archaic* (and of the pre-Socratics) first takes shape.

The rectorial address takes up the crisis of the modern German university as an institution that has lost its authentic vocation and that seeks to recover this vocation by reflecting on its authentic possibilities. Precisely in 1933, with the new beginning of a National Socialist dream of national renewal and rebirth led by Adolf Hitler, Heidegger will reflect on the power of philosophy's 'first beginning' in the early Greeks as a path to a new German future. At root, Heidegger affirms that the contemporary university has lost its connection to originary *Wissenschaft*. Against the prevailing tendency to identify *Wissenschaft* with its institutional forms and practices, Heidegger calls for a rethinking of the essence of science in terms of its original roots, which he then locates in the ancient Greek practice of *philosophia*.

Genuine *Wissenschaft* in this originary sense is not mere evident and self-certain knowledge on the model of science; it has, rather, the character of what Plato in the *Theatetus* (155 d) termed *thaumazein*. *Thaumazein* – or what in the rectorial address Heidegger calls 'the initially awed perseverance of the Greeks in the face of what is' – constitutes the root, the beginning, and the *arche* of genuine *Wissenschaft*.[68] Moreover, it is this archaic-originary experience of philosophical wonder that forms the hidden essence of the German university. Hence, the rectorial address poses an essential question to the German *Volk* at the threshold of revolutionary consciousness in 1933: 'Do we will the essence of the German university or do we not will it?'[69] We can positively answer this question, Heidegger claims,

> only then, if we submit to the power of the beginning of our spiritual-historical existence. This beginning is the awakening of Greek philosophy. That is when, from the culture of one *Volk* and by the power of that *Volk*'s language, Western man rises up for the first time against *the totality of what is* and questions and comprehends it as the being that it is. All science is philosophy, whether it knows it and wills it or not. All science remains bound to that beginning of philosophy and draws from it the strength of its essence, assuming that it still remains at all equal to the beginning.[70]

If the awakening of 1933 is to become essential to the destiny of the German *Volk*, Heidegger declares, it would have to follow its spiritual leaders (*Führer*) in retrieving/recuperating the first beginning of Greek thought. As Heidegger reads it, the *political* revolution of National Socialism is really only a 'first awakening'; what truly matters is the *Volk*'s resolve to prepare itself for 'a second, deeper revolution' – an *ontological* revolution built upon the archaic foundations of pre-Socratic thought.[71] This is the genuine hope animating Heidegger's rectorial address. As Heidegger writes:

> The beginning still *is*. It does not lie *behind us* as something long past, but stands *before* us. [. . .] The beginning has invaded our future; it stands there as a distant injunction that orders us to recover its greatness.[72]

The logic of Heidegger's dream for an ontological revolution of German *Dasein* on the model of the pre-Socratic *arche* follows closely Nietzsche's own discovery/ invention of archaic philosophy in the Greek tragic age. Like Nietzsche, Heidegger envisions this revolution as a *Kulturkampf* that draws its energy from the Heraclitean interpretation of *polemos*. 'Self-assertion', in Heidegger's sense, is intimately bound up with *polemos* – the Nietzschean/Heraclitean principle that beings are generated by the harmonic–contentious play of being that holds sway over all that is. It is this fundamental insight into Greek thinking, that the *arche* rules over all beings, that generates Heidegger's own attitude toward antiquity as the source of German rebirth. 'The beginning is always what is greatest,' Heidegger declares.[73] *Arche* functions in Heidegger's work as the name designating the twofold character of the beginning: on the one hand, it names the origin as such as the source of all emergence or incipience; on the other, it designates the power that comes to presence in the origin itself. *Arche*, in this sense, is the 'ruling origin' of being. *Arche* is not a temporally prior point of origination or a historically discoverable past. Rather, Heidegger contends, the *arche* derives its power from, or rather unfolds its power as, something futural. As Heidegger expresses it in his lectures for the winter semester of 1937–1938:

> The *futural* is the *beginning of all happening*. In the beginning there lies everything sheltered. Even if what has already begun and what has already become, appear to have advanced beyond their beginning, yet the beginning [. . .] remains in power and abides and everything futural comes into confrontation with it. In all authentic history, futurity is decisive [. . .].[74]

In this reading of the *arche,* which underscores that 'the futural is the origin of history,' Heidegger sets up his essential reading of the history of the West under the watchwords of a 'first' and an 'other' beginning.[75] The first Greek beginning in Anaximander, Heraclitus, and Parmenides opened an experience of being as a whole and as such. Yet soon thereafter this experience was molded and formed by an interpretation of truth as 'correctness' in terms of a subject-object metaphysics of presence. Epochally understood, such a transformation of the originary pre-Socratic experience of being by later metaphysicians led to the forgetting of being as 'event' and the encasing of beings into forms of standing presence available for human control and disposition. The nihilistic consequences of such a process have resulted in a metaphysical epoch of technological dominion where being is grasped in terms of beings available for use and mastery. To prepare for a *transition* out of this metaphysical epoch into another kind of comportment towards being requires an attunement towards poetic thinking marked by *Gelassenheit*, a 'releasement toward things' that lets go of voluntaristic dominion and the metaphysics of will. This transition to an 'other' beginning, Heidegger holds, can only take place by way of a thoughtful recovery of the unthought possibilities that lie hidden in the first beginning. The fate of the Western world depends upon just such a recovery of the archaic Greek beginning for a future yet to be unfolded. In this sense, the *arche* does not lie behind us as a romantic source of nostalgic longing; on the contrary, the *arche* demands a revolutionary comportment that breaks with habit, tradition, and 'the domination of the habitual':

> An upheaval is needed, so that the inhabitual and the forward-reaching can open itself and come to power. Revolution – the upheaval of what is habitual – is the authentic relation to the beginning.[76]

Here Heidegger's notion of the archaic functions as a revolutionary overturning of the present order via a leap or *Sprung* from the origin, ahead towards the future; such a movement brings forth a phenomenological understanding of the *arche* as 'originary leap' or *Ur-sprung*.

Yet Heidegger's stance towards the archaic can hardly be understood apart from his own political vision of German homecoming. What authorizes the possibility of such a return to the origin (not for its own sake, but for the sake of a future to come) is the Heideggerian myth of German rootedness in the pre-Socratic *arche* – 'the metaphysics of German existence in its originary bond to the Greeks'.[77] This narrative of homecoming, which Heidegger defines as 'becoming homely in nearness to the origin', takes place not merely as a *philosophical* journey into the archaic past, but also as a *political* excursion into the metaphysics of German national identity.[78] Like so many other German thinkers, Heidegger sought to establish a primordial bond with the ancient Greeks as a way of founding a new idea of German nationhood, a 'secret Germany' in autochthonic kinship with archaic Hellas.[79] In this *völkisch* reading of Graeco-German affinity, Heidegger reprises Fichte's litany that the Germans, like the Greeks, are an *Urvolk* or 'originary people' and that 'the German language is [. . .] as equally originary as the Greek'.[80] In unearthing this primal connection, Heidegger merely follows the fundamental law of Graeco-German autochthony already laid down by Winckelmann, 'to seek the source itself and return to the beginning in order to find truth, pure and unmixed'.[81] Going back to the eighteenth century, the Germans had succeeded in appropriating this myth of originary kinship with the Greeks to sanction their own historico-metaphysical *Sonderweg* or 'special path' in Western history. On the basis of this presumed heritage, the Germans will now claim to be preeminent among nations since, as Wilhelm von Humboldt put it, 'the Germans possess the undeniable merit of having been the first to have truly understood and profoundly experienced Greek culture'.[82] In the teeth of Weimar nihilism – and later during the Second World War – Heidegger will draw on this heritage as a way of establishing Germany's elect status and its 'spiritual mission'.[83] As Heidegger expresses it in his lectures on Heraclitus from the summer semester of 1944, 'the Germans and only the Germans can save the West and its history' (GA, 55: 108).

Heidegger's idea of the archaic, modeled as it is on both Hölderlin and Nietzsche's vision of Hellas, breaks with the Second Humanism of the nineteenth century and its image of 'classical' Greece.[84] Like them, Heidegger employs an image of an 'other' Greece, a more originary and primordial Greece that not merely predates the Periclean age of Athens, but differs from it 'essentially'. At first glance, their visions appear significantly different from Heidegger's, in that both Hölderlin and Nietzsche focus on aesthetics, especially tragedy, as the basis for their judgments, whereas Heidegger focuses on the history of philosophy. Yet with a deeper look we can see how Heidegger's vision of archaic Greece is deeply tragic, formed as it is by the philosopheme of *decline* away from the *arche*. Heidegger underscores this in the early 1930s, claiming that 'the history of Western philosophy is an accelerating decline from its beginning'; indeed, 'directly following its beginning Western philosophy becomes un-Greek and remains so, explicitly or implicitly, until Nietzsche'.[85] This image of philosophy's decline, like Heidegger's diagnosis of the crisis of Western nihilism, is a crisis-narrative of Nietzschean design. What marks the distinctly Heideggerian character of this tragic vision is its attempt to wed pre-Socratic philosophy and a Hölderlinian politics of homecoming. What matters most to Heidegger in his reading of the archaic is to designate Heraclitus, Parmenides, Anaximander, and Sophocles as the sources for finding a path out of the homelessness of modern *Dasein* on the way to finding Germany's hidden, archaic home in being.

This ontological fashioning of German 'destiny' from out of a Greek *arche* would reverberate in the German world even through the Second World War. With the collapse of the National Socialist regime, however, the political configuration of such a myth would also shift. In the postwar years German culture would inevitably confront the failure of the *Sonderweg* vision of Germany's elect status among nations. In the process, the Hölderlinian ideal of a 'secret Germany' promulgated by Stefan George, Norbert von Hellingrath, Ernst Bertram, and other members of the George Circle would come under critical scrutiny. Moreover, the nationalistic appropriation of archaic Greece by National Socialists such as Ernst Krieck, Hans Heyse, Kurt Hildebrandt, Baeumler, Heidegger and others would be exposed as a right-wing fantasy about Graeco-German kinship. What survives of this myth in the postwar world of cultural change and transformation is a wariness about the political uses and abuses of the old German *Griechenbild* under radically different social and pedagogical realities. In the Gymnasium of the late nineteenth century, 46 percent of classroom time was spent on studying the classical languages.[86] One hundred years later, the situation had radically altered: 'The Greek gods, it seems, have come down from Olympus, and Philhellenic neo-humanism has lost its privileged place in German culture'. Nietzsche's dream of a new German Hellas, like Heidegger's of the pre-Socratic *arche* of the future, would be consigned to the long history of a German archaicism whose dream had perished. Winckelmann's Apollo Belvedere had been transformed – and, in the process, dismantled. Our postmodern epoch looks skeptically at Walter F. Otto's divination that 'the gods *are*', preferring to splinter the archaic energy of Dionysus into a multicultural kaleidoscope of difference and diversity.[87] Still, Heidegger's vision of the *archaic* as something more than an 'ideal' or a 'cultural value' bears authentic engagement. For what persists in Heidegger's Graecophilia is a vision of history arched by the future, a vision that understands the hermeneutic power of the *arche* as a *hermaion*, a 'gift from Hermes'.[88] For Heidegger this *hermaion* serves as a Nietzschean 'boundary stone' providing direction for those on a Hölderlinian sojourn through 'the destitute time of the world's night'.[89] If, for Hölderlin, 'the return of the gods is the return of the Greek in the German and as something German', then one can understand Heidegger's Hölderlinian dream of poetic dwelling as the call of the gods harkening us back to the inceptive power of the *arche*.[90] This is the hermeneutic gift of archaic thinking that the beginning offers to all who would heed its call:

We name that which precedes and determines all history, the beginning. Because it does not belong to a past, but lies in advance of that which is to come, that which belongs to the beginning over and over again makes of itself a gift to an epoch.[91]

Notes

1 F. Nietzsche, *Sämtliche Werke: Kritische Studienausgabe*, ed. G. Colli and M. Montinari, 15 vols, Berlin and New York; Munich: Walter de Gruyter; Deutscher Taschenbuch Verlag, 1967–1977 and 1988, vol. 8, 5[1], p. 41. Henceforth cited as KSA, followed by volume, section, and page reference.
2 P. Tillich, *Die sozialistische Entscheidung*, Offenbach: Karl Drott, 1948, p. 18.
3 Liddell-Scott, *A Greek-English Lexicon*, New York: American Book Company, 1897, pp. 227–29.
4 W. Rehm, *Griechentum und Goethezeit: Geschichte eines Glaubens*, Berne: Francke, 1952, p. 367.
5 J. J. Winckelmann, *Kleine Schriften zur Geschichte der Kunst des Altertums*, Leipzig: Insel, 1913, p. 62.
6 Rehm, *Griechentum und Goethezeit*, p. 229.
7 Winckelmann, *Kleine Schriften*, pp. 85–86.
8 The irony here is that Winckelmann's vision of Greek aesthetic perfection was based on a Roman copy. Winckelmann never saw archaic Greek sculpture nor visited Greece.
9 J. W. von Goethe, *Werke* [Hamburger Ausgabe], vol. 12, ed. E. Trunz, Hamburg: Wegner, 1960, pp. 96–129.
10 J. J. Winckelmann, *Geschichte der Kunst des Altertums*, Vienna: Phaidon, 1936, p. 9.
11 There are a number of excellent works dealing with the topic of Graeco-German affinity: see S. L. Marchand, *Down from Olympus: Archaeology and Philhellenism in Germany, 1750–1970*, Princeton: Princeton University Press, 1996; H. Sichtermann, *Kulturgeschichte der klassischen Archäologie*, Munich: Beck, 1996; as well as various articles by Glenn Most, including 'Die Entdeckung der Archaik', in B. Seidensticker and M. Vöhler (eds), *Urgeschichte der Moderne: Die Antike im 20. Jahrhundert*, Stuttgart: Metzler, 2001, pp. 20–39; 'Zur Archäologie der Archaik', *Antike und Abendland* 35, 1989, pp. 1–20; '*Polemos Panton Pater*: Die Vorsokratiker in der Forschung der zwanziger Jahre', in H. Flashar and S. Vogt (eds), *Altertumswissenschaft in den 20er Jahren: Neue Fragen und Impulse*, Stuttgart: Steiner, 1995, pp. 87–114; and 'Philhellenism, Cosmopolitanism, Nationalism', in M. Haagsma, P. den Boer, and E. M. Moormann (eds), *The Impact of Classical Greece on European and National Identities*, Leiden: Brill, 2003, pp. 71–91. For a basic survey, see W. Liefer, *Hellas im deutschen Geistesleben*, Heerenalb: Erdmann, 1963, or (now outdated) E. M. Butler, *The Tyranny of Greece over Germany*, Cambridge: Cambridge University Press, 1935.
12 V. Lambropoulos, *The Rise of Eurocentrism: Anatomy of Interpretation*, Princeton: Princeton University Press, 1993, p. 79.
13 D. t. D. Held, 'Conflict and Repose: Dialectic of the Greek Ideal in Nietzsche and Winckelmann', in Paul Bishop (ed.), *Nietzsche and Antiquity: His Reaction and Response to the Classical Tradition*, Rochester, NY: Boydell and Brewer, 2004, pp. 411–24 (p. 416).
14 Nietzsche, *The Birth of Tragedy*, §2, in *The Birth of Tragedy and other writings*, ed. R. Geuss and R. Speirs, tr. R. Speirs, Cambridge: Cambridge University Press, 1999, p. 22; KSA 1, 34.

15 For a number of helpful works on this topic, see G. Dörr, *Muttermythos und Herrschaftsmythos: Zur Dialektik der Aufklärung um die Jahrhundertwende bei den Kosmikern, Stefan George und in der Frankfurterschule*, Würzburg: Königshausen & Neumann, 2007; C. Horn, 'Die Entdeckung der Archaik' in *Remythisierung und Entmythisierung: Deutschsprachige Antikendramen der klassischen Moderne*, Karlsruhe: Universitätsverlag Karlsruhe, 2007, pp. 67–86; and E. S. Sünderhauf, *Griechensehnsucht und Kulturkritik: Die deutsche Rezeption von Winckelmanns Antikenideal 1840–1945*, Berlin: Akademie-Verlag, 2004.

16 'Encyclopaedie der klass[ischen] Philologie' (1871), §13, in F. Nietzsche, *Werke: Kritische Gesamtausgabe*, ed. G. Colli and M. Montinari, W. Müller-Lauter and K. Pestalozzi, Berlin: Walter de Gruyter, 1967ff., vol. II.3, pp. 339–437 (here: p. 390); subsequently referred to as KGW, followed by volume, section, and page reference. See also A. U. Sommer, *Der Geist der Historie und das Ende des Christentums: Zur 'Waffengenossenschaft' von Friedrich Nietzsche und Franz Overbeck*, Berlin: Akademie-Verlag, 1997, pp. 21–29.

17 *Ecce Homo*, 'The Birth of Tragedy', §2; KSA 6, 311; cf. *Twilight of the Idols*, 'What I owe to the ancients', §4; KSA 6, 158. Max Baeumer – in his 'Nietzsche and the Tradition of the Dionysian', in J. O'Flaherty (ed.), *Studies in Nietzsche and the Classical Tradition*, Chapel Hill: University of North Carolina Press, 1976, pp. 165–89 – claims Nietzsche was not the first to 'discover' this and in a narrow sense he is right. Yet Nietzsche's cultural effect is staggering and is much more lasting than any of his predecessors.

18 See also G. S. Williamson, *The Longing for Myth in Germany: Religion and Aesthetic Culture from Romanticism to Nietzsche*, Chicago: University of Chicago Press, 2004; and R. Schlesier, 'Olympische Religion und Chthonische Religion: Creuzer, K. O. Müller, und die Folgen', in *Kulte, Mythen und Gelehrte: Anthropologie der Antike seit 1800*, Frankfurt am Main: Fischer, 1994, pp. 21–32.

19 F. Creuzer, *Symbolik und Mythologie der alten Völker, besonders der Griechen*, Leipzig: Heyer & Leske, 1819–1823, vol. 1, p. 94; and Baeumer, 'Nietzsche and the Tradition of the Dionysian', pp. 165–89.

20 C. Jamme, *Einführung in die Philosophie des Mythos*, vol. 2, *Neuzeit und Gegenwart*, Darmstadt: Wissenschaftliche Buchgesellschaft, 1991, p. 53.

21 A. Baeumler, 'Einleitung', in J. J. Bachofen, *Der Mythus von Orient und Occident: Eine Metaphysik der alten Welt*, ed. M. Schroeter, Munich: Beck, 1926, p. cvi.

22 Williamson, *The Longing for Myth in Germany*, p. 133.

23 Bachofen writes: 'There is only one, singularly powerful lever that moves all civilization: religion. Every elevation, every declivity of human *Dasein* springs forth from a movement that takes its origin from this highest of realms. Without it there is no side of the old form of life that is comprehensible; the earliest epoch is above all an inscrutable riddle without it' (*Der Mythus von Orient und Occident*, p. 20).

24 Dörr, *Muttermythos und Herrschaftsmythos*, p. 56.

25 J. J. Bachofen, *Die Unsterblichkeitslehre der orphischen Theologie auf den Grabdenkmälern des Altertums*, Basel: Schwabe, 1943, p. 100.

26 For a history of these chthonic interpretations of the archaic, see Baeumler, 'Einleitung', in Bachofen, *Der Mythus von Orient und Occident*, pp. xxiii–ccxliv.

27 G. Most, '100 Years of Fractiousness: Disciplining Polemics in 19th-Century German Classical Scholarship', *Transactions of the American Philological Association*, 127, 1997, 349–61 (here: p. 351).

28 Baeumler, 'Einleitung', in Bachofen, *Der Mythus von Orient und Occident*, pp. clxxxix–cxc, cf. p. 10.

29 J. J. Bachofen, *Griechische Reise*, Heidelberg: Weissbach, 1927, cited in Sichtermann, *Kulturgeschichte der klassischen Archäologie*, p. 226.

30 F. Hölderlin, *Hyperion*, tr. R. Benjamin, Brooklyn, NY: Archipelago Books, 2008, pp. 206–09; Hölderlin, *Sämtliche Werke* [Stuttgarter Ausgabe], vol. 3, *Hyperion*, ed. F. Beißner, Stuttgart: Kohlhammer, 1957, pp. 153–55.
31 H. Ottmann, *Geschichte des politischen Denkens*, vol. 3.3, *Die Neuzeit: Die politischen Strömungen im 19. Jahrhundert*, Stuttgart: Metzler, 2008, p. 233. See also Nietzsche's 'imaginary' letter of 1861 concerning Hölderlin, probably written as a school exercise: 'Nowhere has the yearning for Greece revealed itself in purer tones' (KGW I.2, 338–41 [here: p. 340]).
32 Hölderlin, *Hyperion*, p. 215 [tr. modified]; *Sämtliche Werke*, vol. 3, p. 160.
33 Nietzsche, *The Birth of Tragedy*, §1, p. 14; KSA 1, 25.
34 In *Twilight of the Idols* Nietzsche writes that to understand tragedy means 'to go beyond pity and terror and *to be oneself* the eternal joy in becoming – the joy that even encompasses *joy in destruction*' ('What I owe to the ancients', §5; KSA 6, 160).
35 Nietzsche, *The Birth of Tragedy*, §3, p. 33; KSA 1, 36.
36 Nietzsche, *The Birth of Tragedy*, 'Attempt at a Self-Criticism', esp. §2, pp. 5–11; KSA 1, 13–21.
37 Nietzsche, *The Birth of Tragedy*, §2, pp. 20–21; KSA 1, 33.
38 Nietzsche, *The Birth of Tragedy*, §8–§9, pp. 44–46; KSA 1, 62–65.
39 Nietzsche, KSA 8, 41[67], 593–94.
40 Nietzsche, KSA 8, 3[21], 21; cf. 'We Classicists', tr. W. Arrowsmith, in F. Nietzsche, *Unmodern Observations*, ed. W. Arrowsmith, New Haven and London: Yale University Press, 1990, pp. 305–87 (here: p. 331).
41 KSA 8, 5[47], 53; cf. 'We Classicists', p. 356.
42 Nietzsche, KSA 8, 5[159], 84; cf. 'We Classicists', p. 379. In *The Birth of Tragedy*, Nietzsche writes: 'Alexandrian man is basically a librarian and proof-reader, sacrificing his sight miserably to book-dust and errors' (*The Birth of Tragedy*, §18, p. 88; KSA 1, 120).
43 Nietzsche, KSA 8, 3[70], 34; cf. 'We Classicists', p. 342.
44 Nietzsche, KSA 8, 5[111], 69; cf. 'We Classicists', p. 368.
45 See 'History in the Service and Disservice of Life', Preface, in *Unmodern Observations*, pp. 73–145 (pp. 87–88); KSA 1, 245–46. In his notebooks of 1873, Nietzsche writes: 'The pronouncement of the past is always in the form of an oracle: only as prophet of the future, as one who knows the present, will you be able to interpret it' (KSA 7, 29[96], 676).
46 One of the most penetrating commentators on the idea of Nietzsche and the archaic is Günter Wohlfart; see his *Das spielende Kind: Nietzsche: Postvorsokratiker – Vorpostmoderner*, Essen: Verlag Die Blaue Eule, 1999; and '*Also Sprach Herakleitos*': *Heraklits Fragment B 52 und Nietzsches Heraklit-Rezeption*, Freiburg: Alber, 1991; as well as his 'Nachwort' to Nietzsche, *Die nachgelassenen Fragmente: Eine Auswahl*, Stuttgart: Reclam, 1996, pp. 295–314. Two valuable books on Nietzsche and tragedy are B. von Reibnitz, *Kommentar zu Friedrich Nietzsche, "Die Geburt der Tragödie aus dem Geiste der Musik" (Kapitel 1–12)*, Stuttgart: Metzler, 1992; and M. Landfester, 'Kommentar', in Friedrich Nietzsche, *Die Geburt der Tragödie*, Frankfurt: Insel, 1994, pp. 373–684.
47 Nietzsche, *The Birth of Tragedy*, §10, pp. 52–53; KSA 1, 72–73.
48 Nietzsche, *The Birth of Tragedy*, 'Preface to Richard Wagner', p. 14; KSA 1, 24.
49 For a valuable reading of Nietzsche's aesthetic ontology, see E. Fink, *Nietzsches Philosophie*, Stuttgart: Kohlhammer, 1960, pp. 14–27.
50 Nietzsche, KSA 12, 2[114], 119.
51 Nietzsche, *The Birth of Tragedy*, §5, p. 33; KSA 1, 47.
52 Nietzsche, *Philosophy in the Tragic Age of the Greeks*, tr. M. Cowan, Washington: Gateway, 1962, §6, p. 58; KSA 1, 828.

53 Nietzsche, *The Pre-Platonic Philosophers*, ed. and tr. G. Whitlock, Urbana: University of Illinois Press, 2001, p. 79; KGW II.4, 278.

54 Fink, *Nietzsches Philosophie*, p. 31.

55 Günter Wohlfart offers an important reading of Nietzsche's early work in its significance for the later notions of will to power and eternal recurrence: 'The moment of eternal recurrence is the moment of the aesthetic epiphany of the Apollonian Dionysus, i.e., the un-moral god of art, Dionysus Zagreus [. . .] Art, this "great stimulus of life", is the organon of the philosophy of eternal recurrence' (*Das spielende Kind*, p. 119).

56 Nietzsche, *Philosophy in the Tragic Age of the Greeks*, §7, p. 65; KSA 1, 833.

57 Nietzsche, *Philosophy in the Tragic Age of the Greeks*, §7, p. 62; KSA 1, 830.

58 Nietzsche, *Philosophy in the Tragic Age of the Greeks*, §7, p. 62; KSA 1, 831.

59 See, for example, Nietzsche's comment in section 24 of *The Birth of Tragedy*, where he writes of a 'Dionysian phenomenon, one which reveals to us the playful construction and demolition of the world of individuality as an outpouring of primal pleasure and delight, a process quite similar to Heraclitus the Obscure's comparison of the force that shapes the world to a playing child who sets down stones here, there, and the next place, and who builds up piles of sand only to knock them down again' (*The Birth of Tragedy*, p. 114; KSA 1, 153).

60 Nietzsche, *The Birth of Tragedy*, §15, p. 73; KSA 1, 99.

61 W. F. Otto, *Das Wort der Antike*, ed. Kurt von Fritz, Stuttgart: Klett, 1962, p. 349.

62 A. Baeumler, 'Hellas und Germanien', in *Studien zur deutschen Geistesgeschichte*, Berlin: Junker & Dünnhaupt, 1937, pp. 295–311 (here: pp. 309, 295).

63 Baeumler, 'Nietzsche', in *Studien zur deutschen Geistesgeschichte*, pp. 244–80 (here: pp. 250, 258, 246).

64 Baeumler, 'Antrittsvorlesung in Berlin', in *Männerbund und Wissenschaft*, Berlin: Junker & Dünnhaupt, 1934, pp. 123–38 (here: p. 132).

65 W. F. Otto, *Die Gestalt und das Sein: Gesammelte Abhandlungen über den Mythos und seine Bedeutung für die Menschheit*, Düsseldorf: Diederichs, 1955, p. 94.

66 M. Heidegger, *Zur Bestimmung der Philosophie: 1. Die Idee der Philosophie und das Weltanschauungsproblem (Kriegsnotsemester 1919); 2. Phänomenologie und transzendentale Wertphilosophie (Sommersemester 1919)* [Gesamtausgabe, vol. 56/57], ed. Bernd Heimbüchel, Frankfurt am Main: Klostermann, 1987, pp. 13–17. Henceforth the abbreviation GA is used to Heidegger's *Gesamtausgabe*, Frankfurt am Main: Klostermann, 1976ff.

67 Heidegger, *Phenomenological Interpretations of Aristotle*, Bloomington: Indiana University Press, 2001, p. 128; *Phänomenologische Interpretationen zu Aristoteles: Einführung in die phänomenologische Forschung (Wintersemester 1921/22)* [GA, 61], ed. W. Bröcker and K. Bröcker-Oltmanns, Frankfurt am Main: Klostermann, 1985, p. 170.

68 Heidegger, *Reden und andere Zeugnisse eines Lebensweges (1910–1976)* [GA, 16], ed. H. Heidegger, Frankfurt am Main: Klostermann, 2000, p. 111.

69 Heidegger, *Reden und andere Zeugnisse eines Lebensweges* [GA, 16], p. 116.

70 Heidegger, Rectorial Address ('The Self-Assertion of the German University'), in R. Wolin (ed.), *The Heidegger Controversy*, New York: Columbia University Press, 1991, pp. 29–39 (here: p. 31); *Reden und andere Zeugnisse eines Lebensweges* [GA, 16], pp. 108–09.

71 *Martin Heidegger-Elizabeth Blochmann Briefwechsel, 1918–1969*, ed. J. Storck, Marbach: Deutsche Schillergesellschaft, 1990, pp. 60–61.

72 Heidegger, Rectorial Address, p. 32; *Reden und andere Zeugnisse eines Lebensweges* [GA, 16], pp. 110.

73 Heidegger, *Einführung in die Metaphysik*, Tübingen: Niemeyer, 1953, p. 12.

74 Heidegger, *Grundfragen der Philosophie: Ausgewählte "Probleme" der "Logik" (Wintersemester 1937/38)* [GA, 45], ed. F.-W. von Herrmann, Frankfurt am Main: Klostermann, 1984, p. 36.
75 Heidegger, *Grundfragen der Philosophie* [GA, 45], p. 40.
76 Heidegger, *Grundfragen der Philosophie* [GA, 45], pp. 40–41.
77 *"Mein liebes Seelchen!": Briefe Martin Heideggers an seine Frau Elfride, 1915–1970*, ed. G. Heidegger, Munich: Deutsche Verlags-Anstalt, 2005, p. 186.
78 Heidegger, *Erläuterungen zu Hölderlins Dichtung (1936–1968)* [GA, 4], ed. F.-W. von Herrmann, Frankfurt am Main: Klostermann, 1981, p. 24.
79 Heidegger, *Reden und andere Zeugnisse eines Lebensweges* [GA, 16], p. 290.
80 J. G. Fichte, *Reden an die deutsche Nation*, Hamburg: Meiner, 1978, pp. 106, 72. In his famous '*Spiegel* interview' of 1966 Heidegger likewise 'thinks of the special, inner affinity of the German language with Greek language and thinking' (*Reden und andere Zeugnisse eines Lebensweges* [GA, 16], p. 679).
81 Winckelmann, *Geschichte der Kunst des Altertums*, p. 321.
82 Sichtermann, *Kulturgeschichte der klassischen Archäologie*, p. 157.
83 Heidegger, *Reden und andere Zeugnisse eines Lebensweges* [GA, 16], pp. 107–08, 113–14. For an extended analysis of Heidegger's 'politics of the *Arche*', see C. Bambach, *Heidegger's Roots: Nietzsche, National Socialism, and the Greeks*, Ithaca: Cornell University Press, 2003, which provides a reading of Heidegger's anti-Romanism as a feature of his Graecophilia.
84 According to Heidegger, 'for Hölderlin, the Greek world is never "classical antiquity"' (*Hölderlins Hymne "Der Ister" (Sommersemester 1942)* [GA, 53], ed. W. Biemel, Frankfurt am Main: Klostermann, 1984, p. 67.
85 Heidegger, *Sein und Wahrheit: 1. Die Grundfrage der Philosophie (Sommersemester 1933); 2. Vom Wesen der Wahrheit (Wintersemester 1933/34)* [(GA, 36/37], ed. H. Tietjen, Frankfurt am Main: Klostermann, 2001, pp. 29, 24.
86 Marchand, *Down from Olympus*, pp. 117, 375.
87 K. F. Reinhardt, *Vermächtnis der Antike: Gesammelte Essays zur Philosophie und Geschichtsschreibung*, ed. C. Becker, Göttingen: Vandenhoeck & Ruprecht, 1960, p. 379.
88 W. F. Otto, *Die Götter Griechenlands: Das Bild des Göttlichen im Spiegel des griechischen Geistes*, Frankfurt am Main: Klostermann, 1987, p. 137.
89 Heidegger, *Holzwege* [GA, 5], ed. F.-W. von Herrmann, Frankfurt am Main: Klostermann, 1977, p. 270. In *Philosophy in the Tragic Age of the Greeks* Nietzsche employs the image of *Grenzsteine* (§4 and §9, pp. 46 and 69 [tr. modified]; KSA 1, 818 and 836); cf. the allusion from Hölderlin's 'Brot und Wein', 'und wozu Dichter in dürftige Zeit?' (Hölderlin, *Sämtliche Werke*, vol. 2.1, *Gedichte nach 1800*, ed. F. Beißner, Stuttgart: Kohlhammer, 1951, 'Brod und Wein', pp. 90–95 [here: p. 94]).
90 Rehm, *Griechentum und Goethezeit*, p. 348.
91 Heidegger, *Parmenides (Wintersemester 1942/43)* [GA, 54], ed. M. S. Frings, Frankfurt am Main: Klostermann, 1992, pp. 1–2.

Part II

Jungian approaches:
Context and critique

Chapter 6

Archaic man

C.G. Jung

The word 'archaic' means primal, original. While it is one of the most difficult and thankless of tasks to say anything of importance about the civilized man of today, we are apparently in a more favourable position when it comes to archaic man. In the first case, the speaker finds himself caught in the same presuppositions and is blinded by the same prejudices as those whom he wishes to view from a superior standpoint. In the case of archaic man, however, we are far removed from his world in time, our mental equipment, being more differentiated, is superior to his, so that from this more elevated coign of vantage it is possible for us to survey his world and the meaning it held for him.

With this sentence I have set limits to the subject to be covered in my lecture. Unless I restricted myself to the psychic life of archaic man, I could hardly paint a sufficiently comprehensive picture of him in so small a space. I should like to confine myself to this picture, and shall say nothing about the findings of anthropology. When we speak of man in general, we do not have his anatomy, the shape of his skull, or the colour of his skin in mind, but mean rather his psychic world, his state of consciousness, and his mode of life. Since all this belongs to the subject-matter of psychology, we shall be dealing here chiefly with the psychology of archaic man and with the primitive mentality. Despite this limitation we shall find we have actually widened our theme, because it is not only primitive man whose psychology is archaic. It is the psychology also of modern, civilized man, and not merely of individual 'throw-backs' in modern society. On the contrary, every civilized human being, however high his conscious development, is still an archaic man at the deeper levels of his psyche. Just as the human body connects us with the mammals and displays numerous vestiges of earlier evolutionary stages going back even to the reptilian age, so the human psyche is a product of evolution which, when followed back to its origins, shows countless archaic traits.

When we first come into contact with primitive peoples or read about primitive psychology in scientific works, we cannot fail to be deeply impressed with the

Note: From Jung, C. G., The Collected Works of C. G Jung, Vol. 10. © 1964 Bollingen, 1992 renewed. Reprinted by permission of Princeton University Press and Taylor & Francis Books, UK.

strangeness of archaic man. Lévy-Bruhl himself, an authority in the field of primitive psychology, never wearies of emphasizing the striking difference between
the 'prelogical' state of mind and our own conscious outlook. It seems to him, as
a civilized man, inexplicable that the primitive should disregard the obvious
lessons of experience, should flatly deny the most evident causal connections, and
instead of accounting for things as simply due to chance or on reasonable grounds
of causality, should take his 'collective representations' as being intrinsically
valid. By 'collective representations' Lévy-Bruhl means widely current ideas
whose truth is held to be self-evident from the start, such as the primitive ideas
concerning spirits, witchcraft, the power of medicines, and so forth. While it is
perfectly understandable to us that people die of advanced age or as the result of
diseases that are recognized to be fatal, this is not the case with primitive man.
When old persons die, he does not believe it to be the result of age. He argues that
there are persons who have lived to be much older. Likewise, no one dies as the
result of disease, for there have been other people who recovered from the same
disease, or never contracted it. To him, the real explanation is always magic.
Either a spirit has killed the man, or it was sorcery. Many primitive tribes recognize death in battle as the only natural death. Still others regard even death in
battle as unnatural, holding that the enemy who caused it must either have been a
sorcerer or have used a charmed weapon. This grotesque idea can on occasion
take an even more impressive form. For instance, two anklets were found in the
stomach of a crocodile shot by a European. The natives recognized the anklets as
the property of two women who, some time before, had been devoured by a crocodile. At once the charge of witchcraft was raised; for this quite natural occurrence,
which would never have aroused the suspicions of a European, was given an
unexpected interpretation in the light of one of those presuppositions which Lévy-
Bruhl calls 'collective representations.' The natives said that an unknown sorcerer
had summoned the crocodile, and had bidden it catch the two women and bring
them to him. The crocodile had carried out this command. But what about the
anklets in the beast's stomach? Crocodiles, they explained, never ate people
unless bidden to do so. The crocodile had merely received the anklets from the
sorcerer as a reward.

This story is a perfect example of that capricious way of explaining things
which is characteristic of the 'prelogical' state of mind. We call it prelogical,
because to us such an explanation seems absurdly illogical. But it seems so to us
only because we start from assumptions wholly different from those of primitive
man. If we were as convinced as he is of the existence of sorcerers and other
mysterious powers, instead of believing in so-called natural causes, his inferences
would seem to us perfectly logical. As a matter of fact, primitive man is no more
logical or illogical than we are. Only his presuppositions are different, and that is
what distinguishes him from us. His thinking and his conduct are based on
assumptions quite unlike our own. To all that is in any way out of the ordinary and
that therefore disturbs, frightens or astonishes him, he ascribes what we would
call a supernatural origin. For him, of course, these things are not supernatural,

but belong to his world of experience. We feel we are stating a natural sequence of events when we say: This house was burned down because it was struck by lightning. Primitive man senses an equally natural sequence of events when he says: A sorcerer used the lightning to set fire to this house/There is absolutely nothing in the world of the primitive – provided that it is at all unusual or impressive – that will not be accounted for on essentially similar grounds. But in explaining things in this way he is acting just like ourselves: he does not examine his assumptions. To him it is an unquestionable truth that disease and other ills are caused by spirits or witchcraft, just as for us it is a foregone conclusion that an illness has a natural cause. We would no more put it down to sorcery than he to natural causes. His mental functioning does not differ in any fundamental way from ours. It is, as I have said, his assumptions alone that distinguish him from ourselves.

It is also supposed that primitive man has other feelings than we, and another kind of morality – that he has, so to speak, a 'prelogical' temperament. Undoubtedly he has a different code of morals. When asked about the difference between good and evil, a Negro chieftain declared: 'When I steal my enemy's wives, it is good, when he steals mine, it is bad.' In many regions it is a terrible insult to tread on a person's shadow, and in others it is an unpardonable sin to scrape a sealskin with an iron knife instead of a flint one. But let us be honest. Do we not think it a sin to eat fish with a steel knife, for a man to keep his hat on in a room, or to greet a lady with a cigar in his mouth? With us, as well as with primitives, such things have nothing to do with ethics. There are good and loyal head-hunters, and there are others who piously and conscientiously perform cruel rites, or commit murder from sacred conviction. The primitive is no less prompt than we are to value an ethical attitude. His good is just as good as ours, and his evil is just as bad as ours. Only the forms under which they appear are different; the process of ethical judgment is the same.

It is likewise thought that primitive man has keener senses than we, or that they are somehow different. But his highly refined sense of direction or of hearing and sight is entirely a matter of professional differentation. If he is confronted with things that are outside his experience, he is amazingly slow and clumsy. I once showed some native hunters, who were as keen-sighted as hawks, magazine pictures in which any child of ours would instantly have recognized human figures. But my hunters turned the pictures round and round until one of them, tracing the outline with his finger, finally exclaimed: 'These are white men!' It was hailed by all as a great discovery.

The incredibly accurate sense of direction shown by many primitives is essentially occupational. It is absolutely necessary that they should be able to find their way in forests and in the bush. Even the European, after a short while in Africa, begins to notice things he would never have dreamed of noticing before – and from fear of going hopelessly astray in spite of his compass.

There is nothing to show that primitive man thinks, feels, or perceives in a way fundamentally different from ours. It is relatively unimportant that he has, or seems to have, a smaller area of consciousness than we, and that he has little or no aptitude for concentrated mental activity. This last, it is true, strikes the European as strange. For instance, I could never hold a palaver for longer than two hours, since by that time the natives declared themselves tired. They said it was too diffi-cult, and yet I had asked only quite simple questions in the most desultory way. But these same people were capable of astonishing concentration and endurance when out hunting or on a journey. My letter-carrier, for instance, could run seventy-five miles at a stretch. I saw a woman in her sixth month of pregnancy, carrying a baby on her back and smoking a long pipe of tobacco, dance almost the whole night through round a blazing fire when the temperature was 95°, without collapsing. It cannot be denied that primitives are quite capable of concentrating on things that interest them. If we have to give our attention to uninteresting matters, we soon notice how feeble our powers of concentration are. We are just as dependent as they are on emotional impulses.

It is true that primitives are simpler and more childlike than we, in good and evil alike. This in itself does not impress us as strange. And yet, when we approach the world of archaic man, we have the feeling of something prodigiously strange. As far as I have been able to analyse it, this feeling comes predominantly from the fact that the primary assumptions of archaic man are essentially different from ours, so that he lives in a different world. Until we come to know his presupposi-tions, he is a hard riddle to read; but when we know them, all is relatively simple. We might equally well say that primitive man ceases to be a riddle for us as soon as we get to know our *own* presuppositions.

It is a rational presupposition of ours that everything has a natural and percep-tible cause. We are convinced of this right from the start. Causality is one of our most sacred dogmas. There is no legitimate place in our world for invisible, arbi-trary, and so-called supernatural powers – unless, indeed, we descend with the modern physicist into the obscure, microcosmic world inside the atom, where, it appears, some very curious things happen. But that lies far from the beaten track. We distinctly resent the idea of invisible and arbitrary forces, for it is not so long ago that we made our escape from that frightening world of dreams and supersti-tions, and constructed for ourselves a picture of the cosmos worthy of our rational consciousness – that latest and greatest achievement of man. We are now surrounded by a world that is obedient to rational laws. It is true that we do not know the causes of everything, but in time they will be discovered, and these discoveries will accord with our reasoned expectations. There are, to be sure, also chance occurrences, but they are merely accidental, and we do not doubt that they have a causality of their own. Chance happenings are repellent to the mind that loves order. They disturb the regular, predictable course of events in the most absurd and irritating way. We resent them as much as we resent invisible, arbi-trary forces, for they remind us too much of Satanic imps or of the caprice of a *deus ex machina*. They are the worst enemies of our careful calculations and a continual threat to all our undertakings. Being admittedly contrary to reason, they deserve all our abuse, and yet we should not fail to give them their due. The Arab shows them greater respect than we. He writes on every letter *Insha' allah*, 'If God wills,' for only then will the letter arrive. In spite of our resentment and in spite of the fact that events run true to general laws, it is undeniable that we are always and everywhere exposed to incalculable accidents. And what is more invisible and capricious than chance? What is more unavoidable and more annoying?

If we consider the matter, we could as well say that the causal connection of events according to general laws is a theory which is borne out about half the time, while for the rest the demon of chance holds sway. Chance events certainly have their natural causes, and all too often we must discover to our sorrow how commonplace they are. It is not this causality that annoys us; the irritating thing about chance events is that they have to befall us here and now in an apparently arbitrary way. At least that is how it strikes us, and even the most obdurate rationalist may occasionally be moved to curse them. However we interpret chance makes no difference to its power. The more regulated the conditions of life become, the more chance is excluded and the less we need to protect ourselves against it. But despite this everyone in practice takes precautions against chance occurrences or hopes for them, even though there is nothing about chance in the official credo.

It is our assumption, amounting to a positive conviction, that everything has a 'natural' cause which, at least in theory, is perceptible. Primitive man, on the other hand, assumes that everything is brought about by invisible, arbitrary powers – in other words, that everything is chance. Only he does not call it chance, but intention. Natural causation is to him a mere pretence and not worthy of mention. If three women go to the river to draw water, and a crocodile seizes the one in the middle and pulls her under, our view of things leads us to the verdict that it was pure chance that that particular woman was seized. The fact that the crocodile seized her at all seems to us quite natural, for these beasts do occasionally eat human beings.

For primitive man such an explanation completely obliterates the facts and accounts for no aspect of the whole exciting story. He rightly finds our explanation superficial or even absurd, for according to this view the accident could just as well not have happened and the same explanation would fit that case too – that it was 'pure chance' it did not. The prejudice of the European does not allow him to see how little he is saying when he explains things in that way.

Primitive man expects far more of an explanation. What we call pure chance is for him wilful intention. It was therefore the intention of the crocodile – as everyone could observe – to seize the middle one of the three women. If it had not had this intention it would have taken one of the others. But why did the crocodile have this intention? Ordinarily these creatures do not eat human beings. That is quite correct – as correct as the statement that it does not ordinarily rain in the Sahara. Crocodiles are rather timid animals, easily frightened. Considering their numbers, they kill astonishingly few people, and it is an unexpected and unnatural event when they devour a man. Such an event calls for an explanation. Of his own accord the crocodile would not take a human life. By whom, then, was he ordered to do so?

It is on the facts of the world around him that primitive man bases his verdicts. When the unexpected occurs he is justifiably astonished and wishes to know the specific causes. To this extent he behaves exactly as we do. But he goes further than we. He has one or more theories about the arbitrary power of chance. We say: Pure chance. He says: Calculating intention. He lays the chief stress on the confusing and confused breaks in the chain of causation, which we call chance – on those occurrences that fail to show the neat causal connections which science expects, and that constitute the other half of happenings in general. He has long ago adapted himself to nature in so far as it conforms to general laws; what he fears is unpredictable chance whose power makes him see in it an arbitrary and incalculable agent. Here again he is right. It is quite understandable that every-thing out of the ordinary should frighten him. Anteaters are fairly numerous in the regions south of Mount Elgon where I stayed for some time. The anteater is a shy, nocturnal animal that is rarely seen. If one happens to be seen by day, it is an extraordinary and unnatural event which astonishes the natives as much as the discovery of a brook that occasionally flows uphill would astonish us. If we knew of actual cases in which water suddenly overcame the force of gravity, such a discovery would be exceedingly disquieting. We know that tremendous masses of water surround us, and can easily imagine what would happen if water no longer conformed to gravitational law. This is the situation in which primitive man finds himself with respect to the happenings in his world. He is thoroughly familiar with the habits of anteaters, but when one of them suddenly transgresses the natural order of things it acquires for him an unknown sphere of action. Primitive man is so strongly impressed by things as they are that a transgression of the laws of his world exposes him to incalculable possibilities. It is a portent, an omen, comparable to a comet or an eclipse. Since such an unnatural event as the appear-ance of an anteater by day can have no natural causes, some invisible power must be behind it. And the alarming manifestation of a power which can transgress the natural order obviously calls for extraordinary measures of placation or defence. The neighbouring villages must be aroused, and the anteater must be dug up with their concerted efforts and killed. The oldest maternal uncle of the man who saw the anteater must then sacrifice a bull. The man descends into the sacrificial pit and receives the first piece of the animal's flesh, whereupon the uncle and the other participants in the ceremony also eat. In this way the dangerous caprice of nature is expiated.

As for us, we should certainly be alarmed if water suddenly began to run uphill for unknown reasons, but are not when an anteater is seen by day, or an albino is born, or an eclipse takes place. We know the meaning and sphere of action of such happenings, while primitive man does not. Ordinary events constitute for him a coherent whole in which he and all other creatures are embraced. He is therefore extremely conservative, and does what others have always done. If something happens, at any point, to break the coherence of this whole, he feels there is a rift in his well-ordered world. Then anything may happen – heaven knows what. All occurrences that are in any way striking are at once brought into connection with the unusual event. For instance, a missionary set up a flagstaff in front of his house so that he could fly the Union Jack on Sundays. But this innocent pleasure cost him dear, for when shortly after his revolutionary action a devastating storm broke out, the flagstaff was of course made responsible. This sufficed to start a general uprising against the missionary.

It is the regularity of ordinary occurrences that gives primitive man a sense of security in his world. Every exception seems to him a threatening act of an arbitrary power that must somehow be propitiated. It is not only a momentary interruption of the ordinary course of things, but a portent of other untoward events. This seems absurd to us, inasmuch as we forget how our grandparents and great-grandparents still felt about the world. A calf is born with two heads and five legs. In the next village a cock has laid an egg. An old woman has had a dream, a comet appears in the sky, there is a great fire in the nearest town, and the following year a war breaks out. In this way history was always written from remote antiquity down to the eighteenth century. This concatenation of events, so meaningless to us, is significant and convincing to primitive man. And, contrary to all expectation, he is right to find it so. His powers of observation can be trusted. From age-old experience he knows that such concatenations actually exist. What seems to us a wholly senseless heaping-up of single, haphazard occurrences – because we pay attention only to single events and their particular causes – is for the primitive a completely logical sequence of omens and of happenings indicated by them. It is a fatal outbreak of demonic power showing itself in a thoroughly consistent way.

The calf with two heads and the war are one and the same, for the calf was only an anticipation of the war. Primitive man finds this connection so unquestionable and convincing because the whims of chance seem to him a far more important factor in the happenings of the world than regularity and conformity to law. Thanks to his close attention to the unusual, he discovered long before us that chance events arrange themselves in groups or series. The law of the duplication of cases is known to all doctors engaged in clinical work. An old professor of psychiatry at Würzburg always used to say of a particularly rare clinical case: 'Gentlemen, this case is absolutely unique – tomorrow we shall have another just like it.' I myself often observed the same thing during my eight years' practice in an insane asylum. On one occasion a person was committed for a very rare twilight state of consciousness – the first case of this kind I had ever seen. Within two days we had a similar case, and that was the last. 'Duplication of cases' is a joke with us in the clinics, but it was also the first object of primitive science. A recent investigator has ventured the statement: 'Magic is the science of the jungle.' Astrology and other methods of divination may certainly be called the science of antiquity.

What happens regularly is easily observed because we are prepared for it. Knowledge and skill are needed only in situations where the course of events is interrupted in a way hard to fathom. Generally it is one of the shrewdest and wiliest men of the tribe who is entrusted with the observation of meteorological events. His knowledge must suffice to explain all unusual occurrences, and his art to combat them. He is the scholar, the specialist, the expert on chance, and at the same time the keeper of the archives of the tribe's traditional lore. Surrounded by respect and fear, he enjoys great authority, yet not so great but that his tribe is secretly convinced that the neighbouring tribe has a sorcerer who is stronger than theirs. The best medicine is never to be found close at hand, but as far away as possible. I stayed for some time with a tribe who held their old medicine-man in the greatest awe. Nevertheless he was consulted only for the minor ailments of cattle and men. In all serious cases a foreign authority was called in – a *M'ganga* who was brought at a high fee from Uganda – just as with us.

Chance events occur most often in larger or smaller series or groups. An old and well-tried rule for foretelling the weather is this, that when it has rained for several days it will also rain tomorrow. A proverb says, 'Misfortunes never come singly.' Another has it that 'It never rains but it pours.' This proverbial wisdom is primitive science. The common people still believe it and fear it, but the educated man smiles at it – until something unusual happens to him. I will tell you a disagreeable story. A woman I know was awakened one morning by a peculiar tinkling on her night-table. After looking about her for a while she discovered the cause: the rim of her tumbler had snapped off in a ring about a quarter of an inch wide. This struck her as peculiar, and she rang for another glass. About five minutes later she heard the same tinkling, and again the rim of the glass had broken off. This time she was greatly disquieted, and had a third glass brought. Within twenty minutes the rim broke off again with the same tinkling noise. Three such accidents in immediate succession were too much for her. She gave up her belief in natural causes on the spot and brought out in its place a primitive 'collective representation' – the conviction that an arbitrary power was at work. Something of this sort happens to many modern people – provided they are not too thick-skulled – when they are confronted with events which natural causation fails to explain. We naturally prefer to deny such occurrences. They are unpleasant because they disrupt the orderly course of our world and make anything seem possible, thus proving that the primitive mind in us is not yet dead.

Primitive man's belief in an arbitrary power does not arise out of thin air, as was always supposed, but is grounded in experience. The grouping of chance occurrences justifies what we call his superstition, for there is a real measure of probability that unusual events will coincide in time and place. We must not forget that our experience is apt to leave us in the lurch here. Our observation is inadequate because our point of view leads us to overlook these matters. For instance, it would never seriously occur to us to take the following events as a sequence: in the morning a bird flies into your room, an hour later you witness an accident in the street, in the afternoon a relative dies, in the evening the cook drops the soup tureen, and, on coming home at night, you find that you have lost your key. Primitive man would not have overlooked a single item in this chain of events. Every new link would have confirmed his expectations, and he would be right – much more nearly right than we are willing to admit. His anxious expectations are fully justified and serve a purpose. Such a day is ill-omened, and on it nothing should be undertaken. In our world this would be reprehensible superstition, but in the world of the primitive it is highly appropriate shrewdness. In that world man is far more exposed to accidents than we are in our sheltered and well-regulated existence. When you are in the bush you dare not take too many chances. The European soon comes to appreciate this.

When a Pueblo Indian does not feel in the right mood, he stays away from the men's council. When an ancient Roman stumbled on the threshold as he left his house, he gave up his plans for the day. This seems to us senseless, but under primitive conditions such an omen inclines one at least to be cautious. When I am not in full control of myself, I am hampered in my movements, my attention wanders, I get absent-minded. As a result I knock against something, stumble, drop something, forget something. Under civilized conditions all these are mere trifles, but in the primeval forest they mean mortal danger. I make a false step on a slippery tree-trunk that serves as a bridge over a river teeming with crocodiles. I lose my compass in the high grass. I forget to load my rifle and blunder into a rhinoceros trail in the jungle. I am preoccupied with my thoughts and step on a puff-adder. At nightfall I forget to put on my mosquito-boots in time and eleven days later I die from an onset of tropical malaria. To forget to keep one's mouth shut while bathing is enough to bring on a fatal attack of dysentery. For us accidents of this kind have their recognizable natural cause in a somewhat distracted psychological state, but for the primitive they are objectively conditioned omens, or sorcery.

It may be rather more than a question of inattention, however. In the Kitoshi region south of Mount Elgon, in East Africa, I went on an expedition into the Kabras forest. There, in the thick grass, I nearly stepped on a puff-adder, and only managed to jump away just in time. That afternoon my companion returned from a hunt, deathly pale and trembling in every limb. He had narrowly escaped being bitten by a seven-foot mamba which darted at him from behind a termite hill. He would undoubtedly have been killed had he not been able to wound the brute with a shot at the last moment. At nine o'clock that night our camp was attacked by a pack of ravenous hyenas which had surprised a man in his sleep the day before and torn him to pieces. In spite of the fire they swarmed into the hut of our cook, who fled screaming over the stockade. Thenceforth there were no accidents throughout the whole of our journey. Such a day gave our Negroes food for thought. For us it was a simple multiplication of chance events, but for them the inevitable fulfilment of an omen that had occurred on the first day of our journey into the wilds. It so happened that we had fallen, Ford car, bridge, and all, into a stream we were trying to cross. Our boys had exchanged glances as if to say: 'Well, that's a fine start.' To cap this calamity, a tropical thunderstorm blew up and soaked us so thoroughly that I was prostrated with fever for several days. On the evening of the day when my friend had had such a narrow escape out hunting,

I could not help saying to him as we white men sat looking at one another: 'You know, it seems to me as if the trouble had begun still further back. Do you remember the dream you told me in Zurich just before we left?' At that time he had had a very impressive nightmare. He dreamed he was hunting in Africa, and was suddenly attacked by a huge mamba, so that he woke up with a cry of terror. The dream had disturbed him greatly, and he now confessed to me that he had thought it portended the death of one of us. He had of course assumed that it was my death, because we always hope it is the other fellow. But it was he who later fell ill of a severe malarial fever that brought him to the brink of the grave.

To read of such a conversation in a corner of the world where there are no mambas and no anopheles mosquitoes means very little. One must imagine the velvety blue of a tropical night, the overhanging black masses of gigantic trees standing in the primeval forest, the mysterious voices of the nocturnal spaces, a lonely fire with loaded rifles stacked beside it, mosquito-nets, boiled swamp-water to drink, and above all the conviction expressed by an old Afrikander who knew what he was talking about: 'This isn't man's country – it's God's country.' There man is not king; it is rather nature, the animals, plants, and the microbes. Given the mood that goes with the place, one understands how it is that we found a dawning significance in things that anywhere else would provoke a smile. That is the world of unrestrained capricious powers which primitive man has to deal with every day. The unusual event is no joke to him. He draws his own conclusions. 'This is not a good place,' 'The day is unfavourable' – and who knows what dangers he avoids by following such warnings?

'Magic is the science of the jungle.' The portent brings about an immediate alteration of a course of action, the abandonment of a planned undertaking, a change of psychic attitude. These are all highly expedient reactions in view of the fact that chance occurrences tend to fall into sequences and that primitive man is wholly unconscious of psychic causality. Thanks to our one-sided emphasis on so-called natural causes, we have learned to differentiate what is subjective and psychic from what is objective and 'natural.' For primitive man, on the contrary, the psychic and the objective coalesce in the external world. In the face of something extraordinary it is not he who is astonished, but rather the thing which is astonishing. It is *mana* – endowed with magic power. What we would call the powers of imagination and suggestion seem to him invisible forces which act on him from without. His country is neither a geographical nor a political entity. It is that territory which contains his mythology, his religion, all his thinking and feeling in so far as he is unconscious of these functions. His fear is localized in certain places that are 'not good.' The spirits of the departed inhabit such and such a wood. That cave harbours devils who strangle any man who enters. In yonder mountain lives the great serpent; that hill is the grave of the legendary king; near this spring or rock or tree every woman becomes pregnant; that ford is guarded by snake-demons; this towering tree has a voice that can call certain people. Primitive man is unpsychological. Psychic happenings take place outside him in an objective way. Even the things he dreams about are real to him; that is his only reason for paying attention to dreams. Our Elgonyi porters maintained in all seriousness that they never had dreams – only the medicine-man had them. When I questioned the medicine-man, he declared that he had stopped having dreams when the British entered the land. His father had still had 'big' dreams, he told me, and had known where the herds strayed, where the cows took their calves, and when there was going to be a war or a pestilence. It was now the District Commissioner who knew everything, and they knew nothing. He was as resigned as certain Papuans who believe that the crocodiles have for the most part gone over to the British Government. It happened that a native convict who had escaped from the authorities had been badly mangled by a crocodile while trying to cross a river. They therefore concluded that it must have been a police crocodile. God now speaks in dreams to the British, and not to the medicine-man of the Elgonyi, he told me, because it is the British who have the power. Dream activity has emigrated. Occasionally the souls of the natives wander off too, and the medicine-man catches them in cages as if they were birds, or strange souls come in as immigrants and cause peculiar diseases.

This projection of psychic happenings naturally gives rise to relations between men and men, or between men and animals or things, that to us are inconceivable. A white man shoots a crocodile. At once a crowd of people come running from the nearest village and excitedly demand compensation. They explain that the crocodile was a certain old woman in their village who had died at the moment when the shot was fired. The crocodile was obviously her bush-soul. Another man shot a leopard that was lying in wait for his cattle. Just then a woman died in a neighbouring village. She and the leopard were identical.

Lévy-Bruhl has coined the expression *participation mystique* for these remarkable relationships. It seems to me that the word 'mystical' is not happily chosen. Primitive man does not see anything mystical in these matters, but considers them perfectly natural. It is only we who find them so strange, because we appear to know nothing about the phenomena of psychic dissociation. In reality, however, they occur in us too, not in this naïve but in a rather more civilized form. In daily life it happens all the time that we presume that the psychology of other people is the same as ours. We suppose that what is pleasing or desirable to us is the same to others, and that what seems bad to us must also seem bad to them. It is only recently that our courts of law have nerved themselves to admit the psychological relativity of guilt in pronouncing sentence. The tenet *quod licet Jovi non licet bovi* still rankles in the minds of all unsophisticated people; equality before the law is still a precious achievement. And we still attribute to the other fellow all the evil and inferior qualities that we do not like to recognize in ourselves, and therefore have to criticize and attack him, when all that has happened is that an inferior 'soul' has emigrated from one person to another. The world is still full of *bêtes noires* and scapegoats, just as it formerly teemed with witches and werewolves.

Projection is one of the commonest psychic phenomena. It is the same as *participation mystique*, which Lévy-Bruhl, to his great credit, emphasized as being an especially characteristic feature of primitive man. We merely give it another name, and as a rule deny that we are guilty of it. Everything that is unconscious in ourselves we discover in our neighbour, and we treat him accordingly. We no longer subject him to the test of drinking poison; we do not burn him or put the screws on him; but we injure him by means of moral verdicts pronounced with the deepest conviction. What we combat in him is usually our own inferior side.

The simple truth is that primitive man is somewhat more given to projection than we because of the undifferentiated state of his mind and his consequent inability to critize himself. Everything to him is absolutely objective, and his speech reflects this in a drastic way. With a touch of humour we can picture to ourselves what a leopard-woman is like, just as we do when we call a person a goose, a cow, a hen, a snake, an ox, or an ass. These uncomplimentary epithets are familiar to us all. But when primitive man attributes a bush-soul to a person, the poison of moral judgment is absent. He is too naturalistic for that; he is too much impressed by things as they are and much less prone to pass judgment than we. The Pueblo Indians declared in a matter-of-fact way that I belonged to the Bear Totem – in other words, that I was a bear – because I did not come down a ladder standing up like a man, but bunched up on all fours like a bear. If anyone in Europe said I had a bearish nature this would amount to the same thing, but with a rather different shade of meaning. The theme of the bush-soul, which seems so strange to us when we meet with it among primitives, has become with us a mere figure of speech, like so much else. If we take our metaphors concretely we return to the primitive point of view. For instance, we have the expression 'to handle a patient.' In concrete terms this means 'to lay hands on' a person, 'to work at with the hands,' 'to manipulate.' And this is precisely what the medicine-man does with his patients.

We find the bush-soul hard to understand because we are baffled by such a concrete way of looking at things. We cannot conceive of a 'soul' that splits off completely and takes up its abode in a wild animal. When we describe someone as an ass, we do not mean that he is in every aspect the quadruped called an ass. We mean that he resembles an ass in some particular respect. We split off a bit of his personality or psyche and personify it as an ass. So, too, for primitive man the leopard-woman is a human being, only her bush-soul is a leopard. Since all unconscious psychic life is concrete and objective for him, he supposes that a person describable as a leopard has the soul of a leopard. If the splitting and concretizing go still further, he assumes that the leopard-soul lives in the bush in the form of a real leopard.

These identifications, brought about by projection, create a world in which man is completely contained psychically as well as physically. To a certain extent he coalesces with it. In no way is he master of this world, but only a fragment of it. Primitive man is still far from the glorification of human powers. He does not dream of regarding himself as the lord of creation. In Africa, for instance, his zoological classification does not culminate in *Homo sapiens*, but in the elephant. Next comes the lion, then the python or the crocodile, then man and the lesser creatures. Man is still dovetailed into nature. It never occurs to him that he might be able to rule her; all his efforts are devoted to protecting himself against her dangerous caprices. It is civilized man who strives to dominate nature and therefore devotes his greatest energies to the discovery of natural causes which will give him the key to her secret laboratory. That is why he strongly resents the idea of arbitrary powers and denies them. Their existence would amount to proof that his attempt to dominate nature is futile after all.

Summing up, we may say that the outstanding trait of archaic man is his attitude towards the arbitrary power of chance, which he considers a far more important factor in the world-process than natural causes. It consists on the one hand in the observed tendency of chance occurrences to take place in a series, and on the other in the projection of unconscious psychic contents through *participation mystique*. For archaic man this distinction does not exist, because psychic happenings are projected so completely that they cannot be distinguished from objective, physical events. For him the vagaries of chance are arbitrary and intentional acts, interventions by animate beings. He does not realize that unusual events stir him so deeply only because he invests them with the power of his own astonishment or fear. Here, it is true, we move on treacherous ground. Is a thing beautiful because I attribute beauty to it? Or is it the objective beauty of the thing that compels me to acknowledge it? As we know, great minds have wrestled with the problem whether it is the glorious sun that illuminates the world, or the sunlike human eye. Archaic man believes it to be the sun, and civilized man believes it to be the eye – so far, at any rate, as he reflects at all and does not suffer from the disease of the poets. He must de-psychize nature in order to dominate her; and in order to see his world objectively he must take back all his archaic projections.

In the archaic world everything has soul – the soul of man, or let us say of mankind, the collective unconscious, for the individual has as yet no soul of his own. We must not forget that what the Christian sacrament of baptism purports to do is a landmark of the utmost significance in the psychic development of mankind. Baptism endows the individual with a living soul. I do not mean that the baptismal rite in itself does this, by a unique and magical act. I mean that the idea of baptism lifts man out of his archaic identification with the world and transforms him into a being who stands above it. The fact that mankind has risen to the level of this idea is baptism in the deepest sense, for it means the birth of the spiritual man who transcends nature.

In the psychology of the unconscious it is an axiom that every relatively independent portion of the psyche has the character of personality, that it is personified as soon as it is given an opportunity for independent expression. We find the clearest instances of this in the hallucinations of the insane and in mediumistic communications. Whenever an autonomous component of the psyche is projected, an invisible person comes into being. In this way the spirits arise at an ordinary spiritualistic séance. So too among primitives. If an important psychic component is projected on a human being, he becomes *mana,* extraordinarily effective – a sorcerer, witch, werewolf, or the like. The primitive idea that the medicine-man catches the souls that have wandered away by night and puts them in cages like birds is a striking illustration of this. These projections give the medicine-man his mana, they cause animals, trees, and stones to speak, and because they are his own psychic components they compel the projicient to obey them absolutely. For this reason an insane person is helplessly at the mercy of his voices; they are projections of his own psychic activity whose unconscious subject he is. He is the one who speaks through his voices, just as he is the one who hears, sees, and obeys.

From a psychological point of view, therefore, the primitive theory that the arbitrary power of chance is the outcome of the intentions of spirits and sorcerers is perfectly natural, because it is an unavoidable inference from the facts as primitive man sees them. Let us not delude ourselves in this connection. If we explain our scientific views to an intelligent native he will accuse us of ludicrous superstitiousness and a disgraceful want of logic, for he believes that the world is lighted by the sun and not by the human eye. My friend Mountain Lake, a Pueblo chief, once called me sharply to account because I had made insinuating use of the Augustinian argument: 'Not this sun is our Lord, but he who made this sun.' Pointing to the sun he cried indignantly: 'He who goes there is our father. You can see him. From him comes all light, all life – there is nothing that he has not made.' He became greatly excited, struggled for words, and finally cried out: 'Even a man in the mountains, who goes alone, cannot make his fire without him.' The archaic standpoint could hardly be more beautifully expressed than by these words. The power that rules us is outside, in the external world, and through it alone are we permitted to live. Religious thought keeps alive the archaic state of mind even today, in a time bereft of gods. Untold millions of people still think like this.

Speaking earlier of primitive man's attitude to the arbitrary power of chance, I expressed the view that this attitude serves a purpose and therefore has a meaning. Shall we, for the moment at least, venture the hypothesis that the primitive belief in arbitrary powers is justified by the facts and not merely from a psychological point of view? This sounds alarming, but I have no intention of jumping from the frying-pan into the fire and trying to prove that witchcraft actually works. I merely wish to consider the conclusions to which we shall be led if we follow primitive man in assuming that all light comes from the sun, that things are beautiful in themselves, and that a bit of the human soul is a leopard – in other words, that the mana theory is correct. According to this theory, beauty moves *us*, it is not we who create beauty. A certain person *is* a devil, we have not projected our own evil on him and in this way made a devil out of him. There are people – mana personalities – who are impressive in their own right and in no way thanks to our imagination. The *mana theory* maintains that there is something like a widely distributed power in the external world that produces all those extraordinary effects. Everything that exists acts, otherwise it would not *be*. It can *be* only by virtue of its inherent energy. Being is a field of force. The primitive idea of mana, as you can see, has in it the beginnings of a crude theory of energy.

So far we can easily follow this primitive idea. The difficulty arises when we try to carry its implications further, for they reverse the process of psychic projection of which I have spoken. It is then not my imagination or my awe that makes the medicine-man a sorcerer; on the contrary, he *is* a sorcerer and projects his magical powers on me. Spirits are not hallucinations of my mind, but appear to me of their own volition. Although such statements are logical derivatives of the mana idea, we hesitate to accept them and begin to look around for a comfortable theory of psychic projection. The question is nothing less than this: Does the psychic in general – the soul or spirit or the unconscious – originate in *us*, or is the psyche, in the early stages of conscious evolution, actually outside us in the form of arbitrary powers with intentions of their own, and does it gradually take its place within us in the course of psychic development? Were the split-off 'souls' – or dissociated psychic contents, as we would call them – ever parts of the psyches of individuals, or were they from the beginning psychic entities existing in themselves according to the primitive view as ghosts, ancestral spirits, and the like? Were they only by degrees embodied in man in the course of development, so that they gradually constituted in him that world which we now call the psyche?

This whole idea strikes us as dangerously paradoxical, but, at bottom, it is not altogether inconceivable. Not only the religious instructor but the educator as well assumes that it is possible to implant something psychic in man that was not there before. The power of suggestion and influence is a fact; indeed, the modern behaviourists have extravagant expectations in this respect. The idea of a complex building-up of the psyche is expressed on a primitive level in a variety of forms, for instance in the widespread belief in possession, the incarnation of ancestral spirits, the immigration of souls, and so forth. When someone sneezes, we still say: 'God bless you,' by which is meant: 'I hope your new soul will do you no harm.' When in the course of our own development we feel ourselves achieving a unified personality out of a multitude of contradictory tendencies, we experience something like a complex growing-together of the psyche. Since the human body is built up by heredity out of a multitude of Mendelian units, it does not seem altogether out of the question that the human psyche is similarly put together.

The materialistic views of our day have one tendency which they share with archaic thought: both lead to the conclusion that the individual is a mere resultant. In the first case he is the resultant of natural causes, and in the second of chance occurrences. According to both accounts, human individuality is nothing in its own right, but rather the accidental product of forces contained in the objective environment. This is thoroughly consistent with the archaic view of the world, in which the ordinary individual is never important, but always interchangeable with any other and easily dispensable. By the roundabout way of strict causalism, modern materialism has returned to the standpoint of archaic man. But the materialist is more radical, because he is more systematic. Archaic man has the advantage of being inconsistent: he makes an exception of the mana personality. In the course of history these mana personalities were exalted to the position of divine figures; they became heroes and kings who shared the immortality of the gods by eating the food of eternal youth. This idea of the immortality of the individual and of his imperishable worth can be found on the earliest archaic levels, first of all in the belief in spirits, and then in myths of the age when death had not yet gained entry into the world through human carelessness or folly.

Primitive man is not aware of this contradiction in his views. My Elgonyi porters assured me that they had no idea what would happen to them after death. According to them a man is simply dead, he does not breathe any more, and the corpse is carried into the bush where the hyenas eat it. That is what they think by day, but the night teems with spirits of the dead who bring diseases to cattle and men, who attack and strangle the nocturnal traveller and indulge in other forms of violence. The primitive mind is full of such contradictions. They could worry a European out of his skin, and it would never occur to him that something quite similar is to be found in our civilized midst. We have universities where the very thought of divine intervention is considered beneath dispute, but where theology is a part of the curriculum. A research worker in natural science who thinks it positively obscene to attribute the smallest variation of an animal species to an act of divine arbitrariness may have in another compartment of his mind a full-blown Christian faith which he likes to parade on Sundays. Why should we excite ourselves about primitive inconsistency?

It is impossible to derive any philosophical system from the fundamental thoughts of primitive man. They provide only antinomies, but it is just these that are the inexhaustible source of all spiritual problems in all times and in all civilizations. We may ask whether the 'collective representations' of archaic man are really profound, or do they only seem so? I cannot answer this most difficult of questions, but I would like, in conclusion, to tell you of an observation I made among the mountain tribe of the Elgonyi. I searched and inquired far and wide for traces of religious ideas and ceremonies, and for weeks on end I discovered nothing. The natives let me see everything and were willing to give me any information. I could talk with them without the hindrance of a native interpreter, for many of the old men spoke Swahili. At first they were rather reserved, but once the ice was broken I had the friendliest reception. They knew nothing of religious customs. But I did not give up, and finally, at the end of one of many fruitless palavers, an old man suddenly exclaimed: 'In the morning, when the sun comes up, we go out of the huts, spit in our hands, and hold them up to the sun.' I got them to perform the ceremony for me and describe it exactly. They hold their hands before their faces and spit or blow into them vigorously. Then they turn their hands round and hold the palms towards the sun. I asked them the meaning of what they did – why they blew or spat in their hands. My question was futile. 'That is how it has always been done,' they said. It was impossible to get an explanation, and it became clear to me that they knew only what they did and not why they did it. They see no meaning in their action. They greet the new moon with the same gesture.

Now let us suppose that I am a total stranger in Zurich and have come to this city to explore the customs of the place. First I settle down on the outskirts near some suburban homes, and come into neighbourly contact with their owners. I then say to Messrs. Müller and Meyer: 'Please tell me something about your religious customs.' Both gentlemen are taken aback. They never go to church, know nothing about it, and emphatically deny that they practise any such customs. It is spring, and Easter is approaching. One morning I surprise Mr. Müller at a curious occupation. He is busily running about the garden, hiding coloured eggs and setting up peculiar rabbit idols. I have caught him *in flagrante*. 'Why did you conceal this highly interesting ceremony from me?' I ask him. 'What ceremony?' he retorts. 'This is nothing. Everybody does it at Eastertime.' 'But what is the meaning of these idols and eggs, and why do you hide them?' Mr. Müller is stunned. He does not know, any more than he knows the meaning of the Christmas-tree. And yet he does these things, just like a primitive. Did the distant ancestors of the Elgonyi know any better what they were doing? It is highly improbable. Archaic man everywhere does what he does, and only civilized man knows what he does.

What is the meaning of the Elgonyi ceremony just cited? Clearly it is an offering to the sun, which for these natives is *mungu* – that is, *mana*, or divine – only at the moment of rising. If they have spittle on their hands, this is the substance which, according to primitive belief, contains the personal *mana*, the life-force, the power to heal and to make magic. If they breathe into their hands, breath is wind and spirit – it is *roho*, in Arabic *ruch*, in Hebrew *ruach*, and in Greek *pneuma*. The action means: I offer my living soul to God. It is a wordless, acted prayer, which could equally well be spoken: 'Lord, into thy hands I commend my spirit.'

Does this merely happen so, or was this thought brooded and willed even before man existed? I must leave this question unanswered.

Chapter 7

Jung and Lévy-Bruhl

Robert A. Segal

For his knowledge of primitive peoples, C. G. Jung relied almost wholly on the work of Lucien Lévy-Bruhl (1857–1939), a celebrated French philosopher who in mid-career turned to anthropology. In a series of six books from 1910 on, Lévy-Bruhl asserted that primitive peoples had been misunderstood by modern Westerners.[1] Rather than thinking like moderns, just less rigorously, primitives think wholly differently from moderns. Their thinking differs in two key ways: it is 'mystical', and it is 'prelogical'. By 'mystical', Lévy-Bruhl means that primitive peoples experience the world as identical with themselves rather than, like moderns, as distinct from themselves. Primitive peoples do not merely conceive but also perceive, or experience, the world as one with themselves. Their relationship to the world, including that to fellow human beings, is one of *participation mystique*. By 'prelogical', Lévy-Bruhl means that primitives are indifferent to contradictions rather than, like moderns, attentive to them. The primitive mind deems all things identical with one another yet somehow still distinct – a logical contradiction. A human is simultaneously a tree and still a human being.

Jung accepted unquestioningly Lévy-Bruhl's depiction of the primitive mind. He did so even when he, unlike Lévy-Bruhl, actually journeyed to the field to see 'native' peoples firsthand. Yet Jung in fact misses the difference for Lévy-Bruhl between the mystical and the prelogical aspects of primitive thinking. He conflates prelogical with mystical and thereby misses the more radical aspect of the primitive mind for Lévy-Bruhl.

At the same time Jung alters Lévy-Bruhl's conception of primitive mentality in three ways. First, he psychologizes primitive thinking. Where for Lévy-Bruhl primitive thinking is to be explained sociologically, for Jung it is to be explained psychologically. Primitive peoples think as they do not because they live in society but because they live in unconsciousness. Second, Jung universalizes primitive mentality. Where for Lévy-Bruhl primitive thinking is ever more being

Note: A shorter version of this article, together with its appended 'Reply to Susan Rowland', was first published in *The Journal of Analytical Psychology* 52/5 (November 2007), 635–58 and 667–71. It is reproduced here with kind permission of Wiley-Blackwell and *The Journal of Analytical Psychology*.

replaced by modern thinking, for Jung primitive thinking is the initial psychological state of all human beings. Third, Jung values primitive thinking. Where for Lévy-Bruhl primitive thinking is false, for Jung it is true – once it is recognized as an expression not of how the world works but of how the unconscious works.

Lévy-Bruhl

Lévy-Bruhl's first and most important anthropological work, *How Natives Think*, was originally published in French in 1910 (*Les fonctions mentales dans les sociétés inférieures*) but was translated into English only in 1926 – three years *after* the translation of his second and next most important anthropological work, *Primitive Mentality* (1923), originally published in 1922. In 1985 Princeton University Press published a reprint of the 1926 translation of *How Natives Think* with a new introduction by American anthropologist C. Scott Littleton ('Lucien Lévy-Bruhl and the Concept of Cognitive Relativity', pp. v–lviii).

Lévy-Bruhl never asserts that primitive peoples are inferior to moderns. On the contrary, he means to be defending primitive peoples *against* this charge, made above all by the pioneering British anthropologists E. B. Tylor and J. G. Frazer. Of Tylor and Frazer, he states: 'let us abandon the attempt to refer their mental activity to an inferior variety of our own' (Lévy-Bruhl, *How Natives Think*, p. 76). For Tylor and Frazer, primitives think the way moderns do. They just think less rigorously than moderns. For Tylor and Frazer, the difference between primitive and modern thinking is only of degree. For Lévy-Bruhl, the difference is of kind.

Lévy-Bruhl attributes primitive thinking to culture, not biology. In accord with other twentieth-century anthropologists,[2] he separates culture from race: 'Undoubtedly they [primitives] have the same senses as ours [. . .] and their cerebral structure is like our own. But we have to bear in mind that which their collective representations instil [sic] into all their perceptions' (Lévy-Bruhl, *How Natives Think*, p. 43).[3] By 'collective representations' (*représentations collectives*), a term taken from the French sociologist Émile Durkheim, Lévy-Bruhl means group beliefs, ones inculcated in all members of society. Those beliefs are the same across all primitive societies. Primitive representations, or *conceptions*, shape *perceptions*, or experiences (see Lévy-Bruhl, *How Natives Think*, pp. 43–45, 106).[4]

According to Lévy-Bruhl, primitive peoples believe that all phenomena, including humans and their artifacts, are part of – or 'participate in' – an impersonal sacred, or 'mystic', realm pervading the natural one: 'Primitive man, therefore, lives and acts in an environment of beings and objects, all of which, in addition to the [observable] properties that we recognize them to possess, are endued with mystic attributes' (Lévy-Bruhl, *How Natives Think*, p. 65). To take Lévy-Bruhl's most famous example, the Bororo of Brazil declare themselves red parakeets: 'This does not merely signify that after their death they become araras [parakeets], nor that araras are metamorphosed Bororos [. . .]. It is not a name they give themselves, nor a relationship that they claim. What they desire to express by it is actual identity' (Lévy-Bruhl, *How Natives Think*, p. 77).

Mysticism is the first of the two key characteristics of primitive mentality. The second characteristic, prelogicality, builds on the first one but is more radical, for it violates the law of noncontradiction: the notion that something cannot simultaneously be both itself and something else. The belief that all things are mystically one is itself neither contradictory nor uniquely primitive. As not only a belief but also a practice, mysticism is to be found in many cultures, including the West from ancient times through modern. But the belief that all things simultaneously remain distinct, that 'phenomena can be [. . .] both themselves and something other than themselves', is uniquely primitive (Lévy-Bruhl, *How Natives Think*, p. 76). The Bororo believe that a human is really a parakeet yet still really a human being. They do not believe that a human and a parakeet are identical invisibly while distinct visibly. That belief would merely be a version of mysticism. Rather, the Bororo believe that humans and parakeets are at once identical and separate in the same respects and at the same time. Visibly as well as invisibly, humans and parakeets are at once the same and different.[5]

Lévy-Bruhl does not conclude, as is conventionally said of him, that primitive peoples cannot think logically, as if they are biologically deficient.[6] Instead, he concludes that primitives, ruled as they are by their collective representations, regularly suspend the practice of logic: primitive thought 'is not *antilogical*; it is not *alogical* either. By designating it "prelogical" I merely wish to state that it does not bind itself down, as our thought does, to avoiding contradiction' (Lévy-Bruhl, *How Natives Think*, p. 78).[7] As painstakingly precise a writer as Lévy-Bruhl was, his choice of terms was often almost perversely misleading, and many readers mistook 'pre-logical' for 'illogical'. He thereby seemed to be making primitive peoples even more hopelessly inferior to moderns than Tylor and Frazer had made them – the opposite of his intent. As unacceptable as the term primitive has become, it was used unashamedly in Lévy-Bruhl's day and is still used, albeit far more neutrally, in French.[8]

In arguing relentlessly that primitive thinking differs in nature from modern thinking, Lévy-Bruhl is not, however, arguing that it is equally true. In asserting that, as Littleton puts it, primitive thinking 'must be understood *on its own terms*',[9] he is not asserting that it must be *judged* on those terms. Primitive thinking does make sense in light of its premises: 'The fact that the "*patterns of thought*" are different does not, once the premises have been given, prevent the primitiv" from reasoning like us, and, in this sense, his thought is neither more nor less "logical" than ours' (Lévy-Bruhl, 'A Letter to E. E. Evans-Pritchard', p. 121 [italics in original]). But the premises of primitive thinking are still illogical. Therefore primitive thinking, while logical once given its premises, is illogical because its premises are illogical: something cannot simultaneously be both itself and something else in the same respects. Where for Tylor and to a lesser extent Frazer primitive thinking is rational but still false, for Lévy-Bruhl primitive thinking is irrational and consequently false.

Put another way, Lévy-Bruhl is not a relativist. He is an absolutist. There are several varieties of relativism – conceptual, perceptual, and moral – and none fits Lévy-Bruhl. Conceptual relativism, which is what Littleton and others wrongly ascribe to Lévy-Bruhl,[10] denies the existence of objective criteria for assessing the diversity of *beliefs* about the world. Beliefs can supposedly be evaluated only within a culture, by only its own standards. Lévy-Bruhl is hardly a conceptual relativist: he deems mysticism and prelogicality outright false beliefs about the world. Perceptual relativism denies the possibility of evaluating objectively the diversity of *experiences* of the world. Conceptual relativism allows for common experiences that simply get interpreted differently. Perceptual relativism, which is bolder, maintains that experiences themselves differ. People 'occupy' different worlds, and there is no way to judge the differences. What to one culture is the experience of a god is to another a delusion. Lévy-Bruhl is hardly a perceptual relativist either: he deems the experience of oneness a delusion. Moral relativism, which denies that objective criteria exist for evaluating the undeniable diversity of *values* around the world, is not relevant to Lévy-Bruhl.

While Lévy-Bruhl takes the concept of collective representations from Durkheim, he stresses the differences rather than, like Durkheim, the similarities between primitive and modern representations. For Lévy-Bruhl, primitive representations alone come between primitives and the world. The representations shape perceptions as well as conceptions, so that primitive peoples experience, not merely think, the world one as well as distinct. By contrast, for Lévy-Bruhl, modern representations shape only conceptions, not also perceptions, which convey the world to moderns rather than come between moderns and the world (see Lévy-Bruhl, *How Natives Think*, pp. 375–76).

In a section of *How Natives Think* entitled 'The Transition to the Higher Mental Types' (pp. 361–86), Lévy-Bruhl writes of 'progress' and 'evolution' in cognition, which requires the filtering out of the emotional elements that color primitive perceptions (Lévy-Bruhl, *How Natives Think*, pp. 380–81). Only modern representations have been subjected to 'the test of experience' (Lévy-Bruhl, *How Natives Think*, pp. 380–81). In fact, only 'scientific theorizing' is abstract enough to be free of emotion and therefore of mystical and prelogical proclivities (Lévy-Bruhl, *How Natives Think*, p. 382). The difference between primitive peoples and moderns is not, then, that moderns think wholly logically. It is that primitive peoples think wholly prelogically. For Lévy-Bruhl, the emotional allure of mystical participation makes its disappearance unlikely, and he cites example after example of the retention of prelogical thinking among moderns (see Lévy-Bruhl, *How Natives Think*, pp. 382–83).[11] Conversely, he traces the lessening of mystical ties even among primitive peoples (see Lévy-Bruhl, *How Natives Think*, pp. 365–79). The opposition that he draws is, then, between primitive and modern *thinking*, not between primitives and moderns *themselves*.

Many others no less absolutist than Lévy-Bruhl have been criticized much less severely. The reason, disputed by Littleton (see Littleton, 'Lucien Lévy-Bruhl and the Concept of Cognitive Relativity', pp. xix–xx), is that, despite his undeniably neutral intent, Lévy-Bruhl in fact characterizes primitive mentality much more negatively than even Tylor (*Primitive Culture* [1958] (1913)) and Frazer (*The Golden Bough* [1922]) do.[12] Tylor and Frazer take for granted that primitive peoples recognize not only the law of noncontradiction but most 'modern' distinctions as well: those between appearance and reality, subjectivity and objectivity, supernatural and natural, human and nonhuman, living and dead, individual and group, one time and another, and one space and another. True, for Tylor and Frazer, primitive peoples fail to think sufficiently critically and thereby produce religion rather than science, but not because of any missed distinctions. For Tylor and Frazer, primitive peoples still think, and think logically and systematically. Religion, no less than science, is the product of observation, hypothesis, and generalization, not of acculturation.

For Lévy-Bruhl, primitive peoples do not even have religion (see Lévy-Bruhl, *Primitive Mythology*, pp. 4, 10). What beliefs they do have come from their collective representations and not from any observations of the world, let alone from any rational responses to observations (see Lévy-Bruhl, *Primitive Mythology*, pp. 27–30). Far from thinking rationally, primitive peoples, brainwashed by their mystical and prelogical beliefs, scarcely think at all.

To be sure, for Frazer (*The Golden Bough*, chapters 3–4), the efficacy of magic, which for him constitutes a stage prior to that of religion, does presuppose the failure to make two distinctions: that between the literal and the symbolic – for otherwise a voodoo doll would merely symbolize, not affect, a person – and that between a part and the whole – for otherwise a severed strand of hair would merely have once been part of a person, not still affect that person. Furthermore, magic for Frazer presupposes a spider-like connection among all things, including that between a doll and a person and that between a part and the whole.

Still, Frazer never assumes that in even this stage primitive peoples are oblivious to other distinctions, such as those between appearance and reality and between subjectivity and objectivity. And any distinctions missed are of conception, not of perception, which Frazer, together with Tylor, considers invariant.[13]

For both Tylor and Frazer, even earliest human beings merely conceive, not perceive, the identity of a doll with a person. In Frazer's stage of religion as well as of magic, primitives may get angry at a stone over which they have stumbled, as if the stone had tripped them, but they still experience it as a stone. Both Frazer and Tylor are conceptual absolutists – pre-scientific beliefs are false – but perceptual universalists – all perceptions are the same.

Like Frazer, Lévy-Bruhl himself considers magic a stage of culture prior to that of religion (see Lévy-Bruhl, *Primitive Mythology*, pp. 183–84). But he views magic far more radically. While he, following Frazer, refers to magic as 'sympathetic magic', he stresses less the imitation of the desired effect than participation in the imitation. Where for Frazer imitation means mere imitation, for Lévy-Bruhl imitation means becoming identical with whatever one imitates (see Lévy-Bruhl, *Primitive Mythology*, chapter 5).

In above all *How Natives Think* Lévy-Bruhl explicitly follows Durkheim's fundamental principle that primitive beliefs not only are social, or 'collective', rather than individual in nature but also must therefore be explained socially rather than, as for Tylor and Frazer, individually. To quote Lévy-Bruhl: 'Collective representations are social phenomena [. . .]. [S]ocial phenomena have their own laws, and laws which the analysis of the individual *qua* individual could never reveal' (Lévy-Bruhl, *How Natives Think*, p. 23). The explanation of primitive thinking is thus to be found in sociology rather than in psychology. As the supposed study of the individual in isolation, psychology is where Lévy-Bruhl and Durkheim alike place both Tylor and Frazer. Collective beliefs, modern as well as primitive, result from socialization. Unlike Durkheim, Lévy-Bruhl simply assumes collective representations rather than accounts for them. For him, they are the given.[14] For him, society rather than the mind is the source of primitive beliefs, but he does not, like Durkheim, root the beliefs in social structure. In contrast to both Durkheim and Lévy-Bruhl, Tylor and Frazer ignore any sociological aspect of group beliefs. For them, Robinson Crusoe could as likely have invented myth and religion as the magician or priest of a community.

For Lévy-Bruhl, again following Durkheim,[15] moderns as well as primitive peoples have collective representations; representations are primarily categorizations; without representations, individuals would have no thoughts rather than merely private ones; primitive representations shape perception as well as conception; primitive representations are laden with emotion, modern ones freer of emotion; primitive representations constitute religion, modern ones science; science succeeds religion as the explanation of the world; and religion is false, science true.

Yet Durkheim, in both *The Elementary Forms of the Religious Life* (1912) and a review of Lévy-Bruhl's first book, castigates Lévy-Bruhl for exaggerating the differences between primitives and moderns.[16] For Durkheim, primitives recognize the same categories as moderns and are therefore not prelogical. Indeed, science inherits these categories from primitive religion. Without primitive religion there would be no science, even though science subsequently bests religion. Admittedly, science for Durkheim is more nearly objective than religion. It is critical, unemotional, and testable. But it differs from religion in degree only.[17]

Durkheim does acknowledge that at least primitive religion, by personifying the world, blurs the line between the human and the nonhuman. Clan members, for example, identify themselves with their animal or plant totem. But Durkheim then argues that these connections, though false, spur the search for others and thus even here spur the rise of science (see Durkheim, *The Elementary Forms of the Religious Life*, pp. 269–72). Lévy-Bruhl, by contrast, sees science as the opposite of anything primitive.

Jung's use of Lévy-Bruhl

Jung enlists Lévy-Bruhl in the same way that he enlists Richard Wilhelm, Karl Kerényi, and Paul Radin, who, ironically, was Lévy-Bruhl's nemesis. Jung uses them all for data, which he then psychologizes. What seemingly is about the world is, properly grasped, really about the unconscious. The use to which Jung puts Lévy-Bruhl, whom he actually knew,[18] is typical. That Lévy-Bruhl, following Durkheim, continually insists that the source of primitive thinking is not sociological (see, for example, Lévy-Bruhl, 'A Letter to E. E. Evans-Pritchard', p. 121) is for Jung no impediment. Jung even labels Lévy-Bruhl 'an authority in the field of primitive *psychology*' ('Archaic Man' [1931 in German], para. 106 [italics added]).[19]

Jung cites Lévy-Bruhl's works throughout his writings, beginning with *Psychological Types* (1921).[20] His fullest use of Lévy-Bruhl is in his three essays on primitive peoples: 'Archaic Man', 'Mind and Earth', and 'The Spiritual Problem of Modern Man'.[21] All three were translated and published in 1933 as part of Jung's collection *Modern Man in Search of a Soul*.

Jung relies on Lévy-Bruhl even when he goes to East Africa for five months to encounter primitive peoples for himself (see Burleson, 'Defining the Primitive: Carl Jung's "Bugishu Psychological Expedition"'). He arrives smitten with Lévy-Bruhl's ideas and by no coincidence finds them confirmed everywhere. Lévy-Bruhl was his Baedeker. No encounter impels Jung to question his characterization of primitive peoples. His three essays on primitive peoples all come *after* his travels yet still defer to Lévy-Bruhl. That, as he puts it in *Memories, Dreams, Reflections*, Jung finds in primitive peoples 'a potentiality of life which has been overgrown by civilization' presupposes the accuracy for him of Lévy-Bruhl's depiction of them.[22]

Jung acknowledges a debt to Lévy-Bruhl for the very concept of archetypes: 'Archetypal statements are based upon instinctive preconditions and have nothing to do with reason; they are neither rationally grounded nor can they be banished by rational arguments. They have always been part of the world scene – *représentations collectives*, as Lévy-Bruhl rightly called them' (Jung, *Memories, Dreams, Reflections*, p. 353). That Jung credits Lévy-Bruhl rather than Durkheim with the concept of collective representations means that the likely link between Durkheim's sociological concept and Jung's psychological concept came through Lévy-Bruhl.[23]

Jung's misreading of Lévy-Bruhl

As influential on Jung as Lévy-Bruhl was,[24] Jung failed to recognize the more radical of Lévy-Bruhl's claims about primitive peoples, which is not that they are mystical but that they are prelogical. Even if Lévy-Bruhl comes to give ever more attention to the mystical than to the prelogical side, in his first and best-known book, *How Natives Think*, the prelogical aspect dominates, and 'prelogical' is the term that Jung most often associates with Lévy-Bruhl.[25]

From the outset of 'Archaic Man' Jung invokes the authority of Lévy-Bruhl:

> When we first come into contact with primitive peoples or read about primitive psychology in scientific works, we cannot fail to be deeply impressed with the strangeness of archaic man. Lévy-Bruhl himself, an authority in the

field of primitive psychology, never wearies of emphasizing the striking difference between the 'prelogical' state of mind and our own conscious outlook. It seems to him, as a civilized man, inexplicable that the primitive should disregard the obvious lessons of experience, should flatly deny the most evident causal connections, and instead of accounting for things as simply due to chance or on reasonable grounds of causality, should take his 'collective representations' as being intrinsically valid. By 'collective representations' Lévy-Bruhl means widely current ideas whose truth is held to be self-evident from the start [. . .]. While it is perfectly understandable to us that people die of advanced age or as the result of diseases that are recognized to be fatal, this is not the case with primitive man. When old persons die, he does not believe it to be the result of age. He argues that there are persons who have lived to be much older [. . .]. To him, the real explanation is always magic. Either a spirit has killed the man, or it was sorcery.

(para. 106)

Jung then cites an example of 'prelogical' thinking that he in fact takes (without acknowledgment) from Lévy-Bruhl himself (*Primitive Mentality*, pp. 52–53): that of two anklets found in the stomach of a crocodile shot by a European. The 'natives' recognized that the anklets belonged to two women who had been eaten by a crocodile. But instead of concluding that the crocodile had on its own caught the two and eaten them, the 'natives' concluded that some sorcerer 'had summoned the crocodile, and had bidden it catch the two women and bring them to him'. Why assume sorcery? Because crocodiles never eat persons unless bidden to do so. The anklets were its reward (see para. 106).

Jung calls this story 'a perfect example of that capricious way of explaining things which is characteristic of the "prelogical" state of mind' (para. 107). But it is not. For Lévy-Bruhl, 'prelogical' would mean the assumption of identity yet still distinction between the sorcerer, or witch, and the crocodile. On the one hand the witch *becomes* the crocodile: 'In districts where crocodiles are common [. . .] the witches are believed sometimes to turn into crocodiles, or to enter and actuate them, and so cause their victim's death by catching him' (Mrs H. M. Bentley, quoted by Lévy-Bruhl, *Primitive Mentality*, p. 52). On the other hand the witch and the crocodile remain distinct. But as interpreted by Jung, prelogical means the sheer enlistment of the crocodile by the witch to kill the victim. The witch and the crocodile are merely distinct and not also identical.[26] For Lévy-Bruhl, magic is mystical because the magician and the agent are identical, and magic is prelogical because the two remain distinct. In Jung's rendition of the event, magic is not even mystical, let alone prelogical.

Jung's next example of primitive thinking also turns out to come from Lévy-Bruhl (*Primitive Mentality*, pp. 49–50):

If three women go to the river to draw water, and a crocodile seizes the one in the middle and pulls her under, our view of things leads us to the verdict that it was pure chance that that particular woman was seized. [. . .] [P]rimitive man expects far more of an explanation. What we call pure chance is for him wilful intention. It was therefore the intention of the crocodile – as everyone could observe – to seize the middle one of the three women. If it had not had this intention, it would have taken one of the others. But why did the crocodile have this intention? Ordinarily these creatures do not eat human beings [. . .]. Considering their numbers, they kill astonishingly few people, and it is an unexpected and unnatural event when they devour a man. Such an event calls for an explanation. Of his own accord the crocodile would not take a human life. By whom, then, was he ordered to do so?

(paras. 115, 117)

Here Jung, like Lévy-Bruhl, deems primitive the ascription of unfortunate events to malevolence rather than to chance, but for Lévy-Bruhl, unlike Jung, the malevolence again requires the magician's becoming the crocodile, even while remaining a human being.

Jung's psychologizing of primitive thinking

Midway in 'Archaic Man' Jung turns, without announcement, to the *psychology* of primitive thinking. Where moderns have learned to differentiate what is inside from what is outside, at least about the physical world, primitive peoples do not. They project themselves altogether onto the physical world, which therefore becomes the playing field of divine and semi-divine figures rather than the manifestation of impersonal laws of nature. As Jung puts it, 'For primitive man, [. . .] the psychic and the objective coalesce in the external world. What in fact is internal is projected outwardly and is therefore experienced as external: primitive man is unpsychological. Psychic happenings take place outside him in an objective way' (para. 128).

The consequence of projection, notes Jung, is the identification of humans with everything else in the world. Take his third example of crocodiles and women: 'A white man shoots a crocodile. At once a crowd of people come running from the nearest village and excitedly demand compensation. They explain that the crocodile was a certain old woman in their village who had died at the moment when the shot was fired. The crocodile was obviously her bush-soul' (para. 129). Jung credits Lévy-Bruhl with 'coin[ing] the expression *participation mystique* for these remarkable relationships' (para. 130). While Jung regrets Lévy-Bruhl's choice of the term 'mystical', he certainly accepts Lévy-Bruhl's characterization of the relationship between the crocodile and the woman, and he himself regularly uses the term *participation mystique*.[27] In Jung's two prior examples the identity is between the magician and the crocodile (see Lévy-Bruhl, *Primitive Mentality*, p. 55). Now it is between the crocodile and the old woman.

Jung then asserts that modern projection is the same as primitive mystical participation: 'We suppose that what is pleasing or desirable to us is the same to others, and that what seems bad to us must also seem bad to others' (para. 130). But this modern version of identification is tame, for we are not here identifying *ourselves* with others, only our judgments with those of others. Better, then, is Jung's equation of projection with *participation mystique*: 'projection [. . .] is the same as *participation mystique*, which Lévy-Bruhl, to his great credit, emphasized as being an especially characteristic feature of primitive man' (para. 131). We moderns 'merely give it another name, and as a rule deny that we are guilty of it. Everything that is unconscious in ourselves we discover in our neighbour, and we treat him accordingly' (para. 131). But by *participation mystique* Lévy-Bruhl means the ascription to others of what we *recognize*, not what we *deny*, in ourselves. He means the assumption of outright identity between us and others. And he means identity not merely between us and other human beings but even more between us and the rest of the world, the inanimate world as well as the animate one of the crocodile.[28] Above all, he means the sheer belief in *participation mystique* and not, as for Jung, the source of it. As the source of participation mystique he would give society, not the mind, and the means involved would for him be acculturation, not projection.

Even in 'Archaic Man' Jung takes any *modern* identification of a human being with a leopard to be merely metaphorical.[29] Jung cites the example of the identification of himself with a bear by the Pueblo Indians, whom he visited:

> The Pueblo Indians declared in a matter-of-fact way that I belonged to the Bear Totem – in other words, that I was a bear – because I did not come down a ladder standing up like a man, but bunched up on all fours like a bear. If anyone in Europe said I had a bearish nature this would amount to the same thing, but with a rather different shade of meaning. [. . .] If we take our metaphors concretely we return to the primitive point of view.
>
> (para. 132)

But even if Jung were to take the identification literally, it would still fall short of the primitive mentality for Lévy-Bruhl. For the Pueblos, according to Jung, are not calling him a bear yet still a human being. For Lévy-Bruhl, the witch is at once crocodile and still witch: 'between the wizard and the crocodile the relation is such that the wizard becomes the crocodile, without, however, being actually at once' (Lévy-Bruhl, *Primitive Mentality*, p. 55).

As mild as Jung's characterization of primitive thinking is, he no less than Lévy-Bruhl is prepared to call it false. No more than Lévy-Bruhl is he a conceptual or perceptual relativist:

> As we know, great minds have wrestled with the problem of whether it is the glorious sun that illuminates the world, or the sunlike human eye. Archaic man believes it to be the sun, and civilized man believes it to be the eye [. . .]. He [modern man] must de-psychize nature in order to dominate her; and in order to see his world objectively he must take back all his archaic projections.
>
> (para. 135)

For both Lévy-Bruhl and Jung, primitives and moderns do not merely think different things, as for Tylor and Frazer, but actually think differently. For Lévy-Bruhl, there is mystical thinking, which involves the ascription of mystical identity to the world, and scientific thinking, which sees the world as it is. For Jung, there is 'fantasy' thinking, which is like primary process thinking for Freud, and 'directed', or 'logical', thinking, which is like secondary process thinking for Freud. Where directed thinking is deliberate, organized, and purposeful, fantasy thinking is spontaneous, associative, and directionless: 'What happens when we do not think directedly? Well, our thinking then lacks all leading ideas and the sense of direction emanating from them. We no longer compel our thoughts along a definite track, but let them float, sink or rise according to their specific gravity' (Jung, *Symbols of Transformation*, para. 18).[30] Fantasy thinking 'leads away from reality into fantasies of the past or future' (Jung, *Symbols of Transformation*, para. 19). By contrast, directed thinking turns outward to the world. While Jung certainly does not, like Freud, maintain that fantasy thinking operates by the pleasure principle, he does, like Freud, maintain that directed thinking operates by the reality principle: 'To that extent, directed or logical thinking is reality-thinking, a thinking that is adapted to reality, by means of which we imitate the successiveness of objectively real things, so that the images inside our mind follow one another in the same strictly causal sequence as the events taking place outside it. We also call this "thinking with directed attention" ' (Jung, *Symbols of Transformation*, para. 11).

For Jung, as for Freud, fantasy thinking is found most fully in dreams and myths. Jung even uses the phrase 'mythic thinking' interchangeably with fantasy thinking. Freud and Jung agree that myths go beyond dreams to project fantasy thinking onto the world. Myths transform the outer world into an extension of the inner one. Mythic thinking is thus not merely a way of thinking but a way of thinking about the world – and in turn a way of experiencing the world:

> We move in a world of fantasies which, untroubled by the outward course of things, well up from an inner source to produce an ever-changing succession of plastic or phantasmal forms. [. . .] Everything was conceived anthropomorphically or theriomorphically, in the likeness of man or beast. [. . .] Thus there arose a picture of the universe which was completely removed from reality, but which corresponded exactly to man's subjective fantasies.
>
> (Jung, *Symbols of Transformation*, para. 24)

For Jung, as for Lévy-Bruhl, primitive peoples are ruled entirely by fantasy thinking. Although scarcely absent among moderns, fantasy thinking has been supplemented and considerably supplanted by directed thinking, which is to be found above all in science. Jung accepts the assumption of his day, summed up in Ernst Haeckel's Law of Recapitulation, that the biological development of the individual (ontogeny) duplicates that of the species (phylogeny): 'The supposition that there may also be in psychology a correspondence between ontogenesis and phylogenesis therefore seems justified. If this is so, it would mean that infantile thinking and dream-thinking are simply a recapitulation of earlier evolutionary stages' (Jung, *Symbols of Transformation*, para. 26).[31] Primitives are therefore the counterpart to children and moderns the counterpart to adults: 'These considerations tempt us to draw a parallel between the mythological thinking of ancient man and the similar thinking found in children, primitive peoples, and in dreams' (Jung, *Symbols of Transformation*, para. 26). Just as the child is governed wholly by fantasy thinking and only the adult is guided substantially by directed thinking, so the primitive is governed completely by fantasy thinking and only the modern is guided significantly by directed thinking. Lévy-Bruhl, too, parallels primitive thinking with that of children (see, for example, Lévy-Bruhl, *L'Expérience mystique et les symboles chez les primitives*, p. 16).

For Jung, myths serve primarily to open adults up to their unconscious, from which, in the course of growing up, they have ineluctably become severed. Myths 'compensate or correct, in a meaningful manner, the inevitable one-sidednesses and extravagances of the conscious mind' (Jung, 'The Psychology of the Child Archetype', para. 276).[32] But for Jung it is the ego consciousness of only moderns that is sufficiently developed to be severed from the unconscious: 'Since the differentiated consciousness of civilized man has been granted an effective instrument for the practical realization of its contents through the dynamics of his will, there is all the more danger, the more he trains his will, of his getting lost in one-sidedness and deviating further and further from the laws and roots of his being' (Jung, 'The Psychology of the Child Archetype', para. 276). By contrast, primitive peoples hover so close to unconsciousness that their ego consciousness has barely begun to develop:

> Primitive mentality differs from the civilized chiefly in that the conscious mind is far less developed in scope and intensity. Functions such as thinking, willing, etc. are not yet differentiated; they are pre-conscious, and in the case of thinking, for instance, this shows itself in the circumstances that the primitive does not think *consciously*, but that thoughts *appear*. [. . .] Moreover, he is incapable of any conscious effort of will [. . .]
> (Jung, 'The Psychology of the Child Archetype', para. 260)

Primitive myths are unrecognized projections onto the world. They are 'original revelations of the pre-conscious psyche, involuntary statements about unconscious psychic happenings, and anything but allegories of physical processes' (Jung, 'The Psychology of the Child Archetype', para. 261):

> All the mythologized processes of nature, such as summer and winter, the phases of the moon, the rainy seasons, and so forth, are in no sense allegories of these objective [i.e., external] occurrences; rather they are symbolic expressions of the inner, unconscious drama of the psyche which becomes accessible to man's consciousness by way of projection – that is, mirrored in the events of nature.
> (Jung, 'Archetypes and the Collective Unconscious', para. 7)[33]

Where moderns have withdrawn projections from the physical world and explain the world scientifically, primitives experience the world as an extension of themselves.

In short, the primitive mind is for Jung no less one-sided than the modern one. It is simply one-sidedly unconscious rather than, like the modern one, one-sidedly conscious. But then it, too, needs correction.

Jung's valuing of primitive thinking

Despite Jung's association of primitive peoples with children, he is not denigrating primitives. Nor is he denigrating their creation – myth. Indeed, he castigates Freudians for denigrating primitives and myth by linking them to children:

> The first attempts at myth-making can, of course, be observed in children, whose games of make-believe often contain historical echoes. But one must certainly put a large question-mark after the [Freudian] assertion that myths spring from the 'infantile' psychic life of the race. [. . .] [T]he myth-making and myth-inhabiting man was a grown reality and not a four-year-old child. Myth is certainly not an infantile phantasm, but one of the most important requisites of primitive life.
>
> (Jung, *Symbols of Transformation*, para. 29)[34]

Lévy-Bruhl no more than Jung means to be denigrating primitive peoples by associating them with children.[35] The difference between Jung and Lévy-Bruhl is that for Jung moderns can learn from primitives, where for Lévy-Bruhl they cannot.

For both Jung and Lévy-Bruhl, moderns are modern to the extent that they have rejected primitive thinking. But where Lévy-Bruhl uncompromisingly celebrates the liberation of moderns from primitive thinking, Jung, while likewise celebrating that liberation, simultaneously laments the severance of moderns from their primitive roots. For both Jung and Lévy-Bruhl, intellectual progress comes from exposing primitive mischaracterizations of the external world. Progress comes from seeing the world as it is. The external world is really natural rather than supernatural, impersonal rather than personal. Science properly replaces myth and religion as the explanation of the world. There is no turning back.

For Lévy-Bruhl, the source of the mischaracterizations is false collective beliefs, which are simply to be discarded. For Jung, the source is the unconscious, which projects falsely onto the world but which itself is real and must be cultivated. The recognition of the source of the mischaracterizations helps redirect the focus away from the erroneous object – the external world – and onto the correct one – the unconscious.

For Jung, moderns are more advanced than primitives, but they have become modern only by disconnecting themselves from their primitive roots – their unconscious. Modernity is better than primitivism, but it is not the ideal state. It is a stage along the way. Having disconnected themselves from their unconscious in order to develop their ego consciousness, moderns must now return to their unconscious and reconnect themselves to it. In this respect moderns have much to learn from primitive peoples, though what they really have to learn about is themselves.[36] For Lévy-Bruhl, by contrast to Jung, modernity is the ideal state, and what remains to be done is only the further withdrawal of any lingering primitive representations. Lévy-Bruhl acknowledges that the task will never be done, but he urges its continuation. For him, primitive peoples are to be respected, even when compared with children, but moderns have nothing to learn from them.

Criticisms of Lévy-Bruhl

From the publication of the first of his six books on, Lévy-Bruhl was attacked for his claim that a distinctively primitive mentality exists. It was fieldworkers who, by dint of their first hand knowledge of primitive peoples, rebutted the portrayal of them by Lévy-Bruhl, who, to be sure, based his views on the first hand accounts of other fieldworkers. The English anthropologist E. E. Evans-Pritchard, who was in fact Lévy-Bruhl's firmest defender, sums up the criticism this way:

> Most specialists who are also fieldworkers are agreed that primitive peoples are predominantly interested in practical economic pursuits; gardening, hunting, fishing, care of their cattle, and the manufacture of weapons, utensils, and ornaments, and in their social contacts; the life of household and family and kin, relations with friends and neighbours, with superiors and inferiors, dances and feasts, legal disputes, feuds and warfare. Behaviour of a mystical type in the main is restricted to certain situations in social life. Moreover, it is generally linked up with practical activities in such a way that to describe it by itself, as Lévy-Bruhl has done, deprives it of the meaning it derives from its social situation and its cultural accretions.
>
> (Evans-Pritchard, 'Lévy-Bruhl's Theory of Primitive Mentality' [1934], p. 11)[37]

The classic anthropological rebuttal to Lévy-Bruhl was American anthropologist Paul Radin's *Primitive Man as Philosopher* (1957), originally published in 1927.[38] Like other anthropological critics, Radin denies that primitive peoples miss the distinctions that Lévy-Bruhl declares them to be bereft of: cause and effect, subject and object, natural and supernatural, non-mystical and mystical, individual and group, and literal and symbolic. Yet Radin, unlike other anthropological critics, divides the members of any society, modern and primitive alike, into 'men of action', who may well fail to make some of Lévy-Bruhl's distinctions, and 'thinkers', who do not. By contrast, Lévy-Bruhl insists that the 'average man' as well as the 'cultured, scientific man' differs from primitive man (see Lévy-Bruhl, *The Notebooks on Primitive Mentality*, p. 49). Radin attacks not only Lévy-Bruhl himself but also those who accept his view – not least Jung.[39]

In *Purity and Danger* (1966) the English anthropologist Mary Douglas praises Lévy-Bruhl for having 'first posed all the important questions about primitive cultures and their distinctiveness as a class' (Douglas, *Purity and Danger*, p. 92).[40] She also shares Lévy-Bruhl's preoccupation with categorizations. Yet she is sharply critical of him. She argues that primitive religion, even when used for practical ends, does not work magically. A rain dance, for example, may serve to ask god for rain, but it does not itself effect rain. Douglas also argues that primitive religion not only presupposes the distinction between the literal and the symbolic but itself often serves to symbolize other categorizations that, contrary to Lévy-Bruhl, primitive peoples make – for example, distinctions in time and space. Thus a rain dance may symbolize the difference between the rainy and the dry seasons.[41] Still more, Douglas argues that categories like pollution and cleanness, foods, animals, and parts of the body serve to symbolize an individual's relationship to society, which one's relationship to the cosmos serves to symbolize in turn.[42]

Against Lévy-Bruhl, the French structural anthropologist Claude Lévi-Strauss, in *The Savage Mind* (1966) and elsewhere, similarly argues that primitive peoples think no differently from moderns.[43] They merely focus on the observable, qualitative aspects of phenomena rather than, like moderns, on the unobservable, quantitative ones. Colors and sounds, not mass and length, faze them. Far from pre-scientific, primitive peoples attain a fully scientific knowledge of the world. Theirs is simply a 'science of the concrete' rather than of the abstract. Indeed, even if they do not, like moderns, separate abstractions from concrete cases, they do express abstractions through concrete cases.

Furthermore, their knowledge is basically taxonomic, so that primitive peoples are quite capable of categorizing. In fact, their taxonomies take the form of oppositions, which, as the equivalent for Lévi-Strauss of contradictions, make primitive peoples not only aware of contradictions but also intent on resolving them. Myths most of all evince the austere, rigorous, logic-chopping nature of primitive thinking. Lévi-Strauss reads myths as the equivalent of mathematical puzzles.[44]

Where Radin argues that most persons, modern and primitive alike, are as indifferent to logic as Lévy-Bruhl claims that all primitive peoples and only primitive peoples are, Lévi-Strauss refuses to divide up human beings into types and instead asserts uncompromisingly that primitive peoples collectively are as consummately logical as moderns collectively. Therefore the 'antinomy' claimed by Lévy-Bruhl between 'logical and prelogical mentality' is 'false'. 'The savage mind is logical in the same sense and the same fashion as ours. [. . .] [C]ontrary to Lévy-Bruhl's opinion, its thought proceeds through understanding, not affectivity, with the aid of distinctions and oppositions, not by confusion and participation' (Lévi-Strauss, *The Savage Mind*, p. 268).[45]

In *The Foundations of Primitive Thought* (1979) the Canadian anthropologist C. R. Hallpike 'consider[s] much of [Lévy-Bruhl's] work on primitive thought to be of great value' (Hallpike, *The Foundations of Primitive Thought*, p. 50, n. 4).[46]

But he, like Douglas, then proceeds to lambaste Lévy-Bruhl. Hallpike complains that Lévy-Bruhl, following Durkheim, attributes primitive thinking entirely to socialization and not at all to psychology. For Hallpike, cognitive psychology, by which he means Piagetian psychology, accounts for the 'processes' by which the 'content' of mind is inculcated (Hallpike, *The Foundations of Primitive Thought*, p. 3, n. 3). And especially in primitive societies, those processes involve more than the acquisition of language (see Hallpike, *The Foundations of Primitive Thought*, pp. 69, 484). As admiring of Lévy-Bruhl as Hallpike professes to be, his denial that 'primitive peoples employ a *different* logic from that recognized by Western philosophers' (Hallpike, *The Foundations of Primitive Thought*, p. 488) leaves unclear just what is left of Lévy-Bruhl's views to be admired.

In *The Domestication of the Savage Mind* (1977) the English anthropologist Jack Goody breaks with almost all of his contemporaries in accepting that primitive peoples fail to recognize contradictions and thereby evince a prelogical mentality.[47] But Goody attributes their failure to the absence of writing:

> [I]t is certainly easier to perceive contradictions in writing than it is in speech, partly because one can formalise the statements in a syllogistic manner and partly because writing arrests the flow of oral converse so that one can compare side by side utterances that have been made at different times and at different places. Hence there is some element of justification behind Lévy-Bruhl's distinction between logical and pre-logical mentality, as well as behind his discussion of the law of contradiction.
>
> (Goody, *The Domestication of the Savage Mind*, pp. 11–12)

Goody nevertheless dismisses as 'totally wrong' Lévy-Bruhl's assumption of any distinctively primitive 'style of thinking' (Goody, *The Domestication of the Savage Mind*, p. 12).

The chief defender of Lévy-Bruhl against unfair charges was Evans-Pritchard.[48] Yet even he faults Lévy-Bruhl for deeming primitive thinking 'prelogical'. Where for Lévy-Bruhl primitive magic takes the place of science, for Evans-Pritchard magic supplements proto-science: magic and proto-science coexist. To the Azande, the sheer physical features of a tree explain its ordinary, natural 'behavior'. Witchcraft, Evans-Pritchard's most famous example of supernatural causality, explains only unfortunate events involving the tree: why one day it falls on one person or, to cite his most famous example, why a granary under which Azande are sitting collapses when it does. For Lévy-Bruhl, by contrast, even events so regular and therefore so seemingly natural as birth, disease, and death get attributed to 'magic' – a term that he, unlike Evans-Pritchard and others, uses broadly to encompass all supernatural causes (see Lévy-Bruhl, *How Natives Think*, pp. 293–98).[49]

Lévy-Bruhl grants that primitive peoples must have practical, worldly skills to survive: 'primitive peoples who betray no apparent interest in the most obvious causal connections are quite able to utilize them to procure what is necessary to them, their food, for example, or some special tool' (Lévy-Bruhl, *Primitive Mentality*, p. 443). Similarly,

> The Australian Aborigines, for instance, could never, with the few weapons and implements at their disposal, have developed their mastery over the animals which provide their diet – kangaroos, emus, possums, small marsupials, birds, fish – without becoming very minutely informed about their habitat, their pattern of behaviour, their seasonal migrations, and in general, everything about the way they live. Knowledge of these things may often be a matter of life or death.
>
> (Lévy-Bruhl, *Primitive Mythology*, p. 66)[50]

Typically unfairly, Lévy-Bruhl is charged with overlooking this point – above all by the Polish-born anthropologist Bronislaw Malinowski but also even by Evans-Pritchard.[51] In actuality, Lévy-Bruhl simply distinguishes the quasi-scientific techniques used by primitives from any scientific explanation of those techniques given by primitives. Primitive peoples *explain* the efficacy of their practices either mystically or not at all.[52]

Shifting from the Azande to the Nuer, Evans-Pritchard contests Lévy-Bruhl's most striking evidence of prelogical mentality: statements that, for example, a cucumber is an ox and that twins are birds (see Evans-Pritchard, *Nuer Religion* [1956], ch. 5). Lévy-Bruhl maintains that mystical representations override the senses, so that primitive peoples somehow actually perceive, not just conceive, a cucumber as an ox. Evans-Pritchard denies that they do either. The Nuer, he asserts, are speaking only metaphorically. They are saying that a cucumber is sufficiently like an ox to serve as a substitute for it:

> When a cucumber is used as a sacrificial victim Nuer speak of it as an ox. In doing so they are asserting something rather more than that it takes the place of an ox. They do not, of course, say that cucumbers are oxen, and in speaking of a particular cucumber as an ox in a sacrificial situation they are only indicating that it may be thought of as an ox in that particular situation; and they act accordingly by performing the sacrificial rites as closely as possible to what happens when the victim is an ox.
>
> (Evans-Pritchard, *Nuer Religion*, p. 128)

Likewise the Nuer are saying that a twin is like a bird in certain respects but not that twins are birds (see Evans-Pritchard, *Nuer Religion*, pp. 131–32).

Rationality

Evans-Pritchard's responses to Lévy-Bruhl have inspired a continuing debate on 'rationality' – a debate less in anthropology, where either relativism or universalism smugly reigns, than in philosophy. Notoriously hard to define,[53] the term gets raised about both beliefs and practices. Applied to beliefs, including values, 'rational' usually means noncontradictory. Applied to practices, especially rituals, it typically means efficacious – more precisely, seemingly efficacious in the wake of the beliefs underlying the practices. Rationality has nothing to do with *truth*, as Evans-Pritchard's characterization of Azande witchcraft as rational yet false attests. Nor has it anything to do with *origin*, on which Lévy-Bruhl himself simply defers to Durkheim. Rationality does not, strictly, deal even with *function*. It deals instead with the *meaning* of beliefs and practices. As the terms sometimes get used, rationality is a matter of interpretation, not explanation. To be rational is to be intelligible and coherent.

Since Lévy-Bruhl never maintains that primitive beliefs and practices are, on the basis of their mystical and prelogical premises, either unintelligible or incoherent, there might seem to be no debate. Yet his characterization of primitive peoples as not merely mystical, itself neutral, but also prelogical seems absolutist: he seems to be judging them by modern standards, and to be judging them as irrational.

Praising Evans-Pritchard for rejecting Lévy-Bruhl's assessment of primitive peoples as prelogical, the English philosopher Peter Winch, in 'Understanding a Primitive Society' (1964), castigates him for nevertheless deeming Azande witchcraft both false and at least partly irrational.[54] The issue of truth is not germane here, and Winch himself coincidentally drops it. As for rationality, Evans-Pritchard does note, for example, that despite their belief in an inherited 'witchcraft-substance', the Azande rarely invoke heredity either to convict or to clear the accused. He concludes not, however, that they are illogical but that for social reasons they fail to pursue the contradiction either between conviction and the innocence of one's kin or between acquittal and the guilt of one's kin (see Evans-Pritchard, *Witchcraft, Oracles and Magic among the Azande*, pp. 24–26, 127–28). Social considerations determine when witchcraft is and is not invoked to account for unfortunate events (see Evans-Pritchard, *Witchcraft, Oracles and Magic among the Azande*, pp. 74–78; *A History of Anthropological Thought*, p. 130).

Like the German-born anthropologist Franz Boas and his American students, especially Ruth Benedict and Melville Herskovits, Winch deems ethnocentric not the assumption of rationality worldwide but the denial of it. Winch takes for granted that Azande witchcraft is, in its own terms, rational. He faults Evans-Pritchard for assuming that witchcraft is meant to be an explanation of unfortunate events. Understanding it in its own terms means grasping not just, as Evans-Pritchard does, its logic, or rules, but also the point of those rules. Doing so requires grasping what kind of activity, or Wittgensteinian 'form of life', witchcraft is. Like Douglas on a rain dance, Winch proposes that witchcraft is an existential rather than an explanatory or a technological activity and that its point is therefore less to explain or to control events than to express, for example, powerlessness over one's life.[55]

Winch's 1964 essay and his 1958 book, *The Idea of a Social Science*, have spawned a never-ending debate over the criteria of rationality.[56] For the absolutists, who consider the criteria entirely external and objective, primitive beliefs and practices invariably prove wholly irrational (Lévy-Bruhl), considerably less rational than modern ones but still rational (Tylor, Frazer, Evans-Pritchard [*Witchcraft, Oracles and Magic among the Azande*], Alasdair MacIntyre, Ernest Gellner, I. C. Jarvie, Joseph Agassi), nearly as rational as modern ones (Durkheim, Robin Horton), or as fully rational as modern ones (Radin). For the relativists, who deem the criteria largely or entirely internal and 'context dependent', primitive beliefs and practices invariably prove as fully rational as modern ones (Boas, Benedict, Herskovits, Evans-Pritchard [*Nuer Religion*], Winch, Douglas [*Implicit Meanings*]).[57] For those who consider the criteria both external and internal, primitive beliefs and practices prove more rational than irrational (Steven Lukes). Finally, for the noncognitivists, the issue of rationality is irrelevant: primitive beliefs and practices, like their modern religious counterparts, are primarily expressions of the meaningfulness of life rather than efforts either to explain or to control the world (Edmund Leach, Raymond Firth, John Beattie, Douglas [*Purity and Danger, Natural Symbols*], Susanne Langer).[58] Whatever the position taken, it is almost always measured against Lévy-Bruhl's.

In his posthumously published *Notebooks* (1949 in French, 1975 in English) Lévy-Bruhl does not, as is conventionally assumed, abandon altogether his claim that primitive peoples have a distinctive mentality.[59] He does cede his characterization of primitive peoples as prelogical, but he retains his characterization of them as mystical.[60] The difference between primitive peoples and moderns thus becomes that of the degree of mystical thinking in each:

> Let us rectify what I believed correct in 1910 [the date of *Les Fonctions mentales dans les sociétes inférieures*]: there is not a primitive mentality distinguishable from the other by *two* characteristics which are peculiar to it (mystical and prelogical). There is a mystical mentality which is more marked and more easily observable among primitive peoples than in our own societies, but it is present in every human mind.
>
> (Lévy-Bruhl, *The Notebooks on Primitive Mentality*, pp. 100–01)

True, it was the prelogical aspect of primitive thinking that Lévy-Bruhl had considered the more important, but that aspect rests on the mystical character, on which he never yields.

Lévy-Bruhl does not, like Evans-Pritchard (*Nuer Religion*, ch. 5), grant that the Bororo, in deeming Trumai tribesmen fish, are merely comparing the Trumai with fish. He does, however, grant that the Trumai are fish spiritually, not physically. Their spiritual 'fishness' complements, not contradicts, their physical humanness (see Lévy-Bruhl, *The Notebooks on Primitive Mentality*, pp. 8–10, 136–38). Primitive peoples thus recognize, not miss, at least the distinction between the non-physical and the physical (see Lévy-Bruhl, *The Notebooks on Primitive Mentality*, pp. 5–12, 19–22, 40–42, 45–50, 125–29).

Where others criticize Lévy-Bruhl for making primitive peoples prelogical or outright illogical, Jung would criticize him for making moderns logical. In his stress on the prelogical or illogical nature of moderns, Jung is really less close to Lévy-Bruhl than to the Italian social theorist Vilfredo Pareto, who deems most *modern* behavior 'nonlogical'. Like other critics of Lévy-Bruhl, Jung would contend that Lévy-Bruhl goes too far – but in the depiction of moderns, not primitives.

Eight decades after his death, Lévy-Bruhl continues to be the subject of debate. He still attracts defenders and antagonists alike.[61] Fields such as child development, ethnoscience, and cognitive psychology have turned, or returned, to Lévy-Bruhl's contrast of primitive thinking to modern and also to his association of primitive thinking with children's thinking. Attempts have been made to defend Lévy-Bruhl by linking his views to those of, notably, Jean Piaget, who argues that the thinking of children is different from that of adults and who himself cites Lévy-Bruhl, although with qualification, for parallels between primitive peoples and children.[62] Lévy-Bruhl's influence, whether lauded or bemoaned, has spread beyond anthropology, psychology, and other social sciences to the humanities.[63] To cite but one example, discussions of 'mentalities' hark back to Lévy-Bruhl.

Lévy-Bruhl does not need Jung. But Jung does need Lévy-Bruhl. Because Jung assumes that the development of the individual recapitulates that of the species, and because he depicts the individual as beginning in sheer unconsciousness and only gradually developing consciousness of the external world, he requires a depiction of our forbears as existing in a womb-like state, cut off from the outer world. Lévy-Bruhl supplies that depiction. Tylor and Frazer, who see our forbears as reacting to the same outer world as ours, do not. Whether Jung's history of the psyche can survive the loss of Lévy-Bruhl is the fit subject of another article.

Appendix: Replies to Susan Rowland and Paul Bishop

Reply to Susan Rowland[64]

To begin with, I do not see how Susan Rowland is interpreting Jung's essay 'Archaic Man' in terms other than those of authorial intent. After all, she is continually seeking to figure out what Jung *means* to be saying. Even if 'Jung is quite capable of adopting "other" authorial personas than the Christian logos', each of those personas is for her still intended.

In the history of modern literary criticism, which I am sure Rowland knows better than I, authorial intent was what the New Critics of the 1940s and 1950s opposed. The identification of the meaning of a work with its author's intent was pronounced an outright fallacy by New Critics William Wimsatt and Monroe Beardsley in their 1946 essay 'The Intentional Fallacy'. The shift of focus away from the author peaked in the 1970s and 1980s with the reader-response criticism of Stanley Fish, Roland Barthes, Wolfgang Iser, and the Freudian Norman Holland. The authority of the reader replaced that of the author.

But how 'reader-oriented' is Rowland's own approach? While she maintains that 'the reader is the necessary other person in its [the essay's] exchange of ideas and secreting of meaning', surely the inclusion of the reader is, for her, Jung's intent. The reader is expected to *get*, not to *set*, Jung's meaning. When Rowland asserts that Jung's 'aim is to get the reader to acknowledge the historical and archaic in him/herself', she is clearly analyzing the essay according to its author's purported intent.

There are three possible ways in which a reader can be involved. In the tamest sense the reader is the intended audience of Jung's essay. Doubtless every author involves the reader to this extent. In a bolder sense the reader is the *subject* of Jung's essay: the archaic mind is not just that of primitive peoples but that of moderns as well, so that Jung is writing not only *to* moderns but also *about* them. Rowland is, I think, maintaining that Jung involves the reader in this second sense.

In the third, most radical sense, which is that of reader-response criticism, the reader rather than the author determines the meaning of the text. Hence Barthes' famous dictum: the 'death of the author'. Rowland is clearly not going this far. She is seeking out what Jung intended to say, not what readers take his essay to be saying.

If Rowland and I do differ, it is over what Jung's intent is. But how much do we differ even here? I agree with her that for Jung, in contrast to Lucien Lévy-Bruhl, moderns should learn from their forbears. I thus agree with her statement that for Jung 'we are taught to see ourselves in the imperfect mirror of the cultural other.'

Still, for Jung we do not learn about ourselves by recognizing our projections *onto* primitive peoples. Where on almost any other topic Jung is commendably preoccupied with identifying the projections that we make, in the case of 'archaic man' he never doubts his own rendition – a rendition that comes wholly from Lévy-Bruhl. Rather than learning about ourselves by recognizing our projections onto primitive peoples, we learn about ourselves by applying what we, using Lévy-Bruhl, learn about primitive peoples themselves.

Jung ventures to North and especially East Africa to encounter first hand peoples as far removed from European civilization as possible. Yet they prove transparent, not to mention uniform. Jung never questions his understanding of them, even though his almost exclusive source of information was subjected to the severest criticism in his own day. When, as Rowland stresses, Jung matches up moderns with primitive peoples, he matches up the occasional projections onto the external world made by moderns with the continuous projections onto the external world made by primitive peoples. He never compares modern projections onto primitive peoples with primitive peoples themselves.

Lévy-Bruhl, for his part, scarcely denies similarities between primitive peoples and moderns. He never claims that moderns are wholly free of their primitive roots. On the contrary, he emphasizes the eternal allure of, if not quite prelogical thinking, then at least mystical thinking. And mystical thinking is all that Jung assumes Lévy-Bruhl means by prelogical thinking. Lévy-Bruhl singles out natural science, and really only the most abstract formulations of natural science, as free of emotion and therefore of mystical proclivities. If Jung claims to be breaking with Lévy-Bruhl in spotting the residue of primitiveness among moderns, he is breaking with a proverbial straw man.

Jung does rightly claim to be breaking with Lévy-Bruhl in recognizing not the persistence of mystical thinking among moderns but the utility of its persistence in diagnosing moderns. For Jung, mystical thinking by moderns is revealing – but revealing about moderns, not about the world. To be diagnostic, mystical thinking must be psychologized, which is to say recognized as projective. But moderns alone have the capacity to recognize their own projections, for they alone possess the modern science of analytical psychology. Jung does scan the cultural horizons for forerunners of his brand of psychology, and he finds them in ancient Gnosticism and medieval alchemy. But these movements are Western, not primitive. At the same time they are pre-scientific. Gnostics and alchemists may have discovered the inner, but they never fully sifted it out from the outer. They, too, projected the inner onto the outer. Surely for Jung no primitive persons, for whom everything inside is cast outside, are budding analysts awaiting certification.

Unlike primitive peoples, moderns for Jung have largely withdrawn their projections from the physical world. Moderns thus project themselves less onto animals, plants, or inanimate objects than onto other people – with the apparent exception of primitive peoples! The development of natural science has liberated moderns from their projections onto nature. Jung applauds, not questions, the achievements of science, social and natural alike.

The difference between Jung and Lévy-Bruhl is that Jung, having identified the lingering projective nature of even much modern thinking, wants moderns to tend to the source of those projections: the unconscious. Lévy-Bruhl, while likewise finding in modern thinking lingering mystical and prelogical representations, simply wants those representations dissolved. Like projections for Jung, representations for Lévy-Bruhl come between humans and the world. Like projections, representations shape perceptions as well as conceptions. But they do not stem from the mind and therefore reveal nothing about it. They stem from society and reveal something about it only.

Lévy-Bruhl, not merely Jung, grants that, in light of its premises, primitive thinking is logical. The statement from 'Archaic Man' quoted by Rowland – that primitive man 'is no more logical or illogical than we are' but merely starts from different assumptions (para. 107) – almost parrots Lévy-Bruhl's statement, which I quoted, that 'the fact that the *"patterns of thought"* are different does not, once the premises have been given, prevent the primitive from reasoning like us, and, in this sense, his thought is neither more nor less "logical" than ours' (Lévy-Bruhl, 'A Letter to E. E. Evans-Pritchard', p. 12 [italics in original]). As fully as Jung does Lévy-Bruhl acknowledge that the primitive point of view is, to use Rowland's terms, 'natural' and 'coherent'.

But once the mystical and prelogical *premises* of primitive thinking are denied, then for Lévy-Bruhl primitive thinking is both illogical and false. And so, too,

I claim, for Jung. A look at anthropologist E. E. Evans-Pritchard's classic, *Witchcraft, Oracles and Magic among the Azande* (1937), shows how a set of primitive beliefs can be appreciated as consummately logical, given the premises, yet still be pronounced illogical and false. To make her case, Rowland needs to show more than that for Jung primitive thinking for primitives themselves is logical. She needs to show that for Jung primitive thinking is outright logical and true, true about the external world.

Surely for Jung the external world operates through the impersonal processes formulated as laws of science, not through the decisions of gods. Surely for Jung no mystical ether unites all things. Mysticism for primitive peoples is not just another name for synchronicity. Mysticism means identity, or the effacement of the distinction between entities. Synchronicity means the working 'in sync' of distinct entities. For Jung, the relationship of crocodiles to women is that either of causality rather than identity (the first two examples) or of identity (the third example) but is not that of synchronicity.

I doubt that Jung is nudging moderns to entertain the primitive explanation of his three examples of crocodiles and women. For Jung, primitive peoples systematically misunderstand how the world works and do so exactly because they conflate themselves with it. Moderns have nothing to learn from primitives about why the sun rises or why the seasons change. That for Jung moderns are likewise susceptible to 'magical' accounts of, especially, surprising events does not mean that the accounts of primitives are therefore correct. Both accounts are incorrect. Not once but twice in 'Archaic Man' Jung cites the line, noted by Rowland, that 'Magic is the science of the jungle'. But no sooner does Jung repeat this mantra than he attributes magic to the failure of primitives 'to differentiate what is subjective and psychic from what is "objective" and natural' (para. 128). For Jung, it may be understandable how those in the jungle came to concoct magic as their explanation of the world, but that explanation is still false. Science is the science of reality.

If, to quote Rowland, Jung is not 'simply embracing a science that rejects all magic', he is retaining magic only by psychologizing it. When Rowland declares that 'Jung's science takes magic seriously, which is not to say he takes it on his own terms as magic', she seems to be acknowledging the same. But then she cannot maintain that for Jung magic and science are on a par. Unless Jung uses magic the way he uses science – to explain both the human world and the external one – magic for him is not the equal of science. Rowland ends with the claim that 'to be inside the theatre of Jungian science' is 'to experience magic in the science and the science of magic.' I do not fathom how if magic is to be understood scientifically, which means psychologically, there can be any magic in science. If psychology is the key to magic, magic is the *subject* of science, not a *characteristic* of science.

Rowland reads Jung as stressing the similarities between primitive peoples and moderns. I read him as stressing the differences. To me, it would be odd if Jung, while continually deferring to the 'authority' of Lévy-Bruhl, were downplaying the differences. Lévy-Bruhl is his mentor, not his foil.

I do read Jung as criticizing Lévy-Bruhl, but for Lévy-Bruhl's depiction of moderns, not for his depiction of primitives. Never does Jung challenge Lévy-Bruhl on primitives. And again, Lévy-Bruhl himself never asserts that moderns have shed their primitiveness altogether, so that even Jung's differentiation of himself from Lévy-Bruhl may be excessive.

In 'Archaic Man', and still more in 'The Spiritual Problem of Modern Man',[65] Jung chides moderns for assuming that they have transcended their primitive origins. Moderns are deluded in considering themselves scrupulously rational, wholly conscious, and omnipotent. But Jung is faulting moderns *as* a scientist of the mind – that is, as a fellow modern, not as a primitive.

In sum, I differ with Rowland because I read Jung as offering a distinctly modern antidote to the spiritual problems besetting moderns. That antidote is analytical psychology. In enabling moderns to reconnect themselves with their unconscious, psychology does carry them back to the state of primitive peoples. But it also carries them forward, far beyond that state. As a modern, Jung considers himself superior to his forbears.

Reply to Paul Bishop[66]

Like me and like Susan Rowland, Paul Bishop focuses on Jung's essay 'Archaic Man'. But Bishop ventures far beyond me in presenting the history of the essay, in linking the term 'archaic' to Jung's kindred terms 'primal' and 'primordial', and above all in working out the relationship of the concept 'archaic' to Jung's philosophical positions and to Jung's nonphilosophical notion of synchronicity.

In my reply to Susan Rowland, I differentiate archaic thinking from synchronicity. I think that Bishop, like Rowland and no doubt like the hapless Jung, fails to see the difference. Archaic thinking identifies humans with the world. Synchronicity merely parallels them. Bishop and Rowland may then be right about Jung, but Jung would still be wrong.

On various key points Bishop and I are agreed. First, for Jung, the 'archaic' refers not merely to our forbears but also to a variety of thinking found in moderns as well. The archaic is universal. Second, for Jung, the 'archaic' and the unconscious go hand in hand. Archaic thinking is the way the unconscious thinks. Third, for Jung, 'archaic' is anything but a dismissive term. It identifies a rich and profound part, even the core, of the personality. Fourth, for Jung, the archaic aspect of the personality is not to be superseded but is to be cultivated, though I far more than Bishop emphasize Jung's insistence that the cultivation of the archaic begin with the withdrawal of archaic projections from the external world. Fifth, for Jung, moderns suffer because they are disconnected from the archaic and should learn from their forbears' attention to it.

Bishop and I differ most on the characterization of the 'archaic' by Jung and Lévy-Bruhl. Bishop's essay is slightly skewed since it is only Jung who uses the term. Lévy-Bruhl uses primitive instead. But as Bishop stresses, Jung uses related terms that are central to Lévy-Bruhl and that Jung takes from Lévy-Bruhl:

'prelogical' and *participation mystique*. I argue that Jung means to be following Lévy-Bruhl scrupulously in Lévy-Bruhl's use of these terms. But I argue that Jung misunderstands these terms and thereby mischaracterizes 'how natives think'. Bishop maintains that Jung is not confused and instead uses the terms differently.

Bishop notes that Jung does object to Lévy-Bruhl's use of a third term: 'mystical'. I agree, and I noted the objection in my essay. I also agree that among Jung's reasons for objecting to the term is its association with religion and with occultism, if not also with muddle-headedness. As Bishop notes, Jung was continually scorned as 'mystical' by Freudians. Bishop cites Freud's *History of the Psychoanalytic Movement*. One might also cite Ernest Jones' biography of Freud, in which various Freudians register their dismay at Jung's interest in mysticism – as if the *study* of mysticism meant the *practice* of it. For example, Jones cites Freud's own comment on a letter that Jones had received from the American Freudian James J. Putnam: 'Putnam's letter was very amusing. Yet I fear, if he keeps away from Jung on account of his mysticism and denial of incest, he will shrink back from us [. . .].'[67] In *Freud or Jung?*, a rabidly anti-Jungian book with a title posing the mother of all rhetorical questions, Edward Glover remarks: 'It would be easy to account for these confusions [in Jung's thought] by saying that they are due in part to Jung's doting attitude to alchemy and oriental occultism, in part to a desire to present mystical ideas in a "modern" scientific guise [. . .].'[68]

Bishop grants that Jung is happy to use Lévy-Bruhl's term *participation mystique*. But for Lévy-Bruhl that term is interchangeable with 'mystical'. Therefore unless Bishop can show that by *participation mystique* Jung, contrary to Lévy-Bruhl, means something different from mystical, Jung's preference for *participation mystique* over mystical is irrelevant to the main issue dividing us: the relationship for Lévy-Bruhl between *participation mystique* and prelogical. I maintain that while Lévy-Bruhl uses both terms to characterize the thinking of 'archaic man', the terms are distinct. I maintain that Jung fails to see the distinction and consequently conflates the terms.

Bishop's preoccupation with Jung's rejection of 'mystical' is irrelevant to his defense of Jung unless he, Bishop, can establish that Jung recognizes the difference for Lévy-Bruhl between *participation mystique* and prelogicality. But Bishop does not even attempt to do so. He shows only that Jung dislikes the term 'mystical'. But even if Jung, by rejecting the term 'mystical', is thereby absolved of my charge of conflating 'mystical' with 'prelogical', he is not absolved of my real charge, which is that, whether or not Jung chooses to use Lévy-Bruhl's term 'mystical', he, Jung, conflates what Lévy-Bruhl means by either 'mystical' or *participation mystique* with what Lévy-Bruhl means by prelogical. Bishop never considers what Jung means by 'prelogical', even though Jung himself, not merely Lévy-Bruhl, uses that term, together with *participation mystique*, to characterize 'archaic man' and even though Jung continually credits both terms to Lévy-Bruhl.

Bishop does enlist one distinction I note between *participation mystique* for Jung and *participation mystique* for Lévy-Bruhl. Where for Jung we project onto others what we deny in ourselves, for Lévy-Bruhl we project onto others what we recognize in ourselves. But that difference is one noted by me, not by Jung. Jung himself does not distinguish his usage from Lévy-Bruhl's, and the distinction that I note has no bearing on the relationship of *participation mystique* to prelogicality.

In my essay I cite passage upon passage in which Jung uses Lévy-Bruhl's terms *participation mystique* and prelogical to describe the mentality of 'archaic man'. I show case upon case in which Jung, while lauding Lévy-Bruhl for explicating archaic thinking and even using some of Lévy-Bruhl's own examples to do so, fails to fathom Lévy-Bruhl's distinction between *participation mystique* and prelogical. I show that sometimes Jung calls Lévy-Bruhl-like cases that by Jung's own analysis evince not even *participation mystique*, much less prelogicality. Unless Bishop can prove that Jung either grasps what Lévy-Bruhl means by prelogical or offers his own definition of the term, I am not prepared to grant that Jung knows what he is doing. For me, Jung, while not necessarily mystical, is muddle-headed.

Jung's examples aside, how can one characterize the identity among things that is meant by *participation mystique* with the failure to recognize the law of noncontradiction that is meant by prelogical? There is nothing prelogical or – a term that Lévy-Bruhl himself strongly rejects – illogical in the assumption of identity among all things. Lévy-Bruhl would never contend that St Teresa of Avila thinks prelogically just because she thinks mystically. For this reason Lévy-Bruhl considers prelogicality a far more significant characteristic of archaic thinking than *participation mystique*. What does Bishop's reconstruction of Jung say of Jung's repeated use of prelogical? Nothing. Perhaps Bishop's inattention to this word reveals his own conflation of it with *participation mystique*.

Bishop brilliantly links up *participation mystique* with still budding ideas of Jung's, especially that of synchronicity. Bishop thereby reinforces his claim that by the 'archaic' Jung means more than chronology – that is, means more than our forbears. But however wide-ranging for Jung the link between *participation mystique* and synchronicity is, that link does not connect prelogical to synchronicity. If synchronicity were tied to prelogical, synchronicity would be the claim that what goes on in us is not only identical with what is going in the external world but also distinct from it. I daresay that Jung does not introduce the concept of synchronicity to stress separateness. It is identity only that he stresses, and identity stems from *participation mystique* exclusively and not also from the prelogical.

Notes

1 See L. Lévy-Bruhl, *Primitive Mentality*, London: Allen and Unwin, 1923 [1922 in French]; *The 'Soul' of the Primitive*, London: Allen and Unwin, 1928 [1927 in French]; *Primitive Peoples and the Supernatural*, London: Allen and Unwin, 1935 [1931 in French]; *L'Expérience mystique et les symboles chez les primitives* [not translated], Paris: Alcan, 1938; 'A Letter to E. E. Evans-Pritchard', *British Journal of Sociology* 3, 1952, 117–23; *The Notebooks on Primitive Mentality*, New York: Harper, 1975 [1949

in French]; *Primitive Mythology*, St Lucia: University of Queensland Press, 1983
[1935 in French]; *How Natives Think*, Princeton, NJ: Princeton University Press, 1985
[1926] [1910 in French].

2 See, as the *locus classicus*, F. Boas, *The Mind of Primitive Man*, rev. edn, New York: Macmillan; London: Collier Macmillan, 1965 [1938], pp. 43–44.

3 See also Lévy-Bruhl, *Primitive Mentality*, pp. 21–33; 'A Letter to E. E. Evans-Pritchard', p. 121; *The Notebooks on Primitive Mentality*, p. 49.

4 See also C. J. Throop, 'Minding Experience: An Exploration of the Concept of "Experience" in the Early French Anthropology of Durkheim, Lévy-Bruhl, and Lévi-Strauss', *Journal of the History of the Behavioral Sciences* 39, 1983, 365–82 (pp. 370–75).

5 C. R. Hallpike points out that the law of noncontradiction is not in fact thereby violated: 'But there is nothing *logically* contradictory in men's claim to be *really* parakeets as well as men: a logical contradiction is of the form that something is both *A* and not-*A*, which is quite different from an assertion that something is both *A* and *B*, even if it happens to be physically impossible for anything to be both *A* and *B* simultaneously' (C. R. Hallpike, *The Foundations of Primitive Thought*, Oxford: Clarendon, 1979, p. 488). See also Hallpike, 'Is There a Primitive Mentality?', *Man* 11, 1976, 253–70.

6 See, unfairly, P. Radin, *Primitive Man as Philosopher*, 2nd edn, New York: Dover, 1957 [1st edn 1927].

7 See also Lévy-Bruhl, *Primitive Mythology*, p. 7.

8 See, for example, Claude Lévi-Strauss' *defence* of primitive, even 'savage', thinking: *La pensée sauvage* (Paris: Librairie Plon, 1962). He writes: 'This thirst for objective knowledge is one of the most neglected aspects of the thought of people we call primitive' (C. Lévi-Strauss, *The Savage Mind*, tr. not given, Chicago: University of Chicago Press, 1966, p. 3). And: 'When we make the mistake of thinking that the Savage is governed solely by organic or economic needs [...]' (Lévi-Strauss, *The Savage Mind*, p. 3). Of course, he may be using the terms 'savage' and primitive ironically, as Nathalie Pilard has pointed out to me.

9 C. S. Littleton, 'Lucien Lévy-Bruhl and the Concept of Cognitive Relativity', Introduction to Lévy-Bruhl, *How Natives Think*, pp. v–lviii (p. xiv [italics in original]).

10 Besides Littleton, see, for example, R. Needham, *Belief, Language and Experience*, Oxford, Blackwell: 1972, pp. 181–83.

11 See also Lévy-Bruhl, *Primitive Peoples and the Supernatural*, p. 33.

12 E. B. Tylor, *Primitive Culture*, 5th edn, 2 vols, New York: Harper Torchbooks, 1958 [1913] [1st edn 1871]; J. G. Frazer, *The Golden Bough*, one-vol. abridgment, London: Macmillan, 1922 [1st edn 1890].

13 To be sure, Tylor does state that in dreams, visions, and other abnormal states primitive perceptions differ from modern ones: see Tylor, *Primitive Culture*, vol. 1, 286, 297, 305–06; vol. 2. pp. 29–31, 62, 83–84, 275, 280, 442). But then those states differ from normal primitive ones as well. And surely Tylor allows for abnormal modern states, too.

14 In defense of Lévy-Bruhl against the common charge that he assumes rather than explains collective representations, see R. Horton, 'Lévy-Bruhl, Durkheim and the Scientific Revolution', in R. Horton and R. Finnegan (eds), *Modes of Thought*, London: Faber and Faber, 1973, pp. 249–305 (pp. 255–56); reprinted in R. Horton, *Patterns of Thought in Africa and the West*, Cambridge: Cambridge University Press, 1993, pp. 63–104. Yet Lévy-Bruhl himself acknowledges this failing: see Lévy-Bruhl, 'A Letter to E. E. Evans-Pritchard', p. 118.

15 See E. Durkheim, *The Elementary Forms of the Religious Life*, tr. J. W. Swain, New York: Free Press, 1965 [1915], pp. 21–33, 267–72, 476–96; E. Durkheim and M. Mauss, *Primitive Classification*, ed. and tr. R. Needham, Chicago: University of Chicago Press, 1963.

16 See Durkheim, *The Elementary Forms of the Religious Life*, pp. 267–72; review of
Lévy-Bruhl, *Les Fonctions mentales dans les sociétes inférieures*, and of Durkheim,
Les Formes élémentaires de la vie religieuse, *L'Année Sociologique* 12, 1913,
pp. 33–37, translated and reprinted in W. S. F. Pickering (ed.), *Durkheim on Religion*,
trs J. Redding and W. S. F. Pickering, London and Boston: Routledge and Kegan Paul,
1975, pp. 169–73.

17 See Durkheim, *The Elementary Forms of the Religious Life*, pp. 270, 477, 486, 493;
review of Lévy-Bruhl, *Les Fonctions mentales dans les sociétes inférieures*, and of
Durkheim, *Les Formes élémentaires de la vie religieuse*, in Pickering (ed.), *Durkheim
on Religion*, p. 171. Lévy-Bruhl never explicitly responded to Durkheim's criticisms.
For a reconstruction of Lévy-Bruhl's tacit and not so tacit rejoinders, see D. Merllié,
'Did Lucien Lévy-Bruhl Answer the Objections Made in *Les Formes elémentaires*?',
in N. J. Allen, W. S. F. Pickering, and W. W. Miller (eds), *On Durkheim's 'Elementary
Forms of Religious Life'*, London and New York: Routledge, 1998. See also
W. Schmaus, 'Lévy-Bruhl, Durkheim, and the Positivist Roots of the Sociology of
Knowledge', *Journal of the History of the Behavioral Sciences* 32, 1996, 424–40
(p. 430).

18 See X. de Angulo, 'Comments on a Doctoral Thesis' [1925], in W. B. McGuire and
R. F. C. Hull (eds), *C. G. Jung Speaking*, Princeton, NJ: Princeton University Press,
1977, pp. 205–18 (p. 214).

19 See C. G. Jung, 'Archaic Man' (1931 in German), in C. G. Jung, *Collected Works*, eds
Sir H. Read, M. Fordham, G. Adler, and W. McGuire, 20 vols, London: Routledge and
Kegan Paul, 1953–1983, vol. 10, paras. 104–47. Reproduced in this volume, see
pp. 171–87.

20 See C. G. Jung, *Psychological Types*, in Jung, *Collected Works*, vol. 6, paras. 1–857
(paras. 12, 123, 216, 692, 781).

21 See C. G. Jung, 'Archaic Man' 'Mind and Earth' (1927 in German), in Jung, *Collected
Works*, vol. 10, paras. 49–103; 'The Spiritual Problem of Modern Man' (1928 in
German), in Jung, *Collected Works*, vol. 10, paras. 148–96.

22 See C. G. Jung, *Memories, Dreams, Reflections*, ed. A. Jaffé, New York: Vintage
Books, 1963, p. 246.

23 See S. Shamdasani, *Jung and the Making of Modern Psychology*, Cambridge:
Cambridge University Press, 2003, pp. 288–93, 295–97. While granting that for
Durkheim collective representations were social rather than psychological, Shamdasani
still considers them a prime source of Jung's concept of archetypes and still considers
Lévy-Bruhl the conduit of the concept from Durkheim to Jung.

24 On Jung and Lévy-Bruhl, see I. Progoff, *Jung's Psychology and Its Social Meaning*,
Garden City, NY: Doubleday Anchor Books, 1973 [1953], pp. 146–51, 233–38; de
Angulo, 'Comments on a Doctoral Thesis', pp. 214–15, who reports that Jung lamented
Lévy-Bruhl's posthumously published recantation of his views; M. V. Adams, *The
Multicultural Imagination*, London and Boston: Routledge, 1996, pp. 54–60, who gets
Lévy-Bruhl right; J. Dourley, 'Archetypal Hatred as Social Bond: Strategies for Its
Dissolution', in J. Beebe (ed.), *Terror, Violence and the Impulse to Destroy*, Einsiedeln,
Switzerland: Daimon Verlag, 2003, pp. 135–59, who relies wholly on Jung for knowl-
edge of Lévy-Bruhl and therefore wrongly takes *participation mystique* rather than
prelogical thinking to be the heart of primitive mentality; B. W. Burleson, 'Defining
the Primitive: Carl Jung's "Bugishu Psychological Expedition" ', *Journal of African
Travel-Writing* 3, 1997, 17–30, and *Jung in Africa*, London and New York: Continuum,
2005, who in both works faults Jung for the amateurishness of his fieldwork in East
Africa and for his uncritical acceptance of Lévy-Bruhl's views yet who nevertheless
commends Jung for his insight into the mystical character of primitive peoples – even
though that character is identical with the one depicted by Lévy-Bruhl!

25 On the shift in Lévy-Bruhl's emphasis, see D. L. O'Keefe, *Stolen Lightning*, New York: Continuum, 1982, pp. 86–88.

26 Jung seems closer to the English anthropologist E. E. Evans-Pritchard, for whom the atypicality of an unfortunate event spells magic without mystical identity: see Evans-Pritchard, *Witchcraft, Oracles and Magic among the Azande*, Oxford: Clarendon Press, 1937.

27 John Beebe has explained to me that Jung uses the term *participation mystique* to understand what today is called borderline personality.

28 Elsewhere Jung himself notes this difference: 'Among civilized peoples it [*participation mystique*] occurs between persons, seldom between a person and a thing' (Jung, *Psychological Types*, para. 781).

29 Ironically, this is the way that Evans-Pritchard, in criticism of Lévy-Bruhl, takes *primitive* identification as well.

30 C. G. Jung, *Symbols of Transformation*, in Jung, *Collected Works*, vol. 5.

31 On the rise and fall of Haeckel's Law, see S. J. Gould, *Ontogeny and Phylogeny*, Cambridge, MA: Belknap Press of Harvard University Press, 1977, esp. pp. 1–9, 76–85, 167–206. On Jung's (as well as Freud's) acceptance of this principle, see Gould, *Ontogeny and Phylogeny*, pp. 155–61 (on Freud) and 161–63 (on Jung).

32 C. G. Jung, 'The Psychology of the Child Archetype', in Jung, *Collected Works*, vol. 9/i, paras. 259–305.

33 C. G. Jung, 'Archetypes and the Collective Unconscious', in Jung, *Collected Works*, vol. 9/i, paras.1–86.

34 On Freud's derogatory view of primitive peoples, see C. Brickman, *Aboriginal Populations in the Mind*, New York: Columbia University Press, 2003, chapter 2.

35 On Lévy-Bruhl's wariness of the parallel between primitive peoples and children, see G. Jahoda, 'Piaget and Lévy-Bruhl', *History of Psychology* 3, 2000, 218–38 (pp. 220–21, 228–30).

36 On the trade-off for Jung between Europeans and Africans, see M. Adams, *The Fantasy Principle*, London: Brunner-Routledge, 2004, chapter 7.

37 E. E. Evans-Pritchard, 'Lévy-Bruhl's Theory of Primitive Mentality', *University of Egypt Bulletin of the Faculty of Arts* 2, 1934, 1–36. Evans-Pritchard also cites professional fieldworkers whose findings support Lévy-Bruhl's views: see Evans-Pritchard, 'Lévy-Bruhl's Theory of Primitive Mentality', pp. 15–16. See also Evans-Pritchard, *A History of Anthropological Thought*, ed. A. Singer, New York: Basic Books, 1981, p. 131.

38 P. Radin, *Primitive Man as Philosopher*. See also Radin, *Primitive Religion*, New York: Viking Press, 1937, pp. 60–61, 269–74; *The World of Primitive Man*, New York: Abelard-Schuman, 1953, pp. 49, 319.

39 See Radin, *Primitive Man as Philosopher*, pp. 39, 63; *Social Anthropology*, New York: McGraw-Hill, 1933, pp. 15–16; *Primitive Religion*, p. 61; *The World of Primitive Man*, p. 306.

40 M. Douglas, *Purity and Danger*, Baltimore: Penguin Books, 1970 [1966].

41 See Douglas, *Purity and Danger*, pp. 92–94; *Evans-Pritchard*, Sussex: Harvester Press, 1980, pp. 17–20, 31–35.

42 See, on pollution and cleanness, Douglas, *Purity and Danger*, passim; *Implicit Meanings*, London and Boston: Routledge & Kegan Paul, 1975, chapter 3; on food, Douglas, *Purity and Danger*, chapter 3; *Natural Symbols*, 2nd edn, New York: Vintage Books, 1973 [1st edn 1970], chapter 3; *Implicit Meanings*, chapter 16; *In the Active Voice*, London and Boston: Routledge and Kegan Paul, 1982, chapter 4; on animals, Douglas, *Purity and Danger*, chapter 3; *Implicit Meanings*, chapters 1–2, 16; on parts of the body, Douglas, *Purity and Danger*, chapter 7; *Natural Symbols*, passim; *Implicit Meanings*, chapters 4, 6.

43 See C. Lévi-Strauss, *The Savage Mind*, esp. chapter 1; *Introduction to a Science of Mythology*, trs. J. and D. Weightman, vol. I: *The Raw and the Cooked*, New York: Harper and Row, 1969, pp. 1–32; *Myth and Meaning*, Toronto: University of Toronto Press, 1978.

44 See C. Lévi-Strauss, 'The Structural Study of Myth', *Journal of American Folklore* 68, 1955, 428–44.

45 In defence of Lévy-Bruhl against Radin and and Lévi-Strauss, see R. Pertierra, 'Lévy-Bruhl and Modes of Thought: A Re-appraisal', *Mankind* 14, 1983, 112–26 (pp. 118, 121–23).

46 Hallpike, *The Foundations of Primitive Thought*. See also Hallpike, 'Is There a Primitive Mentality?'.

47 J. Goody, *The Domestication of the Savage Mind*, Cambridge: Cambridge University Press, 1977. See also Goody, 'Evolution and Communication: The Domestication of the Savage Mind', *British Journal of Sociology* 24, 1973, 1–12.

48 See E. E. Evans-Pritchard, 'Lévy-Bruhl's Theory of Primitive Mentality'; *Witchcraft, Oracles and Magic among the Azande*, Oxford: Clarendon Press, 1937; 'Lucien Lévy-Bruhl', *Man* 40, 1939, 24–25; *Nuer Religion*, Oxford: Clarendon Press, 1956, chapter 5; *Theories of Primitive Religion*, Oxford: Clarendon Press, 1965, chapter 4; *A History of Anthropological Thought*, chapter 12.

49 Though Tylor scarcely generalizes from the example, he, too, cites primitive peoples who attribute death to witchcraft exclusively: see Tylor, *Primitive Culture*, vol. 1, p. 138.

50 See also Lévy-Bruhl, *Primitive Mentality*, p. 442; *The 'Soul' of the Primitive*, p. 19; 'A Letter to E. E. Evans-Pritchard', p. 122.

51 See B. Malinowski, 'Magic, Science and Religion' [1925], in Malinowski, *Magic, Science and Religion and Other Essays*, ed. R. Redfield, Garden City, NY: Doubleday Anchor Books, 1954 [1948], pp. 17–92 (pp. 25–26); Evans-Pritchard, *A History of Anthropological Thought*, pp. 127–28.

52 See Lévy-Bruhl, *How Natives Think*, pp. 228–29; *Primitive Mentality*, pp. 442–44; *The 'Soul' of the Primitive*, pp. 19–20; *Primitive Peoples and the Supernatural*, pp. 24, 92.

53 See S. Lukes, 'Some Problems about Rationality' [1967], in B. R. Wilson (ed.), *Rationality*, New York: Harper Torchbooks, 1971, pp. 194–213 (pp. 207–08).

54 P. Winch, 'Understanding a Primitive Society', *American Philosophical Quarterly* 1, 1964, 307–24. Most confusingly, Winch considers only Evans-Pritchard's *Witchcraft, Oracles and Magic among the Azande* and not also his *Nuer Religion*, in which Evans-Pritchard comes close to Winch.

55 Winch's denial that external criteria exist for judging Azande witchcraft seemingly makes him a relativist. But he vehemently denies that he is – both on the grounds that he acknowledges some external criteria (see Winch, 'Understanding a Primitive Society', pp. 308, 318; but see also pp. 315, 317, and see also S. Lukes, 'On the Social Determination of Truth', in Horton and Finnegan (eds), *Modes of Thought*, pp. 230–48 [p. 233]) and, even more, on the grounds, not in fact germane, that the form of life exemplified by witchcraft is itself universal (see Winch, 'Understanding a Primitive Society', p. 322). The criteria that make witchcraft rational are therefore not just idiosyncratic Azande ones, but they are still Azande ones.

56 P. Winch, *The Idea of a Social Science and Its Relation to Philosophy*, London: Routledge and Kegan Paul, 1958.

57 See also M. Douglas, 'World View and the Core', in S. C. Brown (ed.), *Philosophical Disputes in the Social Sciences*, Sussex: Harvester Press; Atlantic Highlands, NJ: Humanities Press, pp. 177–87.

58 On rationality, see Wilson (ed.), *Rationality*; R. Borger and F. Cioffi (eds), *Explanation in the Behavioural Sciences*, Cambridge: Cambridge University Press, 1970,

pp. 167–230; Horton and Finnegan (eds), *Modes of Thought*, pp. 230–305; S. I. Benn and G. W. Mortimore (eds), *Rationality and the Social Sciences*, London and Boston: Routledge and Kegan Paul, 1976; F. R. Dallmayr and T. C. McCarthy (eds), *Understanding and Social Inquiry*, Notre Dame, IN: University of Notre Dame Press, 1977; M. Hollis and S. Lukes (eds), *Rationality and Relativism*, Cambridge, MA: MIT Press, 1982.

59 See, for example, Needham, *Belief, Language and Experience*, pp. 164–66.

60 See Lévy-Bruhl, *The Notebooks on Primitive Mentality*, esp. pp. 37–39, 47–50, 100–01, 126–27. See also Horton, 'Lévy-Bruhl, Durkheim and the Scientific Revolution', pp. 257–58; G. E. R. Lloyd, *Demystifying Mentalities*, Cambridge: Cambridge University Press, 1990, pp. 1–2.

61 In partial defense of Lévy-Bruhl against Evans-Pritchard, see E. Gellner, 'Concepts and Society' [1970], in Wilson (ed.), *Rationality*, pp. 18–49 (pp. 34–39).

62 See J. Piaget, *Structuralism*, New York: Harper Torchbooks, 1971 [1970], pp. 197–98. On Lévy-Bruhl's influence on Piaget, see Jahoda, 'Piaget and Lévy-Bruhl'.

63 On Lévy-Bruhl's continuing influence, or anticipation of subsequent views, see, pro and con, G. E. R. Lloyd, *Polarity and Analogy*, Cambridge: Cambridge University Press, 1966, pp. 3–6; Lloyd, *Demystifying Mentalities*, pp. 1–5, 116–19, 141–42, 144; Needham, *Belief, Language and Experience*, pp. 167–71; Horton, 'Lévy-Bruhl, Durkheim and the Scientific Revolution', pp. 272–73; J. R. Porter, 'Biblical Classics: III. Johs. Pedersen: Israel', *Expository Times* 90/2, November 1978, 36–40; H. Gardner, *The Mind's New Science*, New York: Basic Books, 1985, esp. pp. 223–27, 257–59; Littleton, 'Lucien Lévy-Bruhl and the Concept of Cognitive Relativity', pp. xxi–xliv; Pertierra, 'Lévy-Bruhl and Modes of Thought: A Re-appraisal', pp. 114–15; D. Price-Williams, 'In Search of Mythopoeic Thought', *Ethos* 27, 1999, 25–32.

64 See Susan Rowland, 'Response to Robert A. Segal's "Jung and Lévy-Bruhl" ', *Journal of Analytical Psychology*, 2007, vol. 52, pp. 659–65. Her article is reproduced in this volume, pp. 219–25.

65 C. G. Jung, 'The Spiritual Problem of Modern Man', in Jung, *Collected Works*, vol. 10, paras. 148–96.

66 See P. Bishop, 'The Timeliness and Timelessness of the "Archaic": Analytical Psychology, "Primordial" Thought, Synchronicity', *Journal of Analytical Psychology*, 2008, vol. 53, pp. 501–23. This article is reproduced in this volume, pp. 226–47. Robert A. Segal's reply has been especially written for this volume.

67 E. Jones, *Sigmund Freud*, vol. 2, London: Hogarth Press, 1955, pp. 115–16.

68 E. Glover, *Freud or Jung?*, New York: Meridian Books, 1956 [1949], p. 50.

Chapter 8

Jung 'performing' the archaic

Response to 'Jung and Lévy-Bruhl'

Susan Rowland

This article[1] is a response to Robert A. Segal's excellent 'Jung and Lévy-Bruhl' in this volume.[2] My work differs from his in taking a radically different view of Jung's essay 'Archaic Man'.[3] Whereas Segal examines the intellectual relationship between Jung's psychology and Lévy-Bruhl's anthropology, my approach is to concentrate on the type of writing that results from Jung's attempt to *stage* a confrontation between modern so-called 'sophisticated' and non-western people's so-called 'primitive' attitudes. In particular, I am arguing for a creative and artistic framework to Jung's work here as he tries to use his psychology, a psyche-logos, writing about the psyche, to *enact* his ideas. Indeed, his far-sighted criticism of colonial attitudes (while demonstrating weaknesses) amounts to a crucial postcolonial component to Analytical Psychology.

My starting position here is that this essay is not a challenge to Segal's conclusions. Rather, it is an attempt to demonstrate how different sorts of knowledge can be constructed from the same raw textual material. By drawing upon epistemologies more familiar in the arts, I aim to show that an innately *creative* treatment of writing and reading is *native* to Jung's texts.

To begin then is to take Segal's perceptive analysis of Jung's indebtedness to Lévy-Bruhl, which concentrates on, without exclusively drawing from, the text 'Archaic Man'. Segal makes the very important point that it is often in the *valuation* of so-called primitive psychological traits that Jung differs from figures such as Lévy-Bruhl and Sigmund Freud. Indeed Segal draws on Jung's formulation of two kinds of thinking, 'directed' and 'fantasy', to show what was significant for him in 'primitive' peoples.

Consequently Segal's scholarship is an important addition to understanding Jung in the history of ideas, and even in tracing the development of western constructs of indigenous peoples. He begins to show just how much psychology has been influenced by the physical encounter with 'other' cultures. Segal's conclusion (a triumph of directed thinking!) appears to be the following:

Note: This chapter was first published in *The Journal of Analytical Psychology* 52(5), 2007, 659–65. It is reproduced here with kind permission of Wiley-Blackwell and *The Journal of Analytical Psychology*.

> For both Jung and Lévy-Bruhl, intellectual progress comes from exposing
> 'primitive' mischaracterizations of the external world [. . .] Science properly
> replaces myth and religion as the explanation of the world. There is no
> turning back.
>
> (p. 200, this volume)

And:

> For Jung . . . [m]odernity is better than primitivism, but it is not the ideal
> state. It is a stage along the way.
>
> (pp. 200–01, this volume)

I will take these ideas as an initial basis for my response. First, let me say that I
am not going to disagree with Segal's analysis; rather I am going to supplement
it, to add another view. For my argument is that Segal has admirably analyzed the
'directed thinking' of 'Archaic man'. However, this logical edifice is far from the
total matter of the text. An analysis of the rationality of 'Archaic Man', in my
view, could be effectively supplemented by looking at its 'fantasy thinking'.

Put another way, Jung's essays typically draw on at least as many rhetorical
techniques as they do logical methods. Moreover, rhetoric is not an adornment to
Jung's work. Rather it is a way of netting many voices, many aspects of the
thinking/dreaming psyche, into the debate. These 'voices' are deeply implicated
in the *matter* of the writing. That is, the voices are both means and subject of the
essay. After all, at the end of 'On the Nature of the Psyche', Jung concludes of his
science:

> I fancied I was working along the best scientific lines [. . .] only to discover
> in the end that I had involved myself in a net of reflections which extend far
> beyond natural science and ramify into the fields of philosophy, theology,
> comparative religion, and the human sciences in general.[4]

Such a description was never more true than of 'Archaic Man' where Jung generates different 'points of view' (a phrase often repeated). This signifies perspectives from different parts of the earth and also different constructions of reality.
Our encultured question 'what does Jung really mean by . . .' is frequently frustrated by the lens provided by the many languages of science, magic, divination,
anecdotes, etc. For appealing to authorial intention as the guarantor of logical
coherence in a text is to regard the author as the rational one God of 'his' creation.
Such a practice, consciously monotheistic, constructs truth as singular and
coherent by the very act of casting out all that is 'other'. Jung is quite capable of
adopting 'other' authorial personas than the Christian logos.

Here, the text's circling, repetitive, interwoven comparisons aim at a radical
undermining of just that reliance upon logic or 'directed thinking' as uniquely
valuable. What I am arguing is that 'Archaic man' *performs* its deepest

achievement: it places the reader within its 'net of reflections' and it does so as an ethical act. I am not saying that the essay is against logic or rationality, or that it devalues it. Rather, I am saying that it works to undermine the sole supremacy of directed thinking in favour of re-*placing* it, changing its *place*, into a relationship, a net of reflections with the kind of 'otherness' that the term 'primitive' comes to represent in 'Archaic' man's rhetoric.

Considered another way, while I can just about accept Segal's 'no going back' from modern science, to me the essay is about the necessity of 'looking back'. So the argument is more multi-directional in time as well as in space. Mood, tone, playfulness, characterization are all contributors for, fundamentally, 'Archaic Man' is *dialogical*: the reader is the necessary other person in its exchange of ideas and secreting of meaning. It is the reader whom the essay seeks to deconstruct out of a comfortable sense of complete distinction from, and superiority over, 'primitive, archaic' man.

This is how it is done.

Looking at 'Archaic Man'

In the opening paragraph, we are told twice that we are 'apparently' in a better position to 'overlook' archaic man than our 'civilized' selves. Of course this will prove to be the fundamental in-direction of the essay because Jung is setting up 'different' vantage points to get at his real subject, the modern European! There is also the language of colonial power here in the notion of 'occupying' space to 'overlook', and the ambiguity of 'overlooking' as *not* seeing, should be borne in mind.

Immediately, the essay begins to diminish the distance between these 'points of view'. Jung is an artist of space and time. 'Archaic Man' is a peculiarly lucent demonstration of how his techniques of argument, and the matter itself, turn on constructions of space (now) and time (later). Here Jung introduces Lévy-Bruhl and the 'primitive' belief in magic. He quotes (as Segal says, without attributing it to Lévy-Bruhl), the story of the crocodile with two anklets in its stomach. Here it seems to me significant that he brings in the perspectives of the people caught up in the crocodile incident themselves: 'The natives said . . . the natives maintained' (para. 106). So now we have the viewpoints of the 'civilized' on the natives and that of the natives on this event, to go with the earlier suggestion that 'civilized' man has 'archaic processes' too.

Jung then begins the process of bringing 'primitive' and civilized together. This is a movement that he will never complete in the sense that he will not say they are exactly the *same*. Nor will he value them the same. Rather his aim is to get the reader to acknowledge the historical and archaic in him/herself. So 'primitive' man 'is no more logical or illogical than we are'; 'he' merely starts from different assumptions (para. 107). And the next part of the essay concentrates on the similarities of 'primitive' to 'civilized' man in adaptation to his very different environment.

True, the 'primitive' has 'a smaller area of consciousness than we' but 'his psychic functioning is essentially the same' (para. 107). This illustrates Jung's method of asserting likeness and unlikeness almost simultaneously in the exchange between rhetorical positions: the process of assimilation and differentiation sets up a dialogical reflective understanding of the civilized. We are taught to see ourselves in the imperfect mirror of the cultural other.

The device of drawing together yet refusing to assimilate enables the essay to begin to open up the otherness within 'civilized' man. We are reminded that our rational belief in causality is a 'sacred dogma', suggesting that it, too, is not wholly susceptible to scientific method. In addition, arbitrariness has crept back into modernity in quantum mechanics, and that 'chance' is detested because it reminds us of beliefs we thought we had forgotten in 'Satanic imps or of a caprice of a *deus ex machina*' (para. 113).

This rhetorical strategy of introducing the register of religion and the occult gathers apace when 'chance' is definitively called a 'demon' (para. 114). We accept chance, while not being able to explain it, a superstitious trait of moderns not 'primitives'. By contrast, the 'primitive' has 'intention' instead of chance. If a crocodile takes one woman out of three from a riverbank then he must have 'intended' to do so (para. 117). But who told him to do it?

Jung explains that 'primitives' have a perfectly rational need for religion in order to cope with the violent vagaries of their world. In a couple of paragraphs, the essay presents the 'point of view' of the 'primitive' as natural and coherent even if blaming a flagstaff for a storm appears illogical to the colonial 'civilized' mentality (para. 118). 'Archaic man' therefore encourages the reader to imaginatively engage with the 'primitive' as 'other', but never to wholly identify with that perspective. Here we note that the world of the 'primitive' is one that is primarily marked 'other' by space.

After occupying the 'ground' of the other (no longer 'overlooking'!), the essay begins to excavate the 'ground' of the civilized. It digs back in time to find another way of framing a dialogue between European and indigene. In the so-called 'civilized space of the Europe that colonized Africa, grandparents were wont to point to the birth of two-headed calves as a harbinger of war' (para. 120). We find chances often occur in sequences and here 'primitive' belief 'precedes' modernity. 'Magic is the science of the jungle,' Jung quotes and will again (para. 121). Unlike Segal, I do not take the essay as saying that science merely replaces the magic of the jungle. Rather, I think 'Archaic Man', *as writing*, is suggesting that magic and science need to relate to each other, in a net of *reflections* reflecting back their biases to each other.

Here, we are told that astrology and divination are the science of antiquity, so again the 'other' in space ('primitives' with their magic) re-connects with the other in time (astrology, divination of the European past) without losing distinctiveness from each other. The space-time link is made even more economically when the Pueblo Indian and Ancient Roman stand together as sensible followers of omens: a good reaction to a dangerous world (para. 125). Jung makes a rhetorically powerful move by putting himself on the same argumentative *ground* as the Indian and the Roman by telling of his dangerous trip to Africa in which his friend nearly died, *and they both came to respect omens.*

After establishing that primitive magic is 'expedient', the essay focuses on the relationship of the psyche with the environment in both primitives and moderns. 'Primitive man is unpsychological' because he does not distinguish between self and environment. This Jung claims as psychic projection, saying that Lévy-Bruhl called it *participation mystique* (para. 130). Segal does a valuable job in pointing out the gaps in his argument here: that just assuming others have the same psychic processes is not really projection, and then also that seeing our own evils in others is not really what 'primitives' do.

Yet, Jung's language here is fascinating! He says we see our own evils as *betes noires* and 'scapegoats', which were formerly 'witches and werewolves' (para. 130). Previously he has noted how 'primitives' have bush-souls in an animal, crocodile, leopard. Now he uses the language of animals and the occult of the European imagination, past and present. Again, I suggest, he is aligning the modern occult and irrational past with the other cultural present, not to wholly identify them, but to *develop a real perspective on the past*, to make visible what modernity has obscured. Crucially, modern idiomatic language carries our 'other' consciousness, symbolized by animals in terms such as *bete noire* or scapegoat, into the modern present.

After all, what Segal does not include is Jung's admission that 'primitive' projection *is* different because they are too 'naturalistic' for the moral judgements of modern projection, which is not to say that Jung's analysis is correct. Elsewhere Jung gives more attention to why modernity's projections are so dark. He concludes in the 'Trickster' essay that loss of myth concentrates the projection into dangerous shadow territories.[5] However, here, the difference in projection between modern and 'primitive' is the source of an important ethical and eco-critical in-sight. The kind of projection that moderns employ is all about power. 'Primitive' man does not dream of ruling nature; modern man's science is designed for little else. Arbitrary forces are resented by moderns because they frustrate the dream of controlling the non-human environment. The search for 'natural causes' is about attempting to find the 'key' that will allow modern man to 'dominate' the other as nature (para. 134). Indeed, Jung makes the powerful eco-critical point that 'civilized man must strip nature of all her psychic attributes in order to dominate it' (para. 135).

After a distinctly ethical and ecological realization of modernity's difference from the 'primitive', Jung brings in religion, not science, as the cultural practice making the break from nature. Baptism is a rite marking personhood as transcendence of nature (para. 136). However, as the very next paragraph asserts, the spiritual is not securely *detached* from the world in either 'primitive' or 'modern' perspective. Non-western peoples take ghosts very seriously while Europe is gripped by a fashion for séances.

This return to *likeness* between modern and 'primitive' is strengthened by another emphasis on the 'primitive' point of view. To primitive man the belief that intentions rule chance and sorcerers contact spirits is 'perfectly natural'. In fact the 'primitive' notion of *mana* or a spiritually powerful person is a sort of very basic theory of energy (paras. 138–39). Science may inhere in myth in such a way that it needs the myth to be fully what it is. The stories of spirits and gods are no petty addition to the primitive's science. Rather the myth-story is part of the technology, the action of the science. It is how the science works in the world of this particular culture.

So with the possibilities that 'primitives' have a viable, yet culturally different science, Jung has acquired the standpoint, the *ground*, on which he can challenge the premises of his own scientific and philosophical tradition. As an intellectual descendant of Descartes, Jung habitually assumes a subject/object distinction. Here the psyche is projected from 'inside' onto another. But he has noticed (*pace* Segal) that 'primitives' do not behave like this. To them it appears that the psyche is outside as well as inside. Might this astounding idea have some basis? The psyche may dwell beyond the living body, be present in the world and account for those ethereal ghosts, spirits and powers (para. 140).

In a bold move, the archaic and modern materialist perspectives are brought together in that each holds man to be the accidental product of the environment (para. 142). This is a clever rhetorical strategy in that it links the 'primitive' and modern viewpoints, without homogenizing them and without the argument's persona, the 'Jung', adopting the position wholly. The new factor of contemporary, godless 'materialism' enters the argument as another wedge to simultaneously connect and distance positions in the mind of the reader.

The last few paragraphs of 'Archaic Man' are rich in human voices and cultural traditions. After noting the inconsistencies of both 'primitives' and moderns, Jung depicts his struggles to find a religious ceremony amongst the Elgonyi. A very similar argument is then imagined with a native of Zurich, who was found to observe a weird and superstitious Easter egg ritual (para. 145). Although Jung says that archaic man *does* and modern man is *aware* of what he does, his example of the unsophisticated Zurich citizen does not bear this out. Sometimes the group distinctions are frustrated by the human variances *within* the categories 'modern' and 'primitive'.

'Archaic Man' ends typically as open to the reader's preferences. Jung interprets the Elgonyi sun ceremony as sounding very like a prayer that may be said by any of the monotheisms: ' "Lord, into thy hands I commend my spirit" ' (para. 146). So he ends with the mystery marking a boundary in both sets of cultures: do our religions reflect a pre-existing teleological truth or are they merely artefacts of culture? By ending, 'I must leave this question unanswered', Jung's rhetorical mode of argument is underlined (para. 147).

The reader who has been exposed to the multifaceted (literally here many faced!) points of view of 'Archaic Man' has come to see that 'grounds' for building an argument upon are multiple, and that plurality is part of the argument itself. 'Archaic Man' is an argument about science and culture in the form of a web, a multi-directional dialogue between different positions. It is in effect a peculiarly lucent study of how space and time *figure* our ideas, our reason, and in particular, our perceptions of the 'other'.

Segal's brilliant essay links this topic of the 'primitives' in Jung to intellectual history. My reading of 'Archaic Man' suggests that Jung is not simply embracing a science that rejects all magic. Rather, Jung's web-like writing resurrects some of the many voices of the whole psyche. His figures here are rhetorical *figures* who speak through forms of otherness that include a number of indigenous cultures, the language of the occult (werewolves, seances), homely European folklore (grandparents and their two-headed calf), sorcerers, the naive citizens of Zurich, the crocodile who had intentions, ghosts, the missionary with the flagstaff, etc.

Put theoretically, the aim of such an essay would seem to be to deconstruct European (colonial) superiority by unpicking the simple division of people into modern or 'primitive'. 'Archaic Man' shows that the two groups are distinct yet similar, and in particular, moral superiority cannot be attributed to the moderns. The linking of modern science with the desire to dominate nature is a fascinating insight in this work of 1931. Making such a point in such a context is prescient. Jung, as always, is building a science that can contain the openness to mystery that he finishes on here.

The writing *performs* the model of knowledge that is the foundation of his science that *includes* the other (as magic, occult, other culture), not just rationally *describes* it. So Jung's science takes magic seriously, which is not to say he takes it on its own terms as magic. On the one hand, 'Archaic Man' is a powerful lens for future post-colonial and eco-critical preoccupations. On the other hand, the equation of the indigenous present as 'primitive' and as similar to the European past is a colonial hierarchy of cultural value. Beyond all this is the rhetorical writing that draws the reader into a dialogue with otherness and many voices. To read this writing is to be inside the theatre of Jungian science. It is to experience magic in the science and the science of magic.

Notes

1 S. Rowland, 'Response to Robert A. Segal's "Jung and Lévy-Bruhl" ', *The Journal of Analytical Psychology* 52 (5), 2007, 659–65. Reproduced here with kind permission of Wiley-Blackwell and *The Journal of Analytical Psychology*.

2 See R. Segal, 'Jung and Lévy-Bruhl', *Journal of Analytical Psychology* 52 (5), 2007, 635–58, reproduced in this volume, pp. 188–218.

3 C. G. Jung, 'Archaic Man', in C. G. Jung, *Collected Works*, ed. Sir H. Read, M. Fordham, G. Adler, and W. McGuire, 20 vols, London: Routledge and Kegan Paul, 1953–1983, vol. 10, *Civilization in Transition*, paras. 104–47, pp. 50–73. Originally published in English as 'Archaic Man', in C. G. Jung, 'Archaic Man', in *Modern Man in Search of a Soul*, tr. W. S. Dell and C. F. Baynes, London: Kegan, Paul, Trench, Trubner, 1933, pp. 143-74. Also reproduced in this volume, pp. 171–87.

4 C. G. Jung, 'On the Nature of the Psyche', in Jung, *Collected Works*, vol. 8, *The Structure and Dynamics of the Psyche*, paras. 343–442, pp. 159–234 (here: para. 421, p. 216).

5 C. G. Jung, 'On the Psychology of the Trickster Figure', in *Collected Works,* vol. 9/i, *The Archetypes and the Collective Unconscious*, paras. 456–88, pp. 255–72.

Chapter 9

The archaic: Timeliness and timelessness

Paul Bishop

Introduction

This essay aims to develop the discussion about the status of the archaic in Jung's thought, initiated by Robert A. Segal and Susan Rowland in their contributions to *The Journal of Analytical Psychology*, both of which are reproduced in this volume,[1] and it in turn was first published in *The Journal of Analytical Psychology*.[2] Segal's contribution is an exercise in intellectual history, which both surveys and analyzes Jung's reception of the thought of the French philosopher-turned-anthropologist, Lucien Lévy-Bruhl (1857–1939), with particular reference to Jung's lecture 'Archaic Man'. He discusses with remarkable clarity Jung's appropriation of such key Lévy-Bruhlian concepts as 'prelogical state' (*état prélogique*), 'collective representations' (*représentations collectives*), and 'mystical participation' (*participation mystique*), as well as pointing out that, in his view, Jung misunderstands these terms. In her response, Rowland offers a reading of 'Archaic Man' that concentrates on its self–performativity as a text. ' "Archaic Man" ', she argues, '*performs* its deepest achievement: it places the reader within its "net of reflections" and it does so as an ethical act' (pp. 220–21, this volume). For this reason, she concludes, 'Jung's web–like writing resurrects the many voices of the whole psyche who speak here through forms of otherness', and thereby '*performs* the model of knowledge that is the foundations of his science that *includes* the other (as magic, occult, other culture), not just rationally *describes* it' (p. 225, this volume).

I wanted to take another look at 'Archaic Man', not simply because Segal kindly gave me the opportunity to read his article in draft form, but because the publication of his and Rowland's articles also coincided with a course I was

Note: This article was first published in *The Journal of Analytical Psychology* 53 (4), 2008, 501–23. It is reproduced here with kind permission of Wiley-Blackwell and *The Journal of Analytical Psychology*.

I should like to thank the Erbengemeinschaft C. G. Jung for permission to consult documents held in the C. G. Jung-Archiv of the ETH-Bibliothek, Zurich, as well as Michael Gasser and his colleagues in the ETH Spezialsammlungen for their kind assistance.

giving on various works by Jung, as part of which I and my students had begun to consider precisely the question of the status of the archaic in analytical psychology. Segal and Rowland's articles, together with the discussions in the class, helped to crystallize some of my thoughts on Jung's paper in relation to the question of the archaic. Faced with Segal and Rowland's sophisticated, if different, approaches – one intellectual-historical, one literary-theoretical – my aim in this essay is to relate Jung's lecture on 'Archaic Man' to those other writings where he discusses the 'archaic', the 'primal', the 'primordial', with a view to clarifying this particular text in the light of his work as a whole. Seen from this perspective, this comparatively obscure and little-discussed text by Jung, included in this volume, assumes a more central importance than it has been hitherto granted, since it both summarizes Jung's thought up to the point of its composition *and* points the way to future developments in the unfolding of analytical psychology.

Jung and the Lesezirkel Hottingen

In 1882 Wilfried Treichler and Hans Bodmer founded the Lesezirkel Hottingen, a literary fraternity based in a town close to Zurich that now forms a district of the city.[3] Of the many prestigious speakers who addressed the Lesezirkel one could name Hermann Hesse, Hugo von Hofmannsthal, Rainer Maria Rilke, and Thomas Mann. Among the lecture cycles organized by the Lesezirkel were discussions on 'Ethics and Thought' (Albert Schweitzer); 'For and Against the Spirit' (talks given by Ivan Ilyin, Romano Guardini, and Ludwig Klages); and a series of lectures held over four evenings, under the title 'The Myth of European Man', given by C. G. Jung, Theodor Haecker, C. J. Burckhardt, and finally Rudolf Kassner (1873–1959).[4] In fact, on 22 October 1930 it was Jung who introduced the series,[5] and went on to give the first talk on 'Archaic Man'. Subsequently his lecture was published in an abridged form in the *Europäische Revue*,[6] and later in 1931 it was included, in a revised form, in *Seelenprobleme der Gegenwart*, a collection of papers that appeared in 1931,[7] from which it was translated by W. S. Dell and Cary F. Baynes in *Modern Man in Search of a Soul* (1933).[8] However, the forum in which 'Archaic Man' was first presented helps explain the highly anecdotal flavour of the text, which was delivered in the Aula of Zurich University at quarter past eight on that Wednesday evening.[9]

The themes adduced in the lecture relate to the core of what Jung had been doing since he had begun to formulate an 'analytical' – originally a 'psycho-synthetic' or later a 'complex' –[10] psychology in *Transformations and Symbols of the Libido* (1911–1912). This psychology attends, as does Freud's, to the 'archaic' – or as Jung glosses the term, the 'primal' (*anfänglich*) or 'original' (*ursprünglich*) – aspects of what he calls 'contemporary' (*heutig*) or 'civilized' (*zivilisiert*) humankind.[11] Or more precisely, as Segal points out (p. 188, this volume), Jung contrasts, not so much two different kinds of people – or two different kinds of psychological attitudes, as he did in his great study of typological difference, *Psychological Types* (1921) – as two different kinds of thinking. In 'Archaic

Man' this difference is initially presented as a chronological one: the central opposition is the 'archaic' vs. the 'contemporary', the 'primal' vs. the 'modern', or the 'primitive' vs. the 'civilized' – yet both sets of terms carry with them connotations of value far beyond simply 'the old' or 'the new'.

For almost immediately after this opposition has, in the opening paragraph, been stated, in his second paragraph Jung – to use the jargon – 'deconstructs' it, writing that 'archaic psychology is not just the psychology of primitives, but also the psychology of modern, civilized humans' (para. 105). Already, then, we find a logical disjunction in Jung's use of the term 'archaic'. The 'archaic' is not simply a chronological category, it is also a category of *quality*. (The same applies – as we shall later see – to another central Jungian concept, the 'primordial'.) The 'archaic' is not simply very, very old – it is *essentially timeless*. To be sure, there is *one* particular context in which Jung uses the term 'archaic' in its chronological sense, when he writes that 'all civilized human beings are, irrespective of their level of consciousness, in the deeper levels of their psyche, still archaic humans', using the analogy of how 'the human body connects us with the mammals and displays numerous vestiges of earlier evolutionary stages going back even to the reptilian age' (para. 105).[12] At the same time, he *also* uses the term 'archaic' as a means of describing *a kind of thinking*, a thinking that is 'prelogical' and displays other characteristics, which we shall shortly discuss.[13] In diagrammatic form, we could thus represent Jung's twofold use of the term 'archaic' as follows, where 'archaic (1)' indicates the chronological use of the term in opposition to 'modern' or 'civilized', and 'ARCHAIC (2)' indicates the term used at a higher, richer level – the persistence of the archaic (in the first sense) in the world of today:[14]

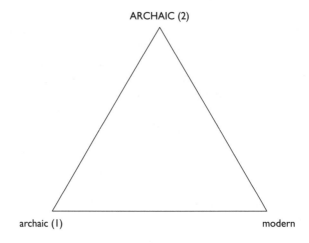

It is in relation to the *quality* of the archaic that Jung finds Lévy-Bruhl most useful – to describe that quality, he uses the terms 'pre-logical' (cf. *état prélogique*), 'pre-individual', i.e., 'collective' (cf. *représentations collectives*), but, interestingly enough, *not* 'mystical'.

Anti-mysticism in Jung's thought

Now, Segal rightly points out that, in their use of the term *participation mystique*, Jung and Lévy-Bruhl mean different things. 'By *participation mystique* Lévy-Bruhl means the ascription to others of what we *recognize* [in ourselves]', Segal observes, 'not what we *deny*' (p. 197, this volume), as appears to be the case when Jung remarks that 'everything that is unconscious in ourselves, we discover in our neighbour, and we treat him accordingly' (para. 131). Even more important, however (or so it seems to me), is Jung's dismissal of the term 'mystical' altogether as a way of describing 'primitive' thought. On this point I disagree with Segal: Jung, it seems to me, does not conflate the 'prelogical' with the 'mystical' (cf. pp. 188, 194 this volume). On the contrary: immediately after he has introduced Lévy-Bruhl's expression *participation mystique* into his text, Jung rejects the very term 'mystical'. 'The word "mystical" seems to me,' Jung writes, 'not happily chosen, because for primitive man it is not a question of anything mystical, but rather of something completely natural' (para. 130).

Why is Jung so particularly touchy about the term 'mystical'? In part, it is because it is one of the labels that, he felt, his opponents were always trying to pin on him. In a footnote in *Transformations and Symbols of the Libido* where he proposes 'psychosynthesis' as a complementary approach to 'psychoanalysis', he complained: 'This time I shall hardly be spared the reproach of mysticism'.[15] Given what Freud was to write shortly afterwards in 'On the History of the Psycho-Analytic Movement' (1914), Jung was right to be worried.[16] In part, it is in line with his own critique of mysticism in *Psychological Types* (1921) as sterile or 'unfruitful'.[17] Significantly, Jung added in a footnote to *The Psychology of the Transference* (1946) that Lévy-Bruhl, who had coined the phrase *participation mystique*, avoided the term 'mystique' in his later publications.[18] And in part, it is because of Jung's likely awareness of the historical use of the term 'mysticism', as when the nineteenth-century Göttingen philologist Karl Otfried Müller (1797–1840) felt the need to defend himself against the charge of 'mysticism', then used as a term to describe the 'idealism' of the Heidelberg classical philologist Friedrich Creuzer (1771–1858) who, in according the pre-eminince he did to the symbol, was felt to have placed an unwarranted emphasis on the irrational, on the religious – on the 'mystical'.[19]

Instead, the 'problem' with Jung's paper is not so much his conflation of the prelogical with the mystical, as the fact that he *appears* to argue for two contradictory positions. What complicates his argument further is his decision, doubtless dictated by the circumstances attendant on its original form as a spoken presentation, to lard his paper with numerous anecdotal examples – the crocodile and the anklets, the missionary and his flagstaff, the calf born with two heads and five legs, Jung's own experiences in the jungle (losing his compass, blundering into a rhinoceros trail, dodging puff-adders), the Elgonyi ritual at sunrise, and the extraordinarily enthusiastic Easter-egg hunting of his neighbour in Zurich – all of which doubtless delighted his audience in the Aula of Zurich University, but risk distracting from what is of central importance in his paper.[20]

The subject–object relation – or, inner nature

This contradiction in Jung's paper can be stated as follows. On the one hand, Jung argues that 'primitive' thinking is, in fact, no different from our own thinking: the only thing that differs is the set of presuppositions on the basis of which we think (paras. 107, 111). This gives Jung an opportunity to make a cultural-relativistic point: '[The primitive] is a hard riddle to read, until we know his presuppositions; then all is relatively simple. We could just as well say: as soon as we know *our* presuppositions, primitive man ceases to be a riddle' (para. 112). So what is prelogical about 'primitive' thinking is merely a question of presuppositions: causal-based, for us, intention-based, for the 'primitive' (paras. 114, 115, 118).

On the other hand, however, 'primitive' thinking *is* different, for otherwise we would not feel what Jung calls 'something prodigiously strange' (*etwas unge-heuer Fremdartiges*) when we 'approach' – or, in fact, 'come into touch with' – 'the world of the archaic human being' (para. 112). Moreover, Jung believes he is able to pin down exactly what is so strange about 'archaic' thought, and to define it. The key to understanding 'archaic' thought lies in the subject–object relation, and more precisely the distinction made between (a) the 'subjective-psychic' and (b) the 'objective-natural' aspects of this relation.

This is a distinction which, Jung argues, 'civilized' human beings are able to make, but 'primitive' humans cannot. (Seen from another perspective, we 'moderns' have merely introduced a distinction which, in reality, is not there – as the 'primitive' knows all too well.) Thus human beings today distinguish sharply between inner and outer reality, between what happens 'inside', in the mind, and what happens 'out there', in the world. In 'primitive' thought, which is a mode of thought to which today we still have access, however, the objective and the psychic coalesce, so that the (artificial) distinction between the inner (the psyche) and the outer (the object) disappears.[21] This aspect of 'archaic' thought, the world-outlook of which Jung conjures up in such colourful terms, from the point of view of the 'primitive' –

> The spirits of the departed inhabit such and such a wood. In yonder mountain lives the great serpent; that hill is the grave of the legendary king; near this spring or rock every woman becomes pregnant; that ford is guarded by snake-demons; this towering tree has a voice that can call certain people,
>
> (para. 128)

– as well as, in a particularly evocative passage, from the point of view of the 'civilized' European who visits, say, Africa –

> The velvety blue of a tropical night, the overhanging black masses of gigantic trees standing in the primeval forest, the mysterious voices of the nocturnal

spaces, a lonely fire with loaded rifles stacked beside it, mosquito-nets, boiled swamp-water to drink, and above all the conviction [. . .]: 'This isn't man's country – it's God's country.'

(para. 127)

– is strongly reminiscent of the approach to nature urged upon us by Goethe in his short, late poem, 'Epirrhema' (1820), which opens with the following lines:

You must, when contemplating Nature,	*Müsset im Naturbetrachten*
Attend to this, in each and every feature:	*Immer eins wie alles achten:*
There's nought outside and nought within,	*Nichts ist drinnen, nichts ist [draußen;*
For she is inside out and outside in.	*Denn was innen, das ist außen.*
Thus will you grasp, with no delay,	*So ergreifet ohne Säumnis*
The holy secret, clear as day.	*Heilig öffentlich Geheimnis.*[22]

Like Goethe, Jung wishes to adopt an 'holistic' approach to nature – an approach intuited, but then mystically re-interpreted, by Jung's contemporary, Rudolf Steiner.[23]

Among the consequences of this 'undifferentiated' consciousness, as Jung refers to it, are a complete lack of self-criticism, exemplified in the 'prelogical' morality that allows a chieftain to assert, 'When I steal my enemy's wives it is good, when he steals mine, it is bad' (para. 108), and a tendency toward projection (see Segal, pp. 196–97, this volume). This projection creates a kind of identity that does not firmly distinguish between subject and object, but conceives of the subject–object relation as a fluid one. For Jung the important thing is that this 'primitive' form of identity is by no means inferior to the 'civilized' one; indeed, it may even, he hints, be superior. For the 'primitive' human is bound up and at one with the world –

Humankind is still part of the order of nature. There is no thought of trying to master nature, instead the highest efforts are devoted to protection from its dangerous caprices [*Zufällen*]

(para. 134)

– whereas the 'civilized' human's approach to the world is governed by an 'instrumental reason', as Theodor W. Adorno (1903–1969) and Max Horkheimer (1895–1973) will call it,[24] that seeks to control, master, and over-power nature, an ambition that ultimately, Jung believes, will turn out to be impossible to achieve:

The civilized human, however, tries to dominate nature, and therefore the strongest striving [*Streben*] is directed to natural causes, which will provide the key to the hidden workshops of nature [*Werkstätten der Natur*].

> For this reason the thought of arbitrary powers and the possibility of their existence are extremely repugnant, for rightly there would be the proof that, in the end, the striving to dominate nature is futile [*vergebliches Streben*].
>
> (para. 134)

Again, the phraseology here is rich in resonances: it emphasizes the quintessentially Goethean term 'striving' (*Streben*), associated with the figure of Faust, whose own approach to nature is fundamentally inquisitive and manipulative;[25] and it echoes the prose text traditionally ascribed to Goethe, 'On Nature', which talks of nature as 'build[ing] always, destroy[ing] always, and her workshop [*ihre Werkstätte*] is beyond our reach'.[26]

That the difference between the 'archaic' and the 'modern' is a fundamental one is further explored by Jung in relation to two, highly telling, instances. First, in an echo of the famous line in a poem by Mörike, 'To a Lamp' – 'Whatever is beautiful appears blessed in itself' (*Was aber schön ist, selig scheint es in ihm selbst*) – which later became the subject of a famous debate between Emil Staiger and Martin Heidegger, then Leo Spitzer,[27] Jung casts this essentially epistemological question into terms of aesthetics: 'Is something beautiful, because I attribute beauty to it? Or does the objective beauty of things compel me to acknowledge it?' (para. 135).[28] Second, in a further allusion to Goethe (who, in turn, is citing Plotinus, and reflecting the view of the Ionian school of pre-Socratic philosophy) –

Were the eye not of the sun,	*Wär nicht das Auge sonnen-* [*haft,*
How could we behold the light?	*Die Sonne könnt' es nie* [*erblicken;*
If God's might and ours were not as one,	*Läg' nicht in uns des Gottes* [*eigne Kraft,*
How could His work enchant our sight?	*Wie könnt' uns Göttliches* [*entzücken?*[29]

– Jung points out how 'great minds have wrestled with the problem, whether the sacred sun illuminates the world, or whether it is the sunlike human eye' (para. 135). At this point we can most clearly see the difference between the 'archaic' and the 'modern': for the 'archaic' point of view believes in the sun (and hence places the objective world at the top of its hierarchy), whereas the 'modern' (or idealist) point of view believes in the eyes (and places its emphasis on the subject side of the subject–object equation).

Jung's solution to this antinomy is a dialectical one: he seeks to discern the object-in-the-subject, to find what is object-ive about the subject-ive. Starting from the axiom that 'every relatively independent part of the psyche [*Seelenteil*] has the character of personality' (para. 137), and the observation that 'in the course of our own development it is from a contradictory multiplicity that the unity of personality gradually emerges' (para. 141), he embarks on a remarkable thought-experiment (on which, because of their respective approaches, neither Segal nor Rowland has an opportunity to comment directly). What if, Jung asks, the archaic point-of-view is *right*? In that case, surely the primitive theory of *mana* – *mana* being a term for a supernatural force that emanates from certain magic individuals, a quasi-divine, primal, vital force –[30] which maintains that 'there exists a universally distributed power that objectively produces extraordi-nary effects' (such as sorcerers, witches, werewolves, and the like); that 'every-thing that exists, acts, or it is not real' (*alles, was ist, wirkt, sondern es nicht wirklich*);[31] that something only *is* by means of its energy; and that Being (*das Seiende*) is a force-field, deserves to be taken seriously (which is not the same as taking it literally)? More specifically, what if, in the terms that Goethe and Plotinus pose the problem, light really *does* come from the sun? What if, to return to Jung's aesthetic example, things actually *are* beautiful, and it is their beauty that moves *us*, not *we* who create their beauty?[32]

In this thought-experiment, Jung wonders whether the object may actually be said to constitute the subject; that is, whether 'the psychic function – the soul, or the spirit, or the unconscious – originates *in me*' or whether 'the psyche is, in the early stages of the formation of consciousness, actually outside us in the form of intentionalities and arbitrary powers, and gradually grows *into us* in the course of psychic development' (para. 140).[33] In the end, Jung concludes that such a view, according to which 'souls' are not split-off parts of the personality, but rather personality itself emerges from the coming-together and embodiment of these 'souls', such that over the historical course of human development they become embodied (literally: *incarnated*) in the individual, until they constitute 'the world [. . .] that we now call psyche' (para. 140), may well be paradoxical, but is never-theless 'not entirely inconceivable' (para. 141).

From this perspective, the psyche is nothing less than our *inner nature*, much in the same way that, in the same year as Jung delivered his paper to the Lesezirkel Hottingen, Freud wrote in *Civilization and its Discontents* (1930) that 'a piece of unconquerable nature' (*ein Stück der unbesiegbaren Natur*) forms 'a part of our own psychical constitution'.[34] Where did Freud find the phrase? It is possible that he took it from Jung who – in his earlier paper on 'Psychological Types' (1923), delivered at the International Congress on Education held in Territet, Switzerland – had remarked that 'the unconscious is the residue of unconquered nature [*der Rest unbezwungener Urnatur*] in us, just as it is also the matrix of our unborn future.'[35] This idea of an inner, primordial nature crops up time and again in the thinking of the first[36] and second[37] generations of the Frankfurt School (who, in their anti-Jungian prejudice, attribute it only to Freud).

Precisely the developmental position that Jung outlines in 'Archaic Man' is explored in further detail a decade or so later in his study of 'The Psychology of the Child Archetype' (1940). Here Jung writes that 'in the last analysis the human body, too, is built of the stuff of the world [*aus dem Stoffe der Welt gemacht*], the very stuff [*an solchem Stoffe*] wherein fantasies become visible; indeed, without it they could not be experienced at all'.[38] To explain what he means, Jung uses here the example of the process of crystallization – if there is no solution in which the crystalline lattice can form, no crystallization can take place –[39] and he takes this argument further, to the point of identifying 'psyche' and 'world':

> The symbols of the self arise in the depths of the body and they express its materiality [*Stofflichkeit*] every bit as much as the structure of the perceiving consciousness. [. . .] The uniqueness of the psyche can never enter wholly into reality, it can only be realized approximately, though it still remains the absolute basis of all consciousness. The deeper 'layers' of the psyche lose their individual uniqueness as they retreat farther and farther into darkness. 'Lower down', that is to say as they approach the autonomous functional systems, they become increasingly collective until they are universalized as extinguished in the body's materiality [*in der Stofflichkeit des Körpers*], i.e., in chemical substances. The body's carbon is simply carbon. Hence 'at bottom' the psyche is simply 'world' [*»Zuunterst« ist daher Psyche überhaupt »Welt«*].[40]

Is this really as materialist a conception of the world, and hence, since they are identical, of the psyche, as it appears to be? Certainly, in 'Archaic Man' Jung seems troubled by the doctrine of materialism. Yet materialism, he speculates, shares with 'archaic' thought the conclusion that, in the end, the individual is maybe no more than 'a mere result', the result of natural causes, in the case of materialism, and the result of chance occurrences, in the case of 'archaic' thought (para. 142). Furthermore, not only has materialism, 'by the roundabout way of strict causalism, returned to the primitive point of view', but it is more systematic, hence more radical. For 'primitive' man makes an exception for the *mana*-personality, and thus, unlike materialism, holds the door open to the divine. Unlike materialism – unless, that is, we consider a possibility left unconsidered in this paper by Jung, who rightly mocks the almost schizophrenic attitude of the modern *Kulturwelt*, in which universities teach both science and theology, and where someone who is a scientist during the week is happy to proclaim his religious faith on Sundays. . . Yet, as Jung has noted, the theory of *mana* has in it the beginnings of a crude theory of energy (para. 139); and, taken as a whole, his paper insinuates the possibility of what might be called a *vitalist-materialist* outlook. Seen in this light, Jung's apparently throwaway remarks about recent discoveries in the field of particle physics assume a greater significance.

Synchronicity and meaning

For it is, so to speak, no coincidence that, in the same year as he wrote 'Archaic Man', Jung also coined the term *synchronicity*.[41] In his memorial address for Richard Wilhelm, given in Munich on 10 May 1930, five months or so before he spoke at the Lesezirkel Hottingen, Jung wrote that the *I Ching* was based, not on the principle of causality, but on a principle he had, albeit tentatively, called 'the *synchronistic* principle'.[42] Although Jung defined synchronicity in his 1935 Tavistock Lectures, with reference to the Chinese concept of *tao*, as 'a peculiar principle [. . .] active in the world, so that things happen somehow together and behave as if they were the same, and yet for us they are not',[43] the major formal exposition of the principle was not until his Eranos conference paper of 1951[44] and his monograph of 1952,[45] following his collaboration with the physicist Wolfgang Pauli (1900–1958).[46] But, if we are to believe the testimony of one of Jung's letters, he had begun considering the possibility of a non-causal principle as early as 1909–1910 and 1912–1913, when he met Einstein and was introduced to the idea of the relativity of space and time.[47]

Thus it is not surprising to find that, at significant moments in 'Archaic Man', Jung gestures towards the concept of synchronicity, particularly in his discussion of the difference between the 'archaic' and the 'modern' world-views in respect of their attitudes to chance (paras. 114, 115, 118). For Jung, to understand the course of events as a causal sequence in accordance with natural laws is a theoretical standpoint which proves itself – but only for about half the time. Over the other half the daimon, chance (*Zufall*), holds sway.[48] By contrast, from the 'primitive' standpoint what we call chance appears as intention (*Absicht*) or deliberate, calculating intent (*berechnender Willkür*) (paras. 115, 117, 118). On this point Jung is quite insistent.[49] In the case of the 'archaic' individual, we find that 'the main emphasis is placed on the other fifty percent of events in the world, thus not on the purely causal connections of science, but on the confusing and confused [*verwirrenden und verworrenen*] breaks in the causal chain [*Durchkreuzungen*[50]], which one calls chance' (para. 118). All the examples Jung gives of apparently chance events, so 'repellent to the mind that loves order', and all too reminiscent of devilish imps or the capricious whim of a *deus ex machina* (para. 113) – when a house burns down, for instance, is it because it was struck by lightning or because a sorcerer used lightning to set fire to the house? (para. 107) – can be interpreted by the 'archaic' mind within a framework of intentionality. Or, to put it another way, they demonstrate that elusive category of 'meaning' that Jung associates with the synchronistic principle.

Within the framework of 'archaic' thought, however, it is important to recognize that synchronicity is not a metaphysical principle, nor even a mystical one. The *mana*-inhabited world of the primitive mind is decidedly immanent: 'invisible, arbitrary forces' (para. 113) may be at work in it, yet these forces are (conceived as being) very much in and of this world. What Jung says about the intention-focused approach of the primitive mind, and the world as it is conceived by such a mind, can fruitfully be compared with the 'figural interpretation' proposed by the German literary critic Erich Auerbach (1892–1957). According to this kind of interpretation, one can read two facts or events, A and B, not simply in terms of a linear relation (A causes B), but in terms of a 'prefigural' relation (A anticipates B, so that the full significance of A is only revealed in and through B). In Auerbach's own words, 'figural interpretation establishes a connection between two events or persons, the first of which signifies not only itself but also the second, while the second encompasses or fulfils the first'.[51] We could equally well describe Auerbach's figural interpretation as a teleological way of reading the world, without a commitment to teleology itself.

Now this, it seems to me, is precisely the kind of relationship Jung is trying to articulate in his notion of 'synchronicity', according to which two events can stand in a 'non-causal', but, or rather hence, 'meaningful' relation to each other.[52] So we might well apply to the phenomenon of synchronistic meaning the following words (originally written in a different context) by the French philosopher Bertrand Vergely, when he says that from 'a life full of correspondences between our interiority and life [. . .] there bursts forth a world of meaning [*sens*], each thing becoming symbolic, the exterior referring to an interior experience.'[53] The idea of seeing the world – or the world revealing itself – as something meaningful is an integral component of Jung's notion of synchronicity, and a correspondence between interior and exterior worlds (psyche and reality) is demonstrated by all the celebrated examples cited by Jung: the rationalistic young woman and the scarabaeid beetle, the fish that pursue Jung throughout his day, or the praying mantis that flew in through the window at the end of Adolf Portmann's Eranos lecture.[54] In its turn, the concept of synchronicity relies on his distinction between 'causal'/'mechanistic' and 'final'/'energic' standpoints.[55] What, in 'primitive', 'archaic' terms, is the world of *mana* is, in ontological terms, a vitalist-materialist vision of the world.

The status of the 'primordial'

Thus read in relation to Jung's later thinking – a reading that is complementary to those offered by Segal and Rowland, not in opposition to them – the lecture 'Archaic Man' gives us good reason to reconsider the status of the 'archaic', the 'primordial' in Jung's thought. Jung enjoys using that exciting German prefix *Ur-*, which immediately opens up a historical dimension to any word when it is placed in front of it: he speaks of the *Urmensch*, of *Urzeit*, of *Urwelt*, *Urwald*, and *Urerlebnis* (see below); in particular, he uses the term *Urbild* when first formulating the idea of 'archetype'. Yet the prefix *Ur-* signifies, not simply that something is old – as, for example, in the case of the *Urzeit* or even the word *ursprünglich* ('original') – but also that something is primary or fundamental: for example, when Goethe talks about the 'primordial phenomenon' or *Urphänomen*, something which, he believed, can be apprehended by a mode of perception he understood as aesthetic.[56]

The chief point to retain, however, is that the 'archaic' and the 'primordial' are not necessarily chronological categories, but can also signify an intensity or authenticity of experience. This synchronic, as well as diachronic (or chronological), dimension to Jung's understanding of the 'archaic' points to the link between the 'archaic' and the concept of synchronicity, the latter being Jung's way of developing a discourse centered on the category of meaning. Without this 'meaning', our world would truly be – to use one of the central categories of existentialist thought – absurd.

Some of the most truly startling passages in Jung's writings rely on the idea of a sudden switch from the present to the past, an unexpected eruption of the primordial into the everyday present. For instance, *Transformations and Symbols of the Libido* opens with a dramatic paragraph comparing the impact on the reader of Freud's explanation of the incest fantasy with the reference to the legend of Oedipus:

> The impression made by this simple reference may be likened to that wholly peculiar feeling which arises in us if, for example, in the noise and tumult of a modern street we should come across an ancient relic – the Corinthian capital of a walled-in column, or a fragment of inscription. Just a moment ago we were given over to the noisy ephemeral life of the present, when something very far away and strange appears to us, which turns our attention to things of another order; a glimpse away from the incoherent multiplicity of the present to a higher coherence in history.[57]

Underneath (or behind, or somewhere obscured by) the 'contemporary', the 'modern', the 'civilized', lies the 'primal', the 'primordial', the 'archaic'. And sometimes the latter, 'something that lives and endures beneath the eternal flux', suddenly intrudes, rhizome-like, into the world of the former, just as, in the prologue to *Memories, Dreams, Reflections*, it is suggested that the only events in Jung's life worthy of narration are 'those when the imperishable world irrupted into this transitory one'.[58]

This idea recurs frequently in Jung's writings. Three years before 'Archaic Man', in 1927, Jung delivered a lecture in Karlsruhe that was subsequently published as 'Analytical Psychology and "Weltanschauung" '. Here he presented analytical psychology as 'a reaction against an exaggerated rationalization of consciousness' which, 'in its endeavour to produce directed processes', thus 'isolates itself from nature' – never a good thing, in Jung's view – 'and so tears humankind from its own natural history'.[59] What happens when we are ripped from our own 'natural history' is portrayed by Jung in terms of an unbearable, claustrophic sense of oppression –

> [Humankind] finds itself [. . .] transplanted into a rationally limited present, stretched over the short span between birth and death. This limitation creates the feeling of randomness [*Zufälligkeit*] and meaninglessness [*Sinnlosigkeit*],[60] and it is this feeling that prevents us from living life with the weight of meaning [*jener Bedeutungsschwere*] that is required for it to be enjoyed to the full [*um völlig ausgeschöpft zu werden*]. [61] Life becomes stale and no longer fully represents the individual. [62] That is why so much unlived life succumbs to the unconscious. One lives just as one walks when one's shoes are too small. That quality of eternity [*Die Ewigkeitsqualität*] [63], which

is so characteristic of the life of primitive humankind, is entirely lacking in our lives. Isolated and hemmed in by our rationalistic walls, we are cut off from the eternity of nature [*die Ewigkeit der Natur*].[64]

– and an oppression from which, thus its hope and its promise, his psychology aims to liberate us:

> Analytical psychology seeks to break through these walls, by digging up again the fantasy-images of the unconscious which the rational understanding has rejected. These images lie beyond the walls, they are part of the *nature in us*,[65] which apparently lies buried in our past and against which we have barricaded ourselves behind the walls of *ratio*. Here arises the conflict with nature, which analytical psychology tries to resolve, not by going 'back to nature' *à la* Rousseau, but, while holding on to the successfully attained modern level of *ratio*, enriching our consciousness with knowledge of the natural spirit [*des natürlichen Geistes*].[66]

In this lecture, Jung beautifully captures the sense of self-alienation when, in a tradition of thought going back to Schiller,[67] he writes:

> Nowhere do we stand closer to the noblest secret of all origins than in the knowledge of our own selves, which we believe we always already know. Yet the immensities of space are better known to us than the depths of the self [*die Tiefen des Selbst*], where – even though we do not understand it – we can listen almost directly to the creativity of Being and Becoming [*das schöpfer-ische Sein und Werden*].[68]

The need for an accommodation with nature, internal, as well as external, receives a yet more vigorous emphasis in his 1932 lecture to the Alsation Pastoral Conference in Strasbourg, published as 'Psychotherapists or the Clergy':

> The opening up of the unconscious means the outbreak [*Ausbruch*] of great spiritual suffering, for it is just as if a flourishing civilization is abandoned to the invasion [*Einbruch*] of barbarian hordes, or if fertile farmland is exposed by the destruction of a protective dam to a havoc-wreaking mountain torrent. Just such a breach [*Durchbruch*] was the World War, which showed as could nothing else how thin the separating wall is that keeps an ordered world apart from ever lurking chaos. But so it is with the individual: behind the rationally ordered world there is nature, violated [*vergewaltigt*] by reason and out for revenge, waiting for the moment when the partition falls, in order to pour herself with destructive consequences into conscious life.[69]

His own therapy, Jung explains in this lecture, is intended to assist and accompany his patient through the experience when – 'it is like an inspired moment'

– 'something Other [*etwas Fremdes*] rises up from the dark realm of the psyche [*Seele*] and confronts him'.[70] This moment, Jung assured his audience, marks 'the beginning of the cure', even though 'this spontaneous activity of the psyche [*diese Eigentätigkeit der Seele*] often becomes intense to the point where an inner voice or visionary images are perceived' – phenomena described as 'a true primordial experience of the spirit [*Urerfahrung des Geistes*]'.[71] For Jung, the primordial is not, as Freud suspected it of being, something infantile and hence problematic; rather, it is the solution to the problem of modernity.

In a short essay from this same period, published in the *Kölner Zeitung* on 7 May 1929 – that is, just over a year before Jung made his appearance at the Lesezirkel Hottingen – Jung contrasts his own position with Freud's. Here he speaks of the libidinal drives (be they erotic, as Freud believed, or power-oriented, as Adler did) as coming into collision with spirit, with *Geist*.[72] Although Jung does not shy away from the fact that he cannot say precisely what *Geist* is, he clearly regards it as a liberating force, as a means of escaping what he called 'the inexorable cycle of biological events' (*die unerbittliche Klammer des biologischen Geschehens*) or, as he strikingly described it, 'the fleshly bond leading back to father and mother or forward to the children that have sprung from our flesh – "incest" with the past and "incest" with the future, the original sin of perpetuation of the "family romance" [*ein "Inzest" mit der Vergangenheit und ein "Inzest" mit der Zukunft, die Erbsunde der Perpetuierung des "Familienromans"*].'[73] From such shackles, Jung wrote in a passage laden with theological and soteriological vocabulary, only *Geist* can free us, and he insisted on the need for an 'authentic-experience-of-the-primordial', for *Urerfahrung*. 'We moderns are faced with the necessity of rediscovering the life of the spirit [*den Geist wieder erleben*]: we must experience it anew for ourselves [*Urerfahrung machen*]', he declared, for 'it is the only way in which to break the spell that binds us to the cycle of biological events.'[74] Thus a return to the 'archaic' and to the 'primordial' is also a way forward to the future: we need, in other words, to (re-) engage with what one might call *le futur archaïque*.[75]

The lecture 'Archaic Man' ends, as Rowland notes, by underlining its own rhetorical mode of argument (p. 224, this volume). After Jung has assimilated the Elgonyi sun ceremony with the monotheism of Christianity, which uses the final words of Christ on the cross, 'Lord, into thy hands I commend my spirit' (para. 146; cf. Luke 23: 46; cf. Ps. 31: 5),[76] he leaves the reader with the following 'unsolved problem': is this coincidence a coincidence – that is, is it by chance that the Elgonyi cult and a Christian prayer are so similar? Or is there something intentional, something 'thought and willed' (*gedacht und gewollt*), something meaningful – something synchronistic – at work here?

Jung took his leave from his audience at the Lesezirkel by offering a final – on reflection, perhaps the most important – distinction between the 'archaic' and the 'modern' modes of being. He told his listeners that 'archaic individuals just do what they do, only civilized individuals know what they do' (para. 145).[77] By placing doing above knowing, Jung is not seeking to lure us into a blind

Aktionismus; rather, he is urging us to reconsider our relation to ourselves and (which turns out, in a way, to amount to the same thing) to the world, and to do so with the ambition of redeeming both Self and World. The articulation of this task constitutes the project of analytical psychology as Jung went on to develop it in the Thirties, Forties, and Fifties; but its first statement, and the foreshadowing of his later work, is to be found in this astonishing talk he gave to the Lesezirkel Hottingen in 1930.

Notes

1 See R. Segal, 'Jung and Lévy-Bruhl', *Journal of Analytical Psychology* 52 (5), 2007, 635–58; 'Response to Susan Rowland', *Journal of Analytical Psychology* 52 (5), 667–71; and S. Rowland, 'Response to Robert A. Segal's "Jung and Lévy-Bruhl" ', *Journal of Analytical Psychology* 52 (5), 2007, 659–65. These articles are reproduced in this volume, pp. 188–218; 219–25.

2 See P. Bishop, 'The timeliness and timelessness of the "archaic": Analytical psychology, "primordial" thought, synchronicity', *Journal of Analytical Psychology* 53 (4), 2008, 501–23; 219–25.

3 C. Ulrich, *Der Lesezirkel Hottingen*, Zurich: Berichthaus, 1981, p. 13. In 1902 Bodmer established within the Lesezirkel Hottingen the Literarischer Club Zürich, which outlived the demise of the former in 1941. And on the anniversary of the death of Schiller, Bodmer founded in 1905 the Schweizerische Schillerstiftung (see H. Bleuler-Waser, *Leben und Taten des Lesezirkels Hottingen: Von seiner Geburt bis zu seinem 25. Altersjahre 1882–1907*, Zurich: Lesezirkel Hottingen, 1907).

4 Ulrich, *Der Lesezirkel Hottingen*, p. 63.

5 The handwritten manuscript and a typescript of Jung's introductory remarks can be found in the C. G. Jung-Archiv of the ETH-Bibliothek, Zurich [Hs 1055: 71b], 'Eröffnung der Vortragsreihe im Hottinger Lesezirkel'. In these remarks, he argued that, whereas in the nineteenth century, 'human knowledge and ability developed to an almost monstrous extent', so that 'the creator became absorbed in his works and the spirit itself went astray and lost itself in the abundance of its forms', the twentieth century was beginning to search for 'the unity that had disappeared in the unbounded multiplicity' – that is, 'the unity of the creative consciousness'. By talking about the characteristics of the 'primitive' mind and its world-view, Jung intended to 'paint the backcloth' for 'the stage on which the forms of the cultural human being [*Culturmensch*]' would, in subsequent lectures, appear.

6 C. G. Jung, 'Der archaische Mensch', *Europäische Revue* 7 (3), 1931, 182–203.

7 C. G. Jung, 'Der archaische Mench', in *Seelenprobleme der Gegenwart* [*Psychologische Abhandlungen*, vol. 3], Zurich: Rascher, 1931, pp. 211–47.

8 C. G. Jung, 'Archaic Man', in *Modern Man in Search of a Soul*, tr. W. S. Dell and C. F. Baynes, London: Kegan, Paul, Trench, Trubner, 1933, pp. 143–74.

9 According to the report of the lecture in the *Neue Zürcher Zeitung*, Jung's lecture was received by his large audience with 'lively applause' ('Der archaische Mensch', *NZZ*, 24 October 1930). Jung's lecture was followed on 19 November 1930 by Theodor Haecker's on 'The Virgilian Man' (*NZZ*, 21 November 1930).

10 In his letter to Freud of 2–12 April 1909, Jung postulates that, if there is a 'psychoanalysis', there must also be a 'psychosynthesis', which 'creates future events according to the same laws' (S. Freud and C. G. Jung, *The Freud/Jung Letters: The Correspondence between Sigmund Freud and C. G. Jung*, ed. W. McGuire, trs. R. F. C. Hull and R. Manheim. Princeton, NJ: Princeton University Press, 1974, p. 216). On Jung's decision in the 1930s to rename it 'complex psychology', see S. Shamdasani, *Jung and the*

Making of Modern Psychology: The Dream of a Science, Cambridge: Cambridge University Press, 2003, pp. 13–14.

11 C. G. Jung, 'Archaic Man' [1931], in C. G. Jung, *Collected Works*, ed. Sir H. Read, M. Fordham, G. Adler, and W. McGuire, 20 vols, London: Routledge and Kegan Paul, 1953–1983, vol. 10, paras. 104–47 (here: para. 104). Henceforth referred to in parenthesis with a paragraph number, whilst other texts by Jung are cited from this edition, referred to as *CW* followed by a volume and paragraph number.

12 There is another sense in which the term is chronological, of course: the sense in which, in Africa or elsewhere, among the Pueblo Indians or the Elgonyi, the archaic is the *contemporary* mode of thought.

13 Earlier, in *Transformations and Symbols of the Libido*, where he used concepts derived from such philosophers as Wolff, Nietzsche, Lotze, Wundt, Jodl, and Mauthner, Jung had developed a distinction between two kinds of thinking, which he called 'directed thinking' (*gerichtetes Denken*) and 'fantasy thinking' (*phantastisches Denken, Träumen, Phantasieren*) (C. G. Jung, *Psychology of the Unconscious: A Study of the Transformations and Symbolisms of the Libido* [1911–1912], tr. B. M. Hinkle, London: Routledge, 1991, paras. 6–57).

14 On this structure of argumentation, known as 'binary synthesis', see E. M. Wilkinson, 'Zur Sprache und Struktur der Ästhetischen Briefe', *Akzente* 6, 1959, 389–418; E. M. Wilkinson and L. A. Willoughby, 'Nachlese zu Schillers Ästhetik: Auf Wegen der Herausgeber', *Jahrbuch der Deutschen Schillergesellschaft* 11, 1967, 374–403; and E. M. Wilkinson and L. A. Willoughby,'Appendix III: Visual Aids', in Friedrich Schiller, *On the Aesthetic Education of Man*, 2nd edn., Oxford: Clarendon Press, 1982, pp. 349–50.

15 Jung, *Psychology of the Unconscious: A Study of the Transformations and Symbolisms of the Libido*, para. 99, n. 17. Compare with Jung's similar defence in 1929, where he writes that 'I am accused of mysticism, but I am not responsible for the fact that, at all times and in all places, humankind has naturally developed religious functions and that the human psyche is thus, from the earliest times [*seit Urzeit*] onwards, saturated and interwoven with religious feelings and ideas' (Jung, 'Freud and Jung: Contrasts' [1929], *CW* 4, paras. 768–84 [here: para. 781]).

16 S. Freud, 'On the History of the Psycho-Analytic Movement' [1914], in *On the History of the Psycho-Analytic Movement, Papers on Metapsychology, and other works* [*Standard Edition of the Collected Works*, vol. 14], eds. J. Strachey and A. Freud, London: Hogarth Press; Institute of Psycho-Analysis, 1957, pp. 1–66 (here: pp. 103–08).

17 Jung, *Psychological Types* [1921], *CW* 6, para. 630.

18 See 'Die Psychologie der Übertragung', in Jung, *Gesammelte Werke*, ed. Lilly Jung-Merker, Elisabeth Ruf, and Leonie Zander, 20 vols, Olten und Freiburg im Breisgau: Walter, 1960–1983, vol. 16, §462, fn. 13 [passage omitted in English translation]. See also the important remarks in a footnote in *Mysterium Coniunctionis*, where Jung claims personal support from Lévy-Bruhl for his interpretation of *état prélogique* in terms of the non-rational, and expresses surprise at Lévy-Bruhl's eventual repudiation of both concepts in his posthumously published notebooks (Jung, *Mysterium coniunctionis* [1955–1956], *CW* 14, para. 336, n. 662).

19 Writing to his parents shortly after his appointment as a professor of philology in Göttingen, Müller wrote on 21 November 1819: 'One has to be very careful here not to be deemed a mystic, for the old bunch of Göttingen professors mixes every possible *Naturphilosophie*, romantic poetry, new theology, higher historical analysis, symbolic mythology etc. in one bowl and then pours it all right down the sink' (C. O. Müller, *Lebensbild in Briefen an seine Eltern, mit dem Tagebuch seiner italienisch-griechischen Reise*, ed. O. and E. Kern, Berlin: Weidmann, 1908, pp. 54–55; cited in J. H. Blok, 'Quests for a Scientific Mythology: F. Creuzer and K. O. Müller on History and Myth', *History and Theory* 33 (3), 1994, 26–52 [here: p. 33]).

242 Jungian approaches: Context and critique

20 As well as being an entertainment factor, Jung's anecdotes make a strategic point: whereas Freud uses contemporary anthropological accounts to write (as in *Totem and Taboo* [1916]) about 'primitive' societies at second hand, Jung actually goes and vists the 'primitive' tribesmen to understand – admire, even – their way of life.

21 This strand of of Jung's thought bears a stronge tinge of Schelling's transcendental idealism; see F. W. J. Schelling, 'Vorlesungen über die Methode des akademischen Studiums' and 'Über das Verhältnis der Naturphilosophie zur Philosophie überhaupt' [1802], in *Werke*, ed. M. Schröter, vol. 3, Munich: Beck, 1958, pp. 229–374 and 526–44 (here: p. 303).

22 J. W. Goethe, *Selected Poems*, ed. C. Middleton, Boston: Suhrkamp/Insel, 1983, pp. 158–58 (tr. C. Middleton). The title of the poem refers to the words spoken by the chorus in ancient Greek tragedy. In his Eranos lecture 'Psychological Aspects of the Mother Archetype' (1938/1954) Jung quotes from this poem, and adds a twist of his own: ' "All that is outside, also is inside", we could say with Goethe. But this "inside", which modern rationalism is so eager to derive from "outside", has its own structure, which precedes all conscious experience as an *a priori*' (Jung, *CW* 9/i, paras. 148–205 [here: para. 187]).

23 For texts presenting Rudolf Steiner's view of Goethe's scientific writings, see R. Steiner, *Nature's Open Secret: Introductions to Goethe's Scientific Writings*, tr. J. Barnes and M. Spiegler, Great Barrington, MA: Anthroposophic Press, 2000. For further discussion of the relation between anthroposophy and Goethean science, see E. Lehrs, *Man or Matter: Introduction to a Spiritual Understanding of Nature on the Basis of Goethe's Method of Training Observation and Thought*, 3rd edn., London: Rudolf Steiner Press, 1985; and E. Koepke, *Goethe, Schiller und die Anthroposophie: Das Geheimnis der Ergänzung*, Stuttgart: Verlag Freies Geistesleben, 2002.

24 See M. Horkheimer and T. W. Adorno, *Dialectic of Enlightenment: Philosophical Fragments* [1944], tr. J. Cumming, New York: Continuum, 1996. For further discussion of the relation between Jung and critical theory, see T. Evers, *Mythos und Emanzipation: Eine kritische Annäherung an C. G. Jung*, Hamburg: Junius, 1987.

25 '[. . .] Grant me a vision of Nature's forces / That bind the world, all its seeds and sources / And innermost life – all this I shall see' (*Daß ich erkenne, was die Welt / Im Innersten zusammenhält, / Schau' alle Wirkenskraft und Samen*) (*Faust I*, ll. 382–84; J. W. Goethe, *Faust: Part One*, tr. D. Luke. Oxford and New York: Oxford University Press, 1987, p. 15).

26 See 'Nature', in J. W. Goethe, *Scientific Studies*, ed. & tr. D. Miller. New York: Suhrkamp, 1988, pp. 3–5 (here: p. 3). For further discussion of this text and its reception, see I. Hermann, 'Goethes Aufsatz *Die Natur* und Freuds weitere philosophisch-psychologische Lektüre aus den Jahren 1880–1900', *Jahrbuch der Psychoanalyse* 7 (1974), 77–100; and H. F. Fullenwider, 'The Goethean Fragment "Die Natur" in English Translation', *Comparative Literature Studies* 23 (1986), 170–77.

27 See E. Staiger, 'Zu einem Vers von Mörike: Ein Briefwechsel mit Martin Heidegger', *Trivium: Schweizerische Vierteljahrsschrift für Literaturwissenschaft* 9 (1), 1951, 1–16; L. Spitzer, 'Wiederum Mörikes Gedicht "Auf eine Lampe" ', *Trivium* 9 (3), 133–47; and A. A. Grugan (tr.), 'The Staiger-Heidegger Correspondence', *Man and World* 14 (1981), 291–307.

28 For the reviewer of Jung's lecture for the *Neue Zürcher Zeitung*, the issue of 'subjectivity–objectivity' was 'the ultimate question at stake not only in contemporary philosophy but also in psychology'; an 'unsolved question', and one with which, unsolved, Jung had allowed his listeners to depart (*NZZ*, 24 October 1930).

29 Thus writes Goethe, citing the famous lines *neque vero oculus unquam videret solem, nisi factus solaris esset* from the sixth tractate of the first *Ennead* in the preface to his *Theory of Colour* (1810) (Goethe 1988, p. 164). Cf. 'To any vision must be brought an

eye adapted to what is to be seen, and having some likeness to it. Never did an eye see the sun unless it had first become sunlike, and never can the Soul have a vision of the First Beauty unless it itself be beautiful' (*Enneads*, 1.6 [1]; in Plotinus, *The Enneads*, tr. Stephen MacKenna, revised B. S. Page. London: Faber and Faber, 1956, p. 64). For discussion of Goethe's relation to Plotinus and neo-Platonic thought, see P. F. Reiff, 'Plotin und die deutsche Romantik', *Euphorion* 19, 1912, 602–12; H. F. Müller, 'Goethe und Plotinos', *Germanisch-romanische Monatsschrift* 7, 1915–1919, 45–60; F. Koch, *Goethe und Plotin*, Leipzig: J. J. Weber, 1925; F. Koch, 'Plotins Schönheitsbegriff und Goethes Kunstschaffen', *Euphorion* 26, 1925, 50–74; H. Schmitz, *Goethes Altersdenken im problemgeschichtlichen Zusammenhang*, Bonn: Bouvier, 1959, pp. 50–104; and V. Hansen, ' "Gleichsam zum erstenmal im Plato gelesen" – Goethes Platonismus', in R. G. Khoury, J. Halfwassen, and F. Musall (eds), *Platonismus im Orient und Okzident: Neuplatonische Denkstrukturen im Judentum, Christentum und Islam*, Heidelberg: Winter, 2005, pp. 233–45.

30 See the entry for *mana* in A. Samuels, B. Shorter, and F. Plaut, *A Critical Dictionary of Jungian Analysis*. London and New York: Routledge, 1986, p. 89. Ernst Cassirer draws attention to Codrington's work on *mana* (E. Cassirer, *The Philosophy of Symbolic Forms*, vol. 2, *Mythical Thought* [1925], tr. R. Manheim. New Haven and London: Yale University Press, 1955, pp. 75–78; E. Cassirer, *An Essay on Man*. New Haven and London: Yale University Press, 1944, p. 96; see R. H. Codrington, *The Melanesians: Studies in their Anthropology and Folklore*. Oxford: Clarendon Press, 1891, pp. 118–19) and, more recently, Bertrand Vergely relates *mana* and the Sanskrit term *manu* to the English 'man' and German *Mann* and *Mensch*, suggesting a perennial identification of the human being and the activity of the mind or spirit (B. Vergely, *Hegel ou la défense de la philosophie*, Toulouse: Milan, 2001, p. 37). For further discussion of Jung's understanding of the *mana*-personality, see part 2, section 4, of 'The Relations between the Ego and the Unconscious' (1928) (*CW* 7, paras. 202–406 [here: paras. 374–406]).

31 Compare with Jung's slogan, *wirklich ist, was wirkt* (see 'The Relations between the Ego and the Unconscious', *CW* 7, para. 353; C. G. Jung. (1960 [1933]), 'The Real and the Surreal' [1933], in *CW* 8, paras. 742–48 [here: para. 742]; *Symbols of Transformation*, *CW* 5 [here: para. 344]; and his letter to H. Haberlandt of 23 April 1952 (*Letters*, ed. G. Adler and A. Jaffé, tr. R. F. C. Hull, vol. 2, London: Routledge & Kegan Paul, 1973, p. 54). Jung's formulation restates the position taken by Hegel in his *Science of Logic*: 'what is actual can act' (*was wirklich ist, kann wirken*) (G. W. F. Hegel, *Science of Logic*, tr. A. V. Miller, Oxford: Oxford University Press, 1969, p. 546).

32 Implicitly this question represents a shift from Jung's position in *Transformations and Symbols of the Libido*, in which he states 'this world is empty to him alone who does not understand how to direct his libido towards objects, and to render them alive and beautiful for himself' (*Psychology of the Unconscious: A Study of the Transformations and Symbolisms of the Libido*, para. 284). Now, in 1930, Jung thinks differently: maybe the world really is beautiful. . . For the conclusion of this aesthetico-existential trajectory in Jung's thought, see the words attributed to him at the end of *Memories, Dreams, Reflections* – 'The world into which we are born is brutal and cruel, and at the same time of divine beauty' (A. Jaffé [ed.], *Memories, Dreams, Reflections of C. G. Jung*, London: Collins; Routledge & Kegan Paul, 1963, p. 391) – in which *beauty* becomes further transmuted into *meaning* (pp. 391–92).

33 Compare with Ludwig Klages's philosophy of 'daimonic' or 'elementary souls' (L. Klages, *Der Geist als Widersacher der Seele* [1929–1932], 6th edn, Bonn: Bouvier Verlag Herbert Grundmann, 1981, pp. 1116–42), centered on what he calls 'the reality of images' (*die Wirklichkeit der Bilder*) (pp. 1143–1248).

34 S. Freud, 'Civilization and its Discontents' [1930], in *The Future of an Illusion, Civilization and its Discontents, and other works* [*Standard Edition of the Collected*

Works, vol. 21], eds. J. Strachey and A. Freud, London: Hogarth Press; Institute of Psycho-Analysis, 1961, pp. 57–146 (here: p. 86).

35 Jung, 'Psychological Types' [1923], in *CW* 6, paras. 883–914 (here: para. 907).

36 In *Dialectic of Enlightenment* (1944), for example, Adorno and Horkheimer speak of the 'remembrance of nature in the subject' (*Eingedenken der Natur im Subjekt*) as the central point at which the Enlightenment is opposed to tyranny. 'By virtue of this remembrance of nature in the subject,' they write, 'in whose fulfillment the unacknowledged truth of all culture lies hidden, enlightenment is universally opposed to domination' (Horkheimer and Adorno, *Dialectic of Enlightenment*, p. 40).

37 In his *Theory of Communicative Action* (1981), Jürgen Habermas (b.1929) also focuses on this key phrase: as part of *his* critique of Adorno and Horkheimer's critique of instrumental reason, Habermas accuses his colleagues of 'follow[ing] the (largely effaced) path that leads back to the origins of instrumental reason, so as to *outdo* the concept of objective reason'; thus their theory of mimesis, Habermas argues, leads them to speak about it 'only as they would about a piece of uncomprehended nature' (J. Habermas, *The Theory of Communicative Action* [1981], tr. T. McCarthy, vol. 1, *Reason and the Rationalization of Society* (Cambridge: Polity Press, 1987, p. 382).

38 Jung, 'The Psychology of the Child Archetype' [1940], *CW* 9/i, paras. 259–305 (here: para. 290).

39 Jung, 'The Psychology of the Child Archetype', *CW* 9/i, para. 290.

40 Jung, 'The Psychology of the Child Archetype', *CW* 9/i, para. 291.

41 For an excellent discussion of Jung's concept of synchronicity, with particular reference to its cultural context, see R. Main, *The Rupture of Time: Synchronicity and Jung's Critique of Modern Western Culture*, Hove and New York: Brunner-Routledge, 2004.

42 Jung, 'Richard Wilhelm: In Memoriam' [1930/1938], *CW* 15, paras. 74–96 (here: para. 81). In fact, in a passage crossed through in the original manuscript of his lecture, Jung made specific reference to the *I Ching*, noting that, in contrast to the European attitude of mind, 'the attitude of Chinese culture derives directly from archaic presuppositions', and recognizing the *I Ching*, 'the quintessence of Chinese thought', as 'nothing other than a textbook of the science of chance and at the same time a methodology of interpreting chance' (C. G. Jung-Archiv, [Hs 1055: 71a], 'Der archaische Mensch').

43 Jung, *The Tavistock Lectures* [1935], *CW* 18, paras. 1–415 (here: para. 70).

44 Jung, 'On Synchronicity' [1951], *CW* 8, paras. 969–97.

45 Jung, 'Synchronicity: An A-Causal Connecting Principle' [1952], *CW* 8, paras. 816–968. Originally published as 'Synchronizität als ein Prinzip akausaler Zusammenhänge', together with Wolfgang Pauli's 'Der Einfluss archetypischer Vorstellungen auf die Bildung naturwissenschaftlicher Theorien bei Kepler', in *Naturerklärung und Psyche* [*Studien aus dem C. G. Jung-Institut*, vol. 4], Zurich: Rascher, 1952, pp. 1–107, and in *The Interpretation of Nature and the Psyche*, tr. R. F. C. Hull, London; New York: Routledge and Kegan Paul; Pantheon [Bollingen Series, vol. 51], 1955, pp. 5–146.

46 See Wolfgang Pauli and C. G. Jung, *Ein Briefwechsel: 1932–1958*, ed. C. A. Meier, Berlin and Heidelberg: Springer, 1992; and C. A. Meier (ed.), *Atom and Archetype: The Pauli/Jung Letters, 1932–1958*, tr. D. Roscoe, London: Routledge, 2001. For further discussion, see H. Atmanspacher, 'The Hidden Side of Wolfgang Pauli: An Eminent Physicist's Extraordinary Encounter with Depth Psychology', *Journal of Consciousness Studies* 3 (2), 1996, 112–26; S. Gieser, *The Innermost Kernel: Depth Psychology and Quantum Physics: Wolfgang Pauli's Dialogue with C. G. Jung*, Berlin and Heidelberg: Springer, 2004; David Lindorff, *Pauli and Jung: The Meeting of Two Great Minds on the Unity of Matter and Spirit*, Wheaton, IL: Quest Books, 2004; Harald Atmanspacher and Hans Primas (eds), *Recasting Reality: Wolfgang Pauli's Philosophical Ideas and Contemporary Science*, Berlin and Heidelberg: Springer,

2009; and Arthur I. Miller, *Deciphering the Cosmic Number: The Strange Friendship of Wolfgang Pauli and Carl Jung*, New York and London: Norton, 2009. I am grateful to Andrew Burnison for providing some of these bibliographical recommendations relating to the Jung–Pauli encounter.

47 See Jung's letter to Carl Seelig of 25 February 1935 (Jung, *Letters*, vol. 2, pp. 108–09).

48 Compare with the first two stanzas of Goethe's poem, 'Primal Words. Orphic' (*Urworte. Orphisch*) (1817–18) (Goethe, *Selected Poems*, pp. 230–33).

49 As Donald Verene has pointed out, Jung's claim that 'for the primitive mind synchronicity is a self-evident fact; consequently there is no such thing as chance' (Jung, 'Synchronicity: An A-Causal Connecting Principle', *CW* 8, para. 941) chimes well with Cassirer's view that for mythical thought, as a symbolic form, there is equally no such thing as chance, so that, 'for the mythical view, it is the swallow that *makes* the summer' (Cassirer, *Mythical Thought*, pp. 43–45; see D. P. Verene, 'Coincidence, historical repetition, and self-knowledge: Jung, Vico, and Joyce', *Journal of Analytical Psychology* 47 (3), 2002, 459–78 [here: p. 464]).

50 Note the ambiguity of *durchkreuzen*, which can mean 'to cross' (*durchfahren*), but also 'to thwart' or 'to frustrate'. In this sense, Jung speaks in *The Psychology of the Transference* (1946) of something that 'thwarts and "crosses" ' (*durchkreuzt*) the individual in a threefold manner: as the shadow, as the Other ('the individual reality of the "You" '), and as the psychic non-ego (i.e., the collective unconscious) (Jung, *CW* 16, paras. 353–539 [here: para. 470]).

51 E. Auerbach, ' "Figura" ' [1944], in *Scenes from the Drama of European Literature: Six Essays*, tr. R. Manheim, Manchester: Manchester University Press, pp. 9–76 (here: p. 53).

52 See Jung, 'Synchronicity: An A-Causal Connecting Principle' (*CW* 8, paras. 816–968). Compare with Verene's argument which relates Jung's view of the *I Ching* as an embodiment of the principle of synchronicity to the principle of repetition, and hence the idea of meaning. The synchronistic view of the world as one in which 'a pattern of events that has been in the past can suddenly and unexpectedly repeat itself' allows Verene to link Jung's idea that 'the self encounters its own reality' through repetition, not simply through causal production, with Vico's conception of 'ideal eternal history' as *corso* and *ricorso*; and the sixty-four hexagrams of the *I Ching* with Vico's 'mental dictionary' and Jung's archetypes of the collective unconscious (Verene, 'Coincidence, historical repetition, and self-knowledge', pp. 474, 468, 469).

53 B. Vergely, *Petite philosophie du bonheur*, Toulouse: Milan, 2002, pp. 13–14. Compare with Vergely's description – without specific reference to Jung, yet entirely compatible with Jung's outlook – of 'the symbolic life'. 'The discovery of the world as a symbolic world,' Vergely writes, 'is a journey which procures an infinite joy', and he goes on: 'The world was silent, insignificant. Now, suddenly, it begins to resonate with thousands of meanings, which in turn bring into motion an entire interior life unknown up this point' (*Petite philosophie du bonheur*, p. 184).

54 For this final case, see G. Quispel, 'Gnosis and Psychology', in R. Segal (ed.), *The Allure of Gnosticism: The Gnostic Experience in Jungian Psychology and Contemporary Culture*, Chicago & La Salla, IL: Open Court, 1955, pp. 10–25 (here: p. 17).

55 See Jung, 'On Psychic Energy' [1928], *CW* 8, paras. 1–130 (here: paras. 2–3).

56 For further discussion of the 'primordial phenomenon' in Goethe's thought, see R. H. Stephenson, *Goethe's Conception of Knowledge and Science*. Edinburgh: Edinburgh University Press, 1995, pp. 13–15, 54–56, 66–71; H. Bortoft, *The Wholeness of Nature: Goethe's Way of Science*, Edinburgh: Floris Books, 1996; and D. Seamon and A. Zajonc (eds), *Goethe's Way of Science: A Phenomenology of Nature*, Albany, NY: State University of New York Press, 1998.

57 Jung, *Psychology of the Unconscious: A Study of the Transformations and Symbolisms of the Libido*, para. 1.

58 Jaffé (ed.), *Memories, Dreams, Reflections of C.G. Jung*, p. 18. For use of the rhizome image elsewhere, see Jung, *Symbols of Transformation*, *CW* 5, p. xxiv; and *Psychology and Religion* (The Terry Lectures) [1938/1940] , *CW* 11, paras. 1–168 (here: para. 37). The image of the rhizome is later found in the work of Gilles Deleuze and Félix Guattari (see their *Milles Plateaux* [1980], or *A Thousand Plateaus*, tr. Brian Massumi, New York and London: Continuum, 2004). In this regard, it is interesting to note the early Deleuze's fascination with Jung in his work on Sacher-Masoch; see C. Kerslake, 'Rebirth through incest: On Deleuze's Early Jungianism', *Angelaki* 9 (1), 2004, 135–57, and in the same volume Kerslake's translation of Deleuze's 'From Sacher-Masoch to Masochism', 125–33.

59 Jung, 'Analytical Psychology and "Weltanschauung" ' [1928/1931], *CW* 8, paras. 689–741 (here: para. 739).

60 Compare with the observation of Joseph Ratzinger, now Benedict XVI: 'I am willed [into being]; not a child of chance [*des Zufalls*] and of necessity, but one of will and of freedom. Thus I am needed, there is a meaning [*einen Sinn*] for me, a task that is intended just for me; there is an idea of me, that I can search for and find and fulfil' (J. Ratzinger, 'Das "Vater unser" sagen dürfen', in *Sich auf Gott verlassen: Erfahrungen mit Gebeten*, ed. R. Walter, Freiburg im Breisgau, Basle, Vienna: Herder, 1987, pp. 71–76 [here: p. 74]).

61 In this sentence, the hedonist imperative intrinsic to Jung's ethics becomes evident: full enjoyment is dependent on a life lived in a meaningful manner.

62 Jung's insistence on the need for totality is not just (one of) the reason(s) it is attractive to the New Age, but is also an indication of the heritage of German classical aesthetics in his thought; compare with Schiller's remark in his essay 'On Bürger's Poetry' (1791) that 'it is almost only poetry that can bring the separated faculties of the soul into union, which occupies the head and the heart, astuteness and wit, reason and imagination in a harmonious alliance, in other words which restores the *entire human* within us' (F. Schiller, *Werke in drei Bänden*, ed. G. Fricke and H. G. Göpfert, vol. 2, Munich: Hanser, 1966, p. 626).

63 Compare with Spinoza's remark that 'we feel and know by experience that we are eternal' (*Ethics*, part 5, proposition 23, scholium; B. Spinoza, 1928). *Selections*, ed. J. Wild, London: Scribner, 1928, p. 385).

64 Jung, 'Analytical Psychology and "Weltanschauung" ', *CW* 8, para. 739.

65 Compare with the passages cited above about 'nature within'.

66 Jung, 'Analytical Psychology and "Weltanschauung" ', *CW* 8, para. 739.

67 For further discussion, see V. Rippere, *Schiller and 'Alienation'*, Berne, Frankfurt am Main, Las Vegas: Peter Lang, 1981; and R. Schacht, *Alienation*, London: George Allen and Unwin, 1971.

68 Jung, 'Analytical Psychology and "Weltanschauung" ', *CW* 8, para. 737.

69 Jung, 'Psychotherapists or the Clergy' [1932], *CW* 11, paras. 488–538 (here: para. 531).

70 Jung, 'Psychotherapists or the Clergy', *CW* 11, para. 534.

71 Jung, 'Psychotherapists or the Clergy', *CW* 11, para. 535.

72 Jung, 'Freud and Jung', *CW* 4, paras. 768–84 (here: para. 776). Compare with Jung's reference above to 'natural spirit' (*natürlicher Geist*) in 'Analytical Psychology and "Weltanschauung" ' (*CW* 8, para. 739).

73 Jung, 'Freud and Jung', *CW* 4, para. 780.

74 Jung, 'Freud and Jung', *CW* 4, para. 780.

75 The name of an antiquarian bookshop in the French town of Bourges, this intriguing expression nicely captures the full chronological ambiguity of Jung's sense of the archaic: at once historical, timeless, and more full of presence than the present.

76 This assimilation is, however, surely compromised by the fact that this phrase is traditionally found in the Church's night office of Compline, not in its morning liturgy of Laudes, as Jung would doubtless have known?

77 In *Wilhelm Meisters Wanderjahre* (1821/1829), one of the characters of Goethe's
    novel – Jarno, now called Montan – insists on the reciprocity of 'thinking' and 'doing'
    when he delivers the celebrated maxim, 'thought and action, action and thought, that is
    the sum of all wisdom' (*Denken und Tun, Tun und Denken, das ist die Summe aller
    Weisheit*) (*Conversations of German Refugees/Wilhelm Meister's Journeyman Years
    or The Renunciants*, ed. J. K. Brown, tr. J. van Heurck, K. Winston [Goethe Edition,
    vol. 10], New York: Suhrkamp Publishers, 1989, p. 280. See also the aphorisms
    collected under the title of 'Thinking and Doing' in the *Maxims and Reflections* as they
    appear in the *Hamburger Ausgabe* of Goethe's works, where Goethe reflects at length
    on the relationship between reflection and experience (Goethe, *Werke*, ed. E. Trunz,
    14 vols, Hamburg: Christian Wegner, 1948–1960; Munich: Beck, 1981, vol. 12,
    pp. 396–417).

Index

abyss 8–11, 25, 131
acumen (*metis*) 59, 61
Adam Kadmon 97–8
Adorno, Theodor W. 21–2, 231, 244nn36–7
aesthetics: aesthetic cosmodicy 153, 156–7; aesthetic intuition 16–17, 37; aesthetic ontology 154, 156; Apollonian-Dionysian 153–4, 157; the archaic as the aesthetic 36; art and the loss of myth 31–2; Nietzsche 17, 18, 152–9; origin of the work of art 26–7; and *participation mystique* 28; and reason 22
Aetna 110
Agassi, Joseph 206
agon (contest, struggle) 58, 59, 60, 132, 157, 158; agonistic word 56, 58–61
alchemy 128, 130, 131–2, 209, 212 *see also* Philosopher's Stone
Alcibiades 109
Alexander the Great 60, 66n33, 89n12
allegory 19, 46nn123–4, 103, 108–9
Ammon, Oracle of 105
Ananke (goddess) 63n4
ananke (necessity) 33, 56, 57, 58
Anaxagoras of Clazomenae 4
Anaximander of Miletus 3, 4, 24, 27, 63n1, 161, 163
Anaximenes of Miletus 4, 24, 29, 63n1, 118
anguish 35, 57, 58, 131, 132, 133, 135
anima 24
anxiety 35 *see also* dread
Aphrodite 112
Apollo 59, 150, 151; Belvedere 148, 154, 163
Apollonian-Dionysian aesthetics 153–4, 157

Apollonian-Dionysian double myth 154, 157
the archaic: as the aesthetic 36 *see also* aesthetics; 'archaic' thinking 163–4, 211, 213 *see also* 'primitive' thinking; as atoms and emptiness 4–5; as enchanting beginning 69–70; as the *factum* 38; in German thought *see* German thought on the archaic; as the ground/*Ungrund/Urgrund* 8–14, 16, 18; and history *see* archaic theories of history; history of the archaic; sublimity of origins; Jungian approaches to *see* Jung, Carl G.; meaning and derivation of the term 3, 69; mytho-phenomenological approach to 69–88; as the One 6, 8, 13–14, 61; origins as an idea of *see* origins/*Ursprung*; pre-conscious forces 114; as the primordial *see* [the] primordial; Romantic archaic 11–17, 149–52; timeliness and timelessness 226–40
archaic man: critique and discussion of Jung's paper 'Archaic Man' 194–7, 207–13, 219–25, 226–40; Jung's lecture/ paper 171–87; and the Pelasgians 103–6
archaic theories of history 55–63; Hesiod's theory of Eris 56, 57–8; humanist tradition 62–3; and ontology as starting point of language problems 61–2; Pindar and the agonistic word 56, 58–61; Thucydides and the need for demythologization 55–7
archetypes 28, 128–9, 194; of place 86–8
Ares 112
Aretino, Leonardo *see* Bruni, Leonardo
Aristophanes 3
Aristotle 3, 5–6, 17, 31, 111, 123